MOUNTAINS OF NORTHERN EUROPE:

Conservation, Management, People and Nature

THE NATURAL HERITAGE OF SCOTLAND

Each year since it was founded in 1992, Scottish Natural Heritage has organised or jointly organised a conference which has focused on a particular aspect of Scotland's natural heritage. The papers read at the conferences, after a process of refereeing and editing, have been brought together as a book. The twelve titles already published in this series are listed below (No. 6 was not based on a conference).

1. *The Islands of Scotland: a Living Marine Heritage*
    Edited by J.M. Baxter and M.B. Usher (1994), 286pp.

2. *Heaths and Moorlands: a Cultural Landscape*
    Edited by D.B.A. Thompson, A.J. Hester and M.B. Usher (1995), 400pp.

3. *Soils, Sustainability and the Natural Heritage*
    Edited by A.G. Taylor, J.E. Gordon and M.B. Usher (1996), 316pp.

4. *Freshwater Quality: Defining the Indefinable?*
    Edited by P.J. Boon and D.L. Howell (1997), 552pp.

5. *Biodiversity in Scotland: Status, Trends and Initiatives*
    Edited by L.V. Fleming, A.C. Newton, J.A. Vickery and M.B. Usher (1997), 309pp.

6. *Land Cover Change: Scotland from the 1940s to the 1980s*
    By E.C. Mackey, M.C. Shewry and G.J. Tudor (1998), 263pp.

7. *Scotland's Living Coastline*
    Edited by J.M. Baxter, K. Duncan, S.M. Atkins and G. Lees (1999), 209pp.

8. *Landscape Character: Perspectives on Management and Change*
    Edited by M.B. Usher (1999), 213pp.

9. *Earth Science and the Natural Heritage: Interactions and Integrated Management*
    Edited by J.E. Gordon and K.F. Leys (2000), 344pp.

10. *Enjoyment and Understanding of the Natural Heritage*
    Edited by Michael B. Usher (2001), 224pp.

11. *The State of Scotland's Environment and Natural Heritage*
    Edited by Michael B. Usher, Edward C. Mackey and James C. Curran (2002), 354pp.

12. *Birds of Prey in a Changing Environment*
    Edited by D.B.A Thompson, S.M. Redpath, A. Fielding, M. Marquiss and C.A.Galbraith (2003), 570pp.

This is the thirteenth book in the series.

MOUNTAINS OF NORTHERN EUROPE:
Conservation, Management, People and Nature

Edited by D.B.A. Thompson, M.F. Price and C.A. Galbraith

EDINBURGH: TSO SCOTLAND

First published in 2005 by The Stationery Office Limited
71 Lothian Road, Edinburgh, EH3 9AZ

Applications for reproduction should be made to Scottish Natural Heritage, 12 Hope
Terrace, Edinburgh EH9 2AS

British Library Cataloguing in Publication Data
A catalogue record for this book is available from the British Library

ISBN 0 11 497319 9

Cover photography: Front cover: Ben More Assynt beyond Ledmore, North West Scotland
(Photo: Laurie Campbell). Back cover top to bottom: Saeter (*seter*) in
Kvam, Norway (Photo: M.F. Price); A Saami with a reindeer (*Rangifer
tarandus*) (Photo: B. Lind); Seeder-feeder process of orographic
enhancement of precipitation in mountains (from Chapter 6 by
D. Fowler & R. Battarbee); and trailing azalea (*Loiseleuria procumbens*)
(Photo: SNH).

Across the world, mountain areas are directly important for at least half of humanity. These areas contribute services to help maintain almost every aspect of our lives. This includes what we drink and eat, the light and heating in our homes, the furniture we use, and the enjoyment we derive from the outdoors, through to maintaining the many facets of local and national traditions, cultures and societies. Across Northern Europe, are some of the world's oldest mountains, notably the Caledonian range, which includes the Kiolen range of Norway and Sweden, and the Scottish Highlands, and further south, there are some of the younger, more rugged mountains such as the Pyrenees and the Alps. Over the 500 million years since these mountains were formed weathering and erosion have shaped the landforms and water courses we see today. Only relatively recently, in the last 10,000 years or so, have human activities influenced mountain areas, contributing to the landscapes, flora and fauna present today. It is sometimes easy to forget just how ancient some mountain areas are, and that they have been influenced by people in only a tiny fraction of the time of their existence.

At the United Nations Conference on Environment and Development (UNCED, referred to as the Earth Summit or Rio Summit) held in Rio de Janeiro in 1992, mountains were the focus of Chapter 13 within 'Agenda 21' (United Nations, 1992). That chapter outlined the need for sustainable mountain development, highlighting the urgency of the action required. It outlined two programme areas: (a) generating and strengthening knowledge on the ecology and sustainable development of mountain ecosystems; and (b) promoting integrated watershed development and alternative livelihood opportunities. Subsequent to this, several important reviews were published, many of them outlining the importance of mountain areas as centres of biodiversity (e.g. Barthlott *et al.*, 1996; Jeník, 1997; Messerli & Ives, 1997; Mishra, 2000; Hallanaro & Pylvänäinen, 2002; Körner & Spehn, 2002; Nagy *et al.*, 2003).

In 1998, the United Nations (UN) General Assembly supported a resolution to establish 2002 as the International Year of Mountains (IYM 2002). The largest number of states ever to support a UN resolution (130 states) supported this one (Baldascini *et al.*, 2002). IYM 2002 had as its aim the promotion of the conservation and sustainable development of mountain regions, thereby ensuring the well-being of mountain and associated lowland communities. Seventy-eight countries established national IYM committees to guide or co-ordinate the activities devised as part of the IYM.

In the UK, the culmination of IYM 2002 was the conference held in Pitlochry during 6-8 November 2002, entitled 'Nature and People: conservation and management in the mountains of Northern Europe'. This was an international event, attended by almost 300 participants, with speakers and delegates drawn from 15 countries. This book provides an account of that conference, with oral and poster papers substantially updated and edited; all of the papers have been refereed independently. The 38 chapters provide a major contribution to our current knowledge of the mountains of Northern Europe, in particular regarding the interactions between people and nature, and the conservation and management activities needed to benefit the natural heritage.

The book focuses on the mountain areas of Northern Europe, with most of the chapters concerned with the UK and Nordic countries. We have divided the book into five parts. Part 1 contains three introductory chapters, and includes the background to activities leading up to IYM 2002, and follow-up actions. The eight chapters in Part 2 detail the many perspectives on mountain environments. These chapters straddle geological, geomorphological, atmospheric, biological and human related aspects of the mountain environment. Land use changes are the focus of Part 3, where nine chapters detail the complex nature of human influences on mountain landscapes and biodiversity. Part 4 deals with management influences, practices and conflicts. Nine chapters range over the gamut of issues connected with establishing and managing National Parks and other important areas for nature. Part 5 of the book explores the prospects for mountain areas. The nine chapters in this section continue the international theme of the book, and reflect on environmental and social changes which have combined to influence the relationship between people and our natural heritage.

The World Summit on Sustainable Development, held in Johannesburg in 2002, focused on the links between environmental sustainability and human well being. The future of mountain areas rests largely on the coming together of these two strands. Their management for agriculture, energy generation, production of food, water, timber and other products, as well as for enjoyment, amenity interests and nature in the broadest sense, requires an integrated approach. This book reveals important lessons and experiences in managing upland areas throughout much of Northern Europe. If there are important lessons to be drawn from these studies, they are that robust science is needed to identify changes, and the causes of change in mountain environments, and that an inclusive approach is needed to address the influences that can be managed. Climate change now presents a massive challenge for managing nature, and of course other interests, in the mountains. We have to strive to ensure that the science, policy and management activities to address climate change mesh effectively, and have the confidence of people who live in or depend on mountains. We hope that this book will appeal to researchers, decision-makers, land managers and the wider public with an interest in mountain area; it is novel in drawing together such a diverse range of interests connected with mountain areas.

Acknowledgements

It is a pleasure to thank many people involved in the conference and in the preparation of the book. The conference was run jointly between Scottish Natural Heritage and the Centre for Mountain Studies, Perth College, UHI (University of the Highlands and Islands) Millennium Institute, and in partnership with The Food and Agriculture Organization of the United Nations and IUCN – The World Conservation Union. Sponsorship was provided by the British Council in Sweden, the Royal Geographical Society/Institute of British Geographers, Perth and Kinross Council, Perthshire Convention Bureau and Scottish Enterprise Tayside. We thank all of these organisations for their support. Perth and Kinross Council are also thanked for valuable assistance in the planning stages, and for helping to organise transport for speakers/delegates from and to Edinburgh Airport. Provost Mike O'Malley kindly hosted the Welcoming Reception for delegates.

The Icelandic Government, in particular Siv Friðleifsdóttir (Minister of the Environment) and Magnús Jóhannesson of the Ministry for the Environment are thanked for providing advice in the lead-up to the conference. The Scottish Executive provided help and support in planning the conference, and particular thanks go to Mike Foulis, Jane Dalgleish, Ian Melville and Ian Bainbridge. We had the honour of having the conference opened by Siv Friðleifsdóttir, and we are delighted that she has written the Foreword for this book. Magnus Magnusson is warmly thanked for arranging the Icelandic Minister's participation – whilst making one of his regular visits to Iceland.

The conference was greatly helped by the efficient chairing of the six sessions, and for this we thank John Markland, Mike Webster, Janet Sprent, Stuart Housden, Bill Heal and Ian Jardine. The speakers made the work of the chairs less onerous than at many conferences. Virtually all of the speakers have contributed chapters to the book, and are thanked for their industry in preparing their talks and then submitting and updating their manuscripts. We also thank the many referees who have helped us in the editorial process.

The Pitlochry Festival Theatre was a superb venue for the conference, and we thank the Festival Theatre staff, particularly Nikki Axford, Sandra Grieve and John Anderson, and all of the other staff (too many to name individually) for their excellent support of the event. The smooth planning and running of the event was eased greatly by the usual diligence of Helen Forster, supported by Sylvia Conway, Paul Robertson, Joyce Garland and Maureen Scott. Jo Newman has been key to the production of the book, and has worked tirelessly with the authors to check details, including website addresses and licence agreements governing the use of maps and other information. The publisher, TSO Scotland, has been meticulous, and we thank Jane McNair and colleagues for assiduous attention to details, and advice.

Des B.A. Thompson and Colin A. Galbraith, Scottish Natural Heritage
Martin F. Price, Centre for Mountain Studies, Perth College, UHI Millennium Institute

August 2005

References

Baldascini, A., Perlis, A. & Romeo, R.L. (2002). *International Year of Mountains: concept paper.* Food and Agriculture Organization of the United Nations, Rome.

Barthlott, W., Lauer, W. & Placke, A. (1996). Global distribution of species diversity in vascular plants: towards a global map of phytodiversity. *Erdkunde,* **50**, 317-27.

Hallanaro, E.-L. & Pylvänäinen, M. (Ed.) (2002). *Nature in Northern Europe – Biodiversity in a Changing Environment.* Nord 2001: 13. Nordic Council of Ministers, Helsinki.

Houghton, J.T., Ding, Y., Griggs, D.J., Noguer, M., van der Linden, P.J., Dai, X., Maskell, K. & Johnson, C.A. (Ed.) (2001). *Climate Change 2001: The Scientific Basis. Contribution of Working Group I to the Third Assessment Report of the Inter-governmental Panel on Climate Change.* Cambridge University Press, Cambridge.

Jeník, J. (1997). The diversity of mountain life. In *Mountains of the World: A Global Priority,* ed. by B. Messerli & J.D. Ives. Parthenon, Carnforth. pp. 199-233.

Körner, C. & Spehn, E.M. (Ed.) (2002). *Mountain Biodiversity: A Global Assessment.* Parthenon, London.

Messerli, B. & Ives, J.D. (Ed.) (1997). *Mountains of the World: A Global Priority.* Parthenon, Carnforth.

Mishra, H. (2000). *Mountain Matters.* Global Environment Facility, Washington DC.

Nagy, L., Grabherr, G., Körner, C.H. & Thompson, D.B.A. (Ed.) (2003). *Alpine Biodiversity in Europe.* Springer-Verlag, Berlin.

United Nations (1992). *Earth Summit: Agenda 29.* The United Nations Programme of Action from Rio. The final text of agreements negotiated by governments at the United Nations Conference on Environment and Development (UNCED), 3-14 June, 1992, Rio de Janeiro, Brazil.

Editorial note

All website addresses cited in chapters have been checked. However, as a matter of principle such addresses are given for information only, and are not viewed as a formal source of published information. Most chapters have cross-references to other chapters in this book and these are given as author(s), chapter number(s) and are not cited in the References.

CONTENTS

Preface   v

Contents   ix

Foreword: Siv Friðleifsdóttir   xii

Contributors   xvi

PART ONE: INTRODUCTION AND CONTEXT   1

1.   Nature and people: conservation and management in the mountains
     of Northern Europe – with mountains in mind
     Magnus Magnusson   5

2.   The International Year of Mountains, 2002: progress and prospects
     Martin F. Price & Thomas Hofer   11

3.   Managing mountain ecosystems to meet human needs: a case for
     applying the ecosystem approach
     Stephen R. Edwards   23

PART TWO: MOUNTAIN ENVIRONMENTS: PERSPECTIVES   39

4.   The nature of mountains: an introduction to science, policy and
     management issues
     D.B.A. Thompson, L. Nagy, S.M. Johnson & P. Robertson   43

5.   Links between geodiversity and biodiversity in European mountains:
     case studies from Sweden, Scotland the the Czech Republic
     Christer Jonasson, John E. Gordon, Milena Kociánová,
     Melanie Josefsson, Igor J. Dvořák & Des. B.A. Thompson   57

6.   Climate change and pollution in the mountains: the nature of change
     David Fowler & Rick Battarbee   71

7.   Contemporary and historical pollutant status of Scottish mountain
     lochs
     M. Kernan, N.l. Rose, L. Camarero, B. Piña & J. Grimalt   89

8. Climate change and effects on Scottish and Welsh montane ecosystems: a conservation perspective
N.E. Ellis & J.E.G. Good 99

9. Modelling future climates in the Scottish Highlands – an approach integrating local climatic variables and regional climate model outputs
John Coll, Stuart W. Gibb & S. John Harrison 103

10. The effects of nitrogen deposition on mountain vegetation: the importance of geology
A.J. Britton, J. Fisher & G. Baillie 121

11. People, recreation and the mountains with reference to the Scottish Highlands
John W. Mackay 127

PART THREE: LAND USE CHANGE 137

12. The mountains of Northern Europe: towards an environmental history for the last ten thousand years
I.G. Simmons 141

13. The use and management of Norwegian mountains reflected in biodiversity values – what are the options for future food production?
E. Gunilla A. Olsson 151

14. Landscape history and biodiversity conservation in the uplands of Norway and Britain: comparisons and contradictions
M.E. Edwards 163

15. Transhumance and vegetation degradation in Iceland before 1900: a historical grazing model
Amanda Thomson 179

16. Terrain sensitivity on high mountain plateaux in the Scottish Highlands: new techniques
Stefan M. Morrocco 185

17. Mapping the inherent erosion risk due to overland flow: a tool to guide land management in the Scottish uplands
Allan Lilly, John Gordon, Monica Petri & Paula Horne 191

18. Mapping downhill skiing's environmental impact: evaluating the potential of remote sensing in Scotland
Will Cadell & K.B. Matthews 197

19. Mountain tourism in Northern Europe: current patterns and
     recent trends
     Peter Fredman & Thomas A. Heberlein   203

20. Some lessons from the ECN, GLORIA and SCANNET networks for
     international environmental monitoring
     Neil Bayfield, Rob Brooker & Linda Turner   213

**PART FOUR: MANAGEMENT INFLUENCES, PRACTICES
AND CONFLICTS**   223

21. Multi-purpose management in the mountains of Northern Europe
     – policies and perspectives
     Jeff Maxwell & Richard Birnie   227

22. Abernethy Forest RSPB Nature Reserve: managing for birds,
     biodiversity and people
     D.J. Beaumont, A. Amphlett & S.D. Housden   239

23. Ownership in the mountains of Northern Scandinavia and its
     influences on management
     Audun Sandberg   251

24. The economic and socio-cultural dimension of the mountains for the
     Swedish Saami reindeer herders
     Nicolas Gunslay   259

25. Public participation for conflict reconciliation in establishing
     Fulufjället National Park, Sweden
     Per Wallsten   263

26. Establishing the Cairngorms National Park: lessons learned and
     challenges ahead
     Murray Ferguson & John A. Forster   275

27. Two models of National Parks: ethical and aesthetic issues in
     management policy for mountain regions
     Alan P. Dougherty   291

28. Community land ownership in Scotland: progress towards sustainable
     development of rural communities?
     Aylwin Pillai   295

29. Tourists, nature and indigenous peoples – conservation and management in the Swedish mountains
Robert Pettersson & Tuomas Vuorio   299

PART FIVE: PROSPECTS   303

30. Natural heritage trends: an upland saga
Eeva-Liisa Hallanaro & Michael B. Usher   307

31. Delimitation of mountain areas for the purpose of developing EU policies
Hallgeir Aalbu   325

32. Frozen opportunities? Local communities and the establishment of Vatnajökull National Park, Iceland
Karl Benediktsson & Guðríður Þorvarðardóttir   335

33. Outdoor education and outdoor recreation in Scotland
Peter Higgins   349

34. SCANNET: a Scandinavian-North European network of terrestrial field bases
O.W. Heal, N. Bayfield, T.V. Callaghan, T.T. Høye, A. Järvinen, M. Johansson, J. Kohler, B. Magnusson, L. Mortensen, S. Neuvonen, M. Rasch & N.R. Saelthun   359

35. Mountains and wilderness: identifying areas for restoration
S.J. Carver   365

36. Priorities for the conservation and management of the natural hertage in Europe's high mountains
Georg Grabherr   371

37. Personal reflections on the conference
Simon Pepper & Eli Moen   377

38. Looking forward from Pitlochry and the International Year of Mountains
Colin A. Galbraith & Martin F. Price   379

Index   381

It was an honour for me to address the Pitlochry conference held on the initiative of Scottish Natural Heritage to mark the International Year of Mountains 2002. I was particularly honoured when my fellow countryman Magnus Magnusson, founder-chairman of Scottish Natural Heritage, invited me in the winter of 2001 to come to this important conference.

My country and Scotland have shared both history and heritage for more than 1,000 years. The Old Icelandic sagas describe how Norsemen settled in the British Isles and how people from these settlements migrated to Iceland. Genetic studies of Icelanders confirm our kinship with the Scots.

It was indeed very proper to dedicate the year 2002 to mountains, and especially to the relevance of mountains for environmental issues and sustainable development. Mountains play a much more extensive role in people's lives than we generally realise. As we know from poetry and other literature, and the visual arts, mountains have always provided creative inspiration, but they also have a variety of practical functions for human life. To a large extent, mountains govern precipitation and the weather; they are the source of our freshwater resources, they contain a large part of the world's biological diversity and genetic resources, they sustain extensive forests, and they provide food for thousands of people who live in mountain regions. Mountains attract millions of tourists throughout all continents of the world. Increased tourism can have a positive impact on mountainous areas, but it must be ensured that such tourism will develop on sustainable principles. So it was doubly apt that the year 2002 was also the International Year of Eco-tourism.

To mark the International Year of Mountains, Iceland organised a variety of events, including a special postage stamp issue and a competition in schools which involved hiking in the mountains, recognising them and writing about them. A nation-wide election was held for the title of 'Iceland's favourite mountain' which was won, with an overwhelming majority, by Mount Herðubreið (Broad-Shoulder). Herðubreið may not be widely known outside Iceland but, although Iceland's mountains are not the biggest in Europe, some of them have still managed to capture people's imagination elsewhere in Europe. In the old days it was commonly believed in northern Europe that the volcano Hekla was the entrance to Hell, from where you could occasionally hear the cries of the grilled souls of the dammed. The English swearword 'heck' is derived from the name of this supposedly hellish mountain – well, that is our story and we are sticking to it! Another Icelandic mountain, Snæfellsjökull, is the starting point in the French author Jules Verne's novel *Journey to the Centre of the Earth*. This classic cone-shaped volcano, which can be seen from Reykjavik on a clear day, is now protected as a part of Iceland's newest National Park which I had the privilege of opening in 2001.

Modern science has found no evidence that you can enter the Earth's centre, or indeed Hell itself, through Icelandic mountains, but mountains remain a central part of the Icelandic identity, and a dominant feature of Iceland's geography. Iceland is largely highland. Some 60% of the country is at an elevation of 300-400 metres, often referred to as the Central Highlands. To all intents and purposes these highlands are uninhabited, yet provide an important foundation for our everyday life. Nevertheless, there has always been a touch of mystique surrounding the Icelandic mountains; we have an old Icelandic expression, when someone is 'away with the fairies', that he or she has 'just come down from the mountains'.

Iceland's highlands and mountains play a major role in the sustainable development of Icelandic society. The highlands provide Iceland's water supply and play a key role in its hydrology. The Central Highlands also represent a resource for renewable energy from glacial rivers and geothermal fields. An important part of Iceland's biological diversity thrives there, with a variety of habitat types and many species which are not found anywhere else in the country. Because of their unusual natural beauty the highlands offer a range of outdoor leisure activities which is rare in Europe. Virtually untouched wilderness is found there, including one of Europe's largest wilderness areas.

The great potential for harnessing renewable energy in Iceland's Central Highlands has received increased attention from the energy sector recently, at the same time as public demand for protecting the Central Highlands has been growing. In 1999, work was launched on drawing up a framework plan for hydro and geothermal utilisation. The first phase of it is now at completion with the assessment and prioritisation of 25 harnessing options. This work aims to put power development and nature conservation in a sensible order of priority. The first nature conservation programme for Iceland as a whole was adopted by the Icelandic Parliament, the *Alþingi*, in 2004. This emphasises conservation of biological diversity, Icelandic natural forms, geological relics and the main habitat types; it also contains plans for establishing the largest National Park in Europe, the Vatnajökull National Park.

Iceland's tourist industry has developed rapidly, and over the past three decades the number of visitors from abroad has grown five-fold to more than 300,000 a year. Highland and mountain travel has grown enormously, both in winter and summer, and sustainable tourism has been gaining a firmer foothold. Growth in highland tourist traffic has been accompanied by more demand for developing accommodation facilities there. At the same time greater awareness and focus on conservation has seen major natural treasures placed under Protection Orders. Today, 22 areas of the highlands are protected, accounting for 16% of Iceland's total highland area.

When the Ministry for the Environment was established in 1990 two issues dominated the dialogue on protection and utilisation of the highlands: one was the lack of planning or co-ordinated administration for the highlands, and the other was the great uncertainty concerning ownership of the area. A systematic approach to resolve the question of the highlands was widely seen as a matter of national importance.

One of the first major tasks at the Ministry for the Environment was to establish a comprehensive plan for the Central Highlands, which involved achieving co-operation with the municipal authorities on harmonising and approving a regional plan for it. Planning work took 7 years and was completed in 1999. The Central Highlands plan addresses all aspects of planning, protection and utilisation of the region until the year 2015, guided by the principles of sustainability.

As a former Minister for Nordic Co-operation, I would also like to add a few words on future co-operation around the North Atlantic Region. The five Nordic countries, plus

Greenland, the Faroe Islands and the Åland Islands, have been engaged in close co-operation on environmental issues for decades in the forum of the Nordic Council and the Nordic Council of Ministers. Lately, the Nordic countries have been taking a growing interest in co-operation with their western neighbours, especially Scotland. In fact the countries in the West Nordic zone have shown interest for some time, as witnessed by three Nordic conferences which have included delegates from the Scottish mainland and islands. The first conference, which was held in Reykjavik in 1999, addressed in general terms the possibilities for co-operation, the others in the Faroe Islands in the year 2001 and in Shetland in the year 2003 focused on the protection of the North Atlantic.

For decades, Iceland has benefited from close contact with Scotland, both on nature conservation issues in the broadest sense and also in the fields of land reclamation and afforestation. Good co-operation has been created between Scottish Natural Heritage and Iceland's Environmental and Food Agency in the form of visits, meetings and talks about the monitoring and conservation of Protected Natural Areas. I trust that Scottish Natural Heritage has relished these contacts as much as we have in Iceland, and that they will continue to grow and prosper.

Mountains must play an increasing role in the future development of conditions for life on earth. People and mountains are in a much more intimate relationship than we tend to recognise; this means that mountain ecosystems are a factor we need to consider much more closely when we decide on action to ensure the sustainable development of our societies. It should be our aim to devise an array of constructive measures which will allow people and nature to live in harmony with the natural environment of our beloved mountains. Therefore, I pay tribute to Scottish Natural Heritage for their initiative in holding the conference in 2002 to put the spotlight on mountains and the related environmental challenges we all face.

Siv Friðleifsdóttir
Icelandic Minister for the
Environment (2002-2004)

CONTRIBUTORS

Hallgeir Aalbu, EuroFutures, Box 415, 10128 Stockholm, Sweden.

Andy Amphlett, RSPB, Abernethy Forest Reserve, Forest Lodge, Nethybridge, Inverness-shire, PH25 3EF.

G. Baillie, The Macaulay Institute, Craigiebuckler, Aberdeen, AB15 8QH.

Rick Battarbee, Environmental Change Research Centre, University College London, 26 Bedford Way, London, WC1H 0AP.

Neil Bayfield, CEH Banchory, Hill of Brathens, Glassel, Banchory, Aberdeenshire, AB31 4BY.

Dave Beaumont, RSPB, Dunedin House, 25 Ravelston Terrace, Edinburgh, EH4 3TP.

Karl Benediktsson, University of Iceland, Department of Geography and Geology, Askja, IS-101 Reykjavík, Iceland.

Richard Birnie, The Macaulay Institute, Craigiebuckler, Aberdeen, AB15 8QH.

Andrea Britton, The Macaulay Institute, Craigiebuckler, Aberdeen, AB15 8QH.

Rob Brooker, CEH Banchory, Hill of Brathens, Glassel, Banchory, Aberdeenshire, AB31 4BY.

Will Cadell, 8326 Toombs Drive, Prince George, BC, V2K 5A3, Canada.

T.V. Callaghan, Abisko Scientific Research Station, Abisko SE 981-07, Sweden.

L. Camarero, Centre d'Estudis Avançats de Blanes – CSIC, C/ Accés Cala St. Francesc 14, 17300 Blanes, Girona, Spain.

Steve Carver, University of Leeds, School of Geography, Leeds, LS2 9JT.

John Coll, Environmental Research Institute, North Highland College, Castle Street, Thurso, Caithness, KW14 7JD.

Alan P. Dougherty, Institute for Environment, Philosophy and Public Policy, Lancaster University, Furness College, Lancaster, LA1 4YG.

Igor J. Dvořák, Krkonoše National Park Administration, Dobroustého 3, 54301 Vrchlabí, Czech Republic.

Mary Edwards, School of Geography, University of Southampton, Highfield, Southampton, SO17 1BJ.

Stephen R. Edwards, IUCN, Rue Mauverney 28, 1196 Gland, Switzerland.

Noranne Ellis, Scottish Natural Heritage, 2 Anderson Place, Edinburgh, EH6 5NP.

Murray Ferguson, Cairngorms National Park Authority, 14 The Square, Grantown on Spey, PH26 3HG.

J. Fisher, The Macaulay Institute, Craigiebuckler, Aberdeen, AB15 8QH.

John A. Forster, John Forster Associates, Dalsack, Finzean, Aberdeenshire, AB31 6ND.

David Fowler, Centre for Ecology & Hydrology, Bush Estate, Penicuik, Midlothian, EH26 0QB.

Peter Fredman, European Tourism Research Institute, Mid-Sweden University, SE-831 25 Östersund, Sweden.

Colin A. Galbraith, Scottish Natural Heritage, 2 Anderson Place, Edinburgh, EH6 5NP.

Stuart Gibb, Environmental Research Institute, North Highland College, UHI Millennium Institute, Castle Street, Thurso, Caithness, KW14 7JD.

J.E.G. Good, Bod Hyfryd, Graiglwyd Road, Penmaenmawr, Conwy, LL34 6ER.

John E. Gordon, Scottish Natural Heritage, 2 Anderson Place, Edinburgh, EH6 5NP.

Georg Grabherr, University of Vienna, Althanstrasse 14, 1090 Vienna, Austria.

J. Grimalt, Department of Environmental Chemistry, CSIC, Jordi Girona 18, 08034-Barcelona, Spain.

Nicolas Gunslay, University of Lapland, Arctic Centre, PO Box 122, 96101 Rovanieme, Finland.

Eeva-Liisa Hallanaro, Päärynäpolku 1, 02710 Espoo, Finland.

S. John Harrison, Department of Geography, University of Dundee, Dundee, DD1 4HN.

O.W. Heal, Easter Hackwood, Dipton Mill Road, Hexham, Northumberland, NE46 1BP.

Thomas Heberlein, University of Wisconsin-Madison, Department of Rural Sociology, Madison-Wisconsin 53706, USA.

Peter Higgins, Outdoor and Environmental Education, School of Education, University of Edinburgh, St. Leonard's Land, Holyrood Road, Edinburgh, EH8 8AQ.

Thomas Hofer, Forestry Officer (Sustainable Mountain Development), Forestry Department, UN Food and Agriculture Organization, Viale delle Terme di Caracalla, 00100 Rome, Italy.

Paula Horne, The Macaulay Institute, Craigiebuckler, Aberdeen, AB15 8QH.

Stuart Housden, RSPB, Dunedin House, 25 Ravelston Terrace, Edinburgh, EH4 3TP.

T.T. Høye, Department of Population Biology, Biological Institute, University of Copenhagen, Universitetsparken 15, DK-2100 Copenhagen, Denmark.

A. Järvinen, Kilpisjärvi Biological Station, PO Box 17, 000 14 University of Helsinki, Finland.

M. Johansson, Abisko Scientific Research Station, S-981 07 Abisko, Sweden.

S.M. Johnson, Scottish Natural Heritage, 2 Anderson Place, Edinburgh, EH6 5NP.

Christer Jonasson, Abisko Scientific Research Station, S-981 07 Abisko, Sweden.

Melanie Josefsson, Swedish Environmental Protection Agency, PO Box 7050, S-750 07 Uppsala, Sweden.

Martin Kernan, Environmental Change Research Centre, University College London, 26 Bedford Way, London, WC1H 0AP.

Milena Kociánová, Krkonoše National Park Administration, Dobrouského 3, 54301 Vrchlabí, Czech Republic.

J. Kohler, Norwegian Polar Institute, 9296 Tromsø, Norway.

Allan Lilly, The Macaulay Institute, Craigiebuckler, Aberdeen, AB15 8QH.

John W. Mackay, University of St. Andrews, Research Centre for Environmental History, St. Katherine's Lodge, The Scores, St. Andrews, Fife, KY16 9AL.

B. Magnusson, Icelandic Institute of Natural History, Hlemmur 3, PO Box 5320, 125 Reykjavik, Iceland.

Magnus Magnusson, Blairskaith House, Balmore-Torrance, Glasgow, G64 4AX.

Keith Matthews, The Macaulay Institute, Cragiebuckler, Aberdeen, AB15 8QH.

Jeff Maxwell, 12 Kingswood Crescent, Kingswells, Aberdeen, AB15 8TE.

Eli Moen, Ministry of The Environment, PO Box 8013 Dep, N-0030 Oslo, Norway.

Steffan Morrocco, School of Geography & Geosciences, University of St. Andrews, St. Andrews, Fife, KY16 9AL.

L. Mortensen, Faroese Geological Survey, Brekkutún 1, 188 Hoyvík, Postbox 3169, 110 Tórshavn, Faroe Islands.

L. Nagy, McConnell Ecological Research, 41 Eildon Street, Edinburgh, EH3 5JX.

S. Neuvonen, Kevo Subarctic Research Institute, University of Turku, FI-20014, Turku, Finland.

E. Gunilla A. Olsson, Department of Biology, Norwegian University of Science and Technology, N-7491 Trondheim, Norway.

Simon Pepper, WWF Scotland, 8 The Square, Aberfeldy, Perthshire, PH15 2DD.

Monica Petri, Scuola Superiore di Studi Universitari e di Perfezionamento S. Anna, Pisa, Italy.

Robert Pettersson, ETOUR (European Tourism Research Institute), Mid-Sweden University, SE-83125 Östersund, Sweden.

Aylwin L. Pillai, University of Aberdeen, School of Law, Taylor Building, Old Aberdeen, Aberdeen, AB24 3UB.

B. Piña, IBMB-CSIC, Jordi Girona 18, 08034-Barcelona, Spain.

Martin Price, Centre for Mountain Studies, Perth College, UHI Millennium Institute, Crieff Road, Perth, PH1 2NX.

M. Rasch, Danish Polar Centre, Strandgade 100H, DK-1401 Copenhagen K, Denmark.

P. Robertson, Scottish Natural Heritage, 2 Anderson Place, Edinburgh, EH6 5NP.

Neil Rose, Environmental Change Research Centre, University College London, 26 Bedford Way, London WC1H 0AP.

N.R. Saelthun, Norwegian Institute for Water Research, PO Box 173, Kjelsaas, 0411 Oslo, Norway.

Audun Sandberg, Bodø University College, N-8049 Bodø, Norway.

I.G. Simmons, Department of Geography, University of Durham, Science Laboratories, South Road, Durham, DH1 3LE.

Des B.A. Thompson, Scottish Natural Heritage, 2 Anderson Place, Edinburgh, EH6 5NP.

Amanda Thomson, Biosystems Dynamics, Centre for Ecology & Hydrology, Bush Estate, Penicuik, Midlothian, EH26 0QB.

Linda Turner, CEH Banchory, Hill of Brathens, Glassel, Banchory, Aberdeenshire, AB31 4BY.

Michael B Usher, School of Biological & Environmental Sciences, University of Stirling, Stirling, FK9 4LA.

Tuomas Vuorio, ETOUR (European Tourism Research Institute), Mid-Sweden University, SE-83125 Östersund, Sweden and Lahti Polytechnic, 15140 Lahti, Finland.

Per Wallsten, Swedish Environmental Protection Agency, S-106 48 Stockholm, Sweden.

Guðríður Þorvarðardóttir, Nature Conservation Division, Environment and Food Agency of Iceland, Suðurlandsbrant 24, IS-108 Reykjavík, Iceland.

PART 1:
Introduction and Context

Saeter, Steinskvanndalen, Kvam, Norway (Photo: Martin Price).

The opening chapters provide three quite different views of mountain areas. Magnus Magnusson delights in reflecting on the cultural and social importance of mountains. As he puts it, "mountains enlarge the landscape of the mind". Magnusson draws on his experiences in Iceland and the Cairngorms to present a lively account of the delicate relationship between nature and people.

Price & Hofer (Chapter 2) provide a full account of the build-up to the International Year of Mountains, 2002 (IYM 2002). Here, we learn that it was not until shortly before the United Nations Conference on Environment and Development (UNCED), held in Rio de Janeiro in 1992, that a global action plan for mountains was devised. In the 'Agenda 21' action plan endorsed in Rio, Chapter 13 dealt with mountains; it was entitled 'Managing Fragile Ecosystems: Sustainable Mountain Development'. This led, after many inter-governmental meetings and non-governmental activities, to the United Nations General Assembly proclaiming, at its 53rd session in 1998, that 2002 would be the IYM. Price & Hofer detail the range of activities spawned by IYM 2002, and note that by the end of 2004 the 'Mountain Partnership' had in its membership 43 countries, 14 inter-governmental organisations and 55 major groups. The work of this partnership ranges from watershed management, sustainable agriculture and rural development to aspects of education, research and policy.

Stephen Edwards (Chapter 3) explores the different roles mountains play in our lives within what he refers to as the 'Ecosystem Approach'. He takes as his definition of the ecosystem that provided under the Convention on Biological Diversity: 'A dynamic complex of plant, animal and micro-organism communities and their non-living environment interacting as a functional unit'. In developing the ecosystem approach, Edwards draws on his experience in working within IUCN – The World Conservation Union. He looks at the relationships between the ecological, policy and social principles governing the management of mountain ecosystems. His conclusion is a theme that emerges throughout much of the book: namely, that interdisciplinary working needs to be promoted to improve the management of mountain ecosystems.

These three chapters set the scene for linking the conservation and management of mountain areas to the interests of nature and people.

1 Nature and people: conservation and management in the mountains of Northern Europe – with mountains in mind

Magnus Magnusson KBE

I feel greatly flattered, in my dotage, to be invited to indulge my growing penchant for anecdotage, and to inflict some thoughts about mountains and mice – or 'People and Nature', if we defer to the end title of this book.

Let me put my cards on the table at once: I am not a professional environmentalist. I am an arrant amateur – a lover of mountains rather than an expert on mountains. My interest in them is that of an awe-struck observer and admirer.

Mountains enlarge the landscape of the mind. Mountains (especially, I like to think, mountains in Iceland) give vivid form to the chaos of creation. Mountains revive the ailing soul with the power and purity of their beauty, while hugging to themselves some of the deepest secrets of the evolution of our planet. And in honour of my compatriot, Siv Friðleifsdóttir, the former Icelandic Minister for the Environment, I want to start by telling you about a mountain in Iceland (which didn't feature in the mountains popularity poll in Iceland she mentioned in her Foreword). She will not need me to remind her about the mountain in the north of Iceland which was given literary fame by our Nobel Prize-winner, Halldór Laxness, in his splendid novel, *Íslandsklukkan* – 'Iceland's Bell'. In that novel, the main character comments on the fact that mountains in Iceland can have different names, depending upon which side of them you live:

> *There is a mountain in the Kinn, in the north, which is called Bakrángi (Back-ridge) when it is looked at from the east, Ógaungufjall (Hard-to-walk Hill) if one looks at it from the west, but Galti (Goatfell) when seen from the sea in Skjálfandi Bay.*

Halldór Laxness was using this observation as a metaphor for 'truth' – it all depends on where you stand when you look at things like mountains. Mountain truth, you see, has many sides. And that is the text on which I want to base my homily.

When we regard this wild country of rock and scree we call mountains, everything depends on perspectives and perceptions. In Scotland, these perceptions have changed radically in little more than a couple of centuries. Back in the early 1700s, our mountain land was considered forbidding and unattractive – imagine that! To a jaundiced observer like David Burt, a Hanoverian officer in General Wade's entourage in the aftermath of the 1715 Rising, the magic of mountains was harshly absent:

> *The summits of the Highest Mountains are mostly destitute of Earth, and the huge naked Rocks, being just above the heath, produce the disagreeable Appearance of a scabbed Head.*

Magnusson, M. (2005). Nature and people: conservation and management in the mountains of Northern Europe – with mountains in mind. In *Mountains of Northern Europe: Conservation, Management, People and Nature,* ed. by D.B.A. Thompson, M.F. Price & C.A. Galbraith. TSO Scotland, Edinburgh. pp. 5-10.

In 1726, Daniel Defoe, in his *Tour through the Whole Island of Great Britain*, concluded that heath or heather was the common product of a barren land, 'a foil to the beauty of the rest of England'! Fifty years later, that quintessential Englishman, Dr Samuel Johnson, in his *Journey to the Western Islands of Scotland*, referred to the chaos of the hills as 'matter incapable of form or usefulness: dismissed by nature from her care ... quickened only with one sullen power of useless vegetation'.

Within a hundred years, however, this 'one sullen power of useless vegetation' had become a source of very useful pleasure to others, as the Victorians applied their single-minded entrepreneurial drive to Scotland's moorlands and began to manage them specifically for the raising and cropping of red grouse – and only red grouse. Earlier inhabitants had reduced the original woodland cover by slash and burn; the Victorians only burned – and through controlled burning they transformed chaos into quiltwork and moorland into mosaic.

Indeed, to this day the legacy of our uplands is coloured by these Victorian preoccupations. We have an enormously fragmented estate pattern, we have vast ranges of bare and deer-dotted moorlands which reinforce the idea of scenic grandeur which visitors associate with Scotland – whereas to the ecologist that grandeur hides a perilous biological impoverishment; hillsides once clothed in Caledonian wildwood are now scoured by erosion; straths and valleys once verdant are now bitten to the quick by alien beasts – not only sheep but also the rabbits introduced in Norman times and the red deer which have been allowed to multiply, to the considerable detriment of their habitat.

However, a more romantic regard for the wild places was meanwhile being born in the hearts and minds of poets such as Wordsworth. In Scotland itself, Walter Scott's writings heralded an important change – a link between wild land and patriotism which epitomised his historically romantic ardour in the *Lay of the Last Minstrel* (1805):

> *Oh Caledonia! Stern and wild,*
> *Meet nurse for a poetic child!*
> *Land of brown heath and shaggy wood,*
> *Land of the Mountain and the Flood,*
> *Land of my Sires! What Mortal Hand*
> *Can e'er untie the filial band*
> *That knits me to thy rugged strand!*

And so it goes on to this day. Any reader could, I am sure, quote me a hundred favourite snatches of poetry which encapsulate and illuminate the magnificent immensity of mountains. My own personal favourite is a poem by Norman MacCaig, in his 1973 collection *The White Bird*, entitled 'Moment Musical in Assynt':

> *A mountain is a sort of music: theme*
> *And counter theme displaced in air amongst*
> *Their own variations.*
> *Wagnerian Devil signed the Coigach score;*
> *And God was Mozart when he wrote Cul Mor.*

You climb a trio when you climb Cul Beag.
Stac Polly – there's a rondo in seven sharps,
Neat as a trivet.
And Quinag, rallentando in the haze,
Is one long tune extending phrase by phrase.

I listen with my eyes and see through that
Mellifluous din of shapes my masterpiece
Of masterpieces:
One sandstone chord that holds up time in space –
Sforzando Suilven reared on his ground bass.

Ornithology, as my good friend Des Thompson of Scottish Natural Heritage (SNH) would rather die than say, is for the birds. But orogeny is not just for literary aesthetes; it is just as much for the environmentalist, the ecologist, the recreationist, the mountaineer and the amateur, in the strictest sense of the term, of the great outdoors – in fact, for everyone.

We should note that as the Victorian enthusiasm for Scotland's mountains rose, so did the desire of the *hoi polloi*, the non-elite, to be allowed to share more fully in this immense natural resource: not just to write about it, but to use it, to enjoy it, to be inspired by it, to be fulfilled by it. In the middle of the 19th century, just after the formation of the Scottish Rights of Way Society, there was a celebrated dispute about access down an old drove road through Glen Tilt; the protagonists were the Duke of Atholl (one of Scotland's major sporting landlords) and an Edinburgh professor leading an excursion of science students. The upshot, after umpteen lawsuits, was that such rights of way became firmly established. Later that century, from 1884 onwards, there were repeated attempts to bring in a parliamentary Access to the Mountains Bill which would have opened the hills to everyone.

This coincided with a great increase in the recreational use of mountains by middle-class town-dwellers. The first number of the *Scottish Mountaineering Club Journal* in 1890 declared that:

... the love of scenery and of the hills is implanted in the heart of every Scot as part of his very birthright ... On the tops we seem to breathe something else than air ... and look down on every side upon a scene untainted by work of man, just as it came fresh from the Creator's hand.

It was not just the allegedly unemotional Scots who felt this quasi-mystical attitude to mountains. In the Alps, according to Horace Bénédict de Saussure in his *Voyages dans les Alpes*:

The soul is uplifted, the powers of intelligence seem to widen, and in the midst of this majestic silence one seems to hear the voice of Nature and to become the confidante of its most secret workings.

Simultaneously, the skiing fraternity was beginning to find its voice. Members of the Scottish Ski Club, founded in Edinburgh in 1907, saw mountains as an obstacle course to be conquered. As one early enthusiast wrote:

I glory in the victory over self and Nature ... The greatest of all joys of ski-ing is the sense of limitless speed, the unfettered rush through the air at breakneck speed ... Man is alone, gloriously alone against the inanimate universe ... He alone is Man, for whose enjoyment and use Nature exists.

Fighting talk, you might well think. Concomitant with this macho attitude, and corrective to it, was the growth of ecology as a scientific discipline, especially in Scotland. Incidentally, I am constantly intrigued by the close etymological kinship between the words 'ecology' and 'economics', both derived from the Greek word *oikos* meaning 'house' or 'home', and both embodying the concept of prudent housekeeping of our natural heritage.

The early ecologists concentrated on producing systematic descriptions of Britain's vegetation. Only much later did ecology begin to wrestle with the question of the impact of people on nature in Scotland; and, inexorably, attention began to be focussed more and more on the great mountains in the Cairngorms.

And that brings me to the nub of the problem addressed in this book. For me the Cairngorms, which all the winds of the world have made their assembly hall, epitomise the problem of how best to reconcile the conflicting demands of nature and people. Nature and people? Nature for people? People for Nature?

Christopher Smout, the Historiographer Royal in Scotland and a former Deputy Chairman of Scottish Natural Heritage, has splendidly chronicled the story of *The Highlands and the Roots of Green Consciousness, 1750-1990* in an SNH Occasional Paper (Smout, 1993). Let me fast-forward to the 1990s, where Chris Smout left off, using the 'Pause' button only to recall that it was as early as the 1920s that political and popular agitation began for the Cairngorms to be purchased as a National Park ('this Caledonian forest should be a National Sanctuary – a sort of serpentless Eden', opined the nationalist Erskine of Marr). Ramsay Macdonald, the then Prime Minister, approved the idea – although he had reservations about the development of winter sports which might leave 'the Cairngorms spotted with miscellaneous erections'. Prophetic words, indeed. He appointed the Addison Committee to investigate the possibility of British National Parks. As we all know, nothing came of that – in Scotland, at least.

So let's jump in at the start of the 1990s, the years which will always be categorised in my own memory as the first SNH decade. For eight of these years it was my privilege to be chairman of SNH. At the inception of SNH, Scottish Ministers were concerned that its natural heritage functions should be strongly connected to the social context within which SNH works; so there was a need to make strong links between those who use, those who depend upon, those who live among, and those who enjoy the natural heritage. We were all quite clear that this would be the best way forward for its protection. I know that this is no less true today, and I know also that it is an approach shared by the other nations represented at the Pitlochry conference in seeking to take care of the natural environment.

One of my early tasks, even before SNH came into being, was to lead a government Working Party consisting of men and women representing key interests in the Cairngorms. My role was to bring the differing and conflicting interests together – practical social science, rather than natural science, if you like. At that time, Ministers wanted to find solutions which built on existing structures – and this, in effect, excluded the possibility of a National Park. That was our brief – essentially, 'to test the voluntary principle to destruction'.

But it wasn't destroyed. Simply by working together patiently, debating the issues over an extended period, holding meetings all around the area, and listening to what others had to say, we emerged with an outcome based on a cooperative future. Indeed, our report was titled '*Common Sense and Sustainability*' (Cairngorms Working Party, 1992) to reflect the need to find sensible and practical ways forward, agreed by consensus (and therefore likely to be supported) and which also gave us a long-term vision.

Two members of the Working Party (John Hunt of the Royal Soociety for the Protection of Birds (RSPB) and the veteran mountaineer Eric Langmuir) had other views, and asked for their dissenting report to be published. They felt that a National Park was the only possible solution for conflict. Well, time may well have proved them right. We now have a Cairngorms National Park, established in September 2003.

Looking back over the years of partnership working which followed the lead given by the Cairngorms Working Party, I now feel that this period of continued cooperative working was an essential part of a long process of forging changed attitudes. We are still in that process. While I am certain that the Cairngorms National Park, Scotland's second, will be able to take stronger responsibility for delivering new initiatives, its work will still depend upon partnership working and on continuing to build a strong consensus with people and their needs, both locally and nationally.

Landscapes like those of the Cairngorms are harsh and wild in character, and can make for difficult living. Their supreme beauty will not have been foremost in the minds of those who in the past had sheer survival as their priority, but we know that these are places to which their people have had, and continue to have, a deep and close emotional attachment. Today, it is mainly those of us who live the softer life who find these places special for their landscapes, their nature and the challenges of open-air recreation, and for the sense of fulfilment and self-development these activities contribute – not just to the individual but to the national well-being.

So where are we now in the relationship between nature and people in these mountain areas? One can read the runes in different ways. People in mountain areas are now much less directly dependent on nature or the use of natural resources, and are much more dependent on the rest of society, both in the provision of modern services and in public subvention to maintain most of the primary land uses. Both agriculture and forestry rely heavily upon public sector support, while private subsidy can be important, too, on land owned and used for hunting. We have in a very real way crossed a threshold, at which the primary land uses (or the direct use of nature) are no longer the main basis for production profitable enough to maintain lifestyles at the standards expected in modern living. That is no criticism – it is a reality and, indeed, a very welcome reality, that standards of living have increased so much for those living in the remoter areas.

There has been in Scotland – and I know in other parts of the north-western seaboard of Europe – a huge commitment, indeed at times heroic efforts, to overcome many kinds of disadvantage. In Scotland much of this has been driven through public sector intervention, by development agencies at the national and local levels, and by much innovation and investment, to find ways of regenerating local economic enterprise, all designed to bring employment and income levels up to modern expectations, and to draw these societies into the mainstream of economic activity.

But it has not been easy to address the considerable disadvantages still attached to being on the periphery of the wider commercial marketplace. The efforts towards economic repositioning of remote communities have had some success but they have been hard-won, at times with as many retreats as advances, simply because chasing these goals involves catching up with an economic world which does not stand still. And it has at times involved engaging with some of the most globally competitive parts of industry. Thus major developments to establish paper-pulp making and new aluminium smelting failed here in Scotland in the 1970s when exposed to the harsh global economics of those trades. Likewise, our efforts to create a domestic softwood industry in the uplands have struggled against adverse global pricing. Tourism, seen by many as an economic lifeline, is also in a highly competitive global market, and other successes in creating local employment, such as fish-farming, have created serious environmental problems as a by-product. Currently, renewable energy development is hotly debated.

But I am an optimist by nature. I fully recognise the immense difficulties involved in that hoped-for reconciliation of Nature and People which is the *raison d'être* not just of this book but also of the International Year of Mountains in 2002.

There are, as we well know, no ready solutions to this great Nature and People debate – we would not need to deliberate these matters at length if there were. I go back to my challenging experience with the Cairngorms Working Party, which taught me above all that there is a need to continue to deliberate these issues. In doing so we should aim to make sure that:

- we are sharing aspirations about longer-term goals. If there is lack of awareness of what different people seek, we can hardly be surprised if misunderstanding arises;
- we should be clear about the facts, because rational debate and the creation of an informed consensus need clear and balanced information; and
- we need to explore what values are important to different people. Sharing values is not an easily-attained goal, but if we can achieve it, the way forward becomes clearer. Sharing values is not so much a matter of conversion on some hilly Damascene road as a widening of our perspectives to acknowledge the validity of the values held by others.

If I am allowed to sneak in a last quotation (I only quote others, as Montaigne once observed, the better to express myself), allow me to remind you of this apothegm from the *Notebooks* of William Blake:

> *Great things are done when men and mountains meet;*
> *This is not done by jostling in the street.*

Keep thinking; keep talking; and, above all, keep hoping: that is the only advice I have to offer.

References

Cairngorms Working Party (1992). *Common Sense and Sustainability: A Partnership for the Cairngorms.* The Scottish Office, Edinburgh.

Smout, C. (1993). The Highlands and the Roots of Green Consciousness, 1750-1990. SNH Occasional Paper No. 1. Scottish Natural Heritage, Perth.

2 The International Year of Mountains, 2002: progress and prospects

Martin F. Price & Thomas Hofer

Summary

1. In 1992, the inclusion of a chapter on sustainable mountain development (SMD) in 'Agenda 21' took mountain issues into the global debate on environment and development. During the 1990s, regional inter-governmental consultations on SMD were held around the world. In 1996, the Government of Kyrgyzstan proposed an International Year of Mountains (IYM); this proposal was endorsed by the UN General Assembly in 1998.

2. The IYM 2002, aimed to "promote the conservation and sustainable development of mountain regions, thereby ensuring the well-being of mountain and lowland communities." Progress should be measured in relation to this objective, recognising the need for long-term, sustainable outcomes. The remarkable number of meetings, at local to global levels, on diverse themes relating to mountains are important because they brought together many people who would otherwise never have met, leading to increased understanding. Diverse media have featured mountain issues, raising the awareness of a massive number of people.

3. Seventy-eight countries established national IYM committees, often with representation from government, non-governmental organisation, and private sectors. These Committees have been cross-sectoral, recognising the need to address the challenges of mountain areas holistically. Many of these structures may disappear, yet all have provided opportunities for dialogue; some will continue, leading to national strategies for sustainable mountain development (SMD), new laws and policies supporting SMD, and to concrete action on the ground. Many scientific programmes have been initiated or accelerated by the IYM. At the international level, the success of many of these initiatives should be assured through the International Partnership for Sustainable Development in Mountain Regions, launched at the World Summit on Sustainable Development in Johannesburg in 2002.

4. The momentum, enthusiasm, and attention created by the IYM present unique opportunities to launch and realise initiatives in favour of mountain regions, their inhabitants, and their resources, with positive impacts reaching far into the future.

Price, M.F. & Hofer, T. (2005). The International Year of Mountains, 2002: progress and prospects. In *Mountains of Northern Europe: Conservation, Management, People and Nature,* ed. by D.B.A. Thompson, M.F. Price & C.A. Galbraith. TSO Scotland, Edinburgh. pp. 11-22.

2.1 Introduction

Mountains occupy 24% of the global land surface (Kapos *et al.*, 2000) and host 12% of the global population (Huddleston *et al.*, 2003). A further 14% of the global population lives adjacent to mountain areas (Meybeck *et al.*, 2001); mountain areas include not only remote, poor and disadvantaged people and communities, but also wealthy tourist resorts and urban centres within and close to the mountains – including mega-cities such as Mexico City and Jakarta. As sources of water, energy, agricultural and forest products, and as centres of biological and cultural diversity, traditional knowledge, religion, recreation, and tourism, mountains are important for at least half of humanity (Messerli & Ives, 1997).

2.2 From the Rio Earth Summit to the International Year of Mountains

Over the past decade, recognition of the diverse values of the world's mountains has been reflected in a global shift of perception and action. Mountains have moved from a topic of interest amongst a relatively small number of scientists, development experts, and decision-makers – as well as mountaineers – to having an important role on the global stage of environment and development debate. The United Nations Conference on Environment and Development (UNCED), held in Rio de Janeiro in 1992, presented a unique opportunity to move mountains into this role, through the inclusion of a specific chapter on mountains in 'Agenda 21', the plan for action endorsed in Rio by the Heads of State or Government of most of the world's nations (Price, 1998; Stone, 2002). Chapter 13 of 'Agenda 21' is entitled 'Managing fragile ecosystems: sustainable mountain development'. Its inclusion in 'Agenda 21' placed mountains on a comparable footing with other major issues such as climate change, tropical deforestation, and desertification.

At the global level, formal implementation of Chapter 13 began in 1993, when the UN Inter-agency Committee on Sustainable Development appointed the Food and Agriculture Organization of the United Nations (FAO) as Task Manager for Chapter 13. In this role, FAO has convened an *ad hoc* Inter-Agency Group on Mountains (IAGM) which, in spite of its name, involves more than UN agencies. From the beginning, FAO recognised that diverse actors are involved in processes relating to the sustainable development of mountain areas. Consequently, FAO invited a number of non-govermental organisations (NGOs) to join the group, and they have participated in all seven meetings to date. Among the recommendations made by the IAGM at its first meeting was that national governments and NGOs should become directly involved in the implementation of Chapter 13. A key means to this end was a series of regional inter-governmental consultations, bringing together governments within the African, Asia-Pacific, European, and Latin America/Caribbean regions in 1994 to 1996. In total, representatives of 62 countries and the European Union attended these meetings (Price, 1999).

Parallel to this inter-governmental process, a non-governmental process took place. Its importance was underlined by the IAGM, recognising that the process that had led to Chapter 13 – in contrast to many other chapters of Agenda 21 – had been driven by a relatively small number of academics and development experts, mainly from industrialised countries. In 1995, a global NGO consultation held in Lima, Peru, brought together 110 participants from 40 countries. This meeting led to the establishment of the Mountain Forum, "a global network for information exchange, mutual support, and advocacy for equitable and ecologically sustainable mountain development and conservation." The Mountain Forum has subsequently been organised through both global and regional structures, and had over 4,600 individual and 400 organisational members in more than 130 countries at the beginning of 2005. Key means of information sharing include: 15 discussion lists, electronic conferences, and an interactive website (www.mtnforum.org) with membership services, calendar of events, on-line library, and links to other networks (Taylor, 2000). The Mountain Forum is also a key member of the Mountain Partnership (see below).

In the five years following Rio, other related activities took place in various nations around the world. A number of countries established national-level or sub-national institutions concerned with the sustainable development of their mountain areas. Others, particularly in Europe, developed laws and policies effectively to this end (Price, 1999; Villeneuve et al., 2002). It was in this context of a gathering international momentum of support for mountain areas that the participants in the international conference 'Mountain Research – Challenges for the 21st Century', held in Bishkek, Kyrgyzstan in 1996, proposed that sustainable mountain development should be the theme of a forthcoming international year. This idea was proposed to the UN Economic and Social Council (ECOSOC) by the Kyrgyz Ambassador to the UN in 1997, resulting in a resolution, co-sponsored by 44 member countries, requesting the Secretary-General to undertake an exploratory process. At its following session, ECOSOC adopted a resolution, co-sponsored by 105 member countries, recommending to the General Assembly that 2002 should be declared the International Year of Mountains (IYM). The outcome was that the UN General Assembly proclaimed, at its 53rd session in 1998, in a resolution sponsored by 130 countries, that 2002 would be the IYM.

2.3 The International Year of Mountains: objectives and activities

The mission statement of the IYM, developed by FAO in its role as Lead Agency for the Year, was to "promote the conservation and sustainable development of mountain regions, thereby ensuring the well-being of mountain and lowland communities". As stated in the concept paper for the IYM, it "should provide an opportunity to initiate processes that eventually advance the development of mountain communities, and act as a 'springboard' or catalyst for long-term, sustained, and concrete action" (FAO, 2000). Progress should be measured in relation to these objectives, and in particular to the degree of success in consolidating and building on existing efforts to improve the quality of life in mountain regions and protecting mountain environments for generations to come. As much as possible, such processes should take place 'on the ground', with full involvement of mountain people.

The IYM represented a unique opportunity to raise awareness, across society as a whole, of the manifold values of mountain regions and the fragility of their resources, building on the IYM motto 'We are all mountain people'. Around the world, diverse media – postage stamps, newspapers, magazines, radio, television, internet - featured mountain issues. Many books on mountain issues were published. A major source of information was the IYM website (www.mountains2002.org) developed by FAO as part of its global communications programme. All these means raised the awareness of an uncountable number of people with regard to the diverse values of mountains at all scales: an investment in their future, as the

IYM must not be regarded as a 'one-off', but as a unique year in the vital process of fostering sustainable mountain development, which "concerns both mountain regions and populations living downstream or otherwise dependent on these regions in various ways" (Price & Kim, 1999).

2.3.1 National committees

During the planning of the IYM, it was recognised that one measure of success would be the extent to which it contributed to establishing effective programmes, projects and policies. While this requires participation at all levels from villages to international organisations, the greatest efforts need to come from those working at the national level to achieve sustainable mountain development. This includes governments, businesses, environmental groups, social and cultural organisations, and research institutions. National organisations are well placed to understand local needs and priorities and support community development initiatives. They can also act as a bridge between international agencies and local community development groups who are working on the ground to bring change. Consequently, a major emphasis of FAO, in its capacity as Lead Agency for IYM, was to support the development and activities of national committees for the IYM.

By the end of 2002, 78 countries had established such national committees or similar mechanisms. There were countries, including Norway, where governments designated a national focal point; in others, different mechanisms were created to ensure the coordination of IYM activities. This was the case in Scotland, where a group of environmental and recreational NGOs developed a proposal for a Project Officer, whose position was largely funded by Scottish Natural Heritage, a government agency, with additional support from the NGOs themselves. To support these various structures, FAO developed a number of resources which were made available on the IYM website and on CD.

While most national IYM committees were led by a government agency, many included representatives of mountain people, grass-roots organisations, NGOs, civil society institutions, the private sector, research institutions, UN agencies, national government agencies, and decentralised authorities. In some countries, the national IYM committee has been the first national mechanism for the sustainable development of mountains and the first opportunity to implement a holistic approach to mountains.

National-level activities made a significant contribution to the success of the IYM and ensured that a growing network of governments, organizations, major groups and individuals around the world know that mountains are vital to life. Activities generating awareness of the importance of mountains and the need to protect them included school competitions, nationwide TV and radio programmes, and diverse meetings and other events (see section 2.3.2). Countries issued commemorative stamps, phone cards and books. Thirty national committees established IYM websites to widen dialogue, exchange information and promote country-level action.

Many of these national committees are currently in the process of institutional transformation into more permanent structures to work on sustainable mountain development in a long-term capacity. A number of countries have embarked on developing and implementing sustainable mountain development strategies, policies and laws designed to respond to the specific needs, priorities and conditions of their respective mountain areas. FAO is providing seed money and technical support to many of these processes.

2.3.2 National meetings and other events

As with any International Year, the IYM was marked by a large number of meetings and other events, on almost every theme relating to mountains that one could imagine. The calendar of events on the IYM website documented the diversity: there were meetings devoted to mountain women, children, water, mining, war, forests, biodiversity, arts ... All of them were important because they have brought together many people who would otherwise never have met, leading to increased understanding both of issues and of others' viewpoints. There were also many other events, including lectures, film festivals, concerts, exhibits, book launchings, mountain climbs and hikes, and projects to repair areas damaged by recreational over-use. Figure 2.1 shows the geographical distribution of meetings included in the calendar of events – and there were many others.

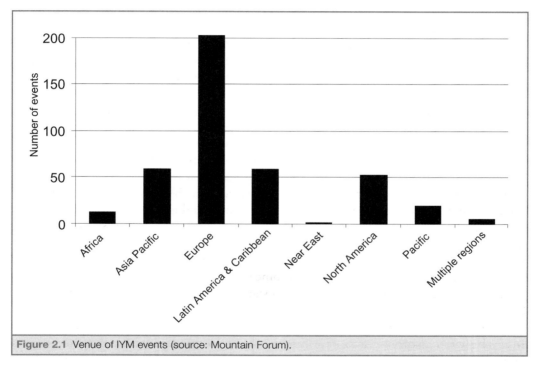

Figure 2.1 Venue of IYM events (source: Mountain Forum).

Taking just a few examples from different parts of the world, events and initiatives included the following.

- In Austria, a chain of fires was lit from the eastern frontier of the Alps to the Swiss border at the summer solstice.
- Bhutan established *in situ* orchid and rhododendron gardens as educational, recreational and conservation centres.
- Cuba developed a programme of fact-finding expeditions to the mountain sources of 295 rivers, not only to create public awareness of the source of freshwater but also to evaluate its conservation status, meet the needs of local populations, and develop recommendations for the care and protection of river sources.

- Ghana organized a durbar of chiefs and mountain people on a specially designated National Mountains Day, which drew attention to improper farming practices and the indiscriminate exploitation of bushmeat.
- In Italy, 'The Olympic Games of Mountain Cheeses' in November 2002 was the first competition-exhibition on a world scale exclusively reserved for mountain cheeses.
- In Lesotho, a national symposium on mountain ecosystems attracted rural community representatives who began initiatives in biodiversity conservation and ecotourism.
- Korea conducted a clean-up campaign in many mountainous areas with the participation of about 60,000 forest officials, members of the Korean Alpine Club, and local communities.
- In Mexico, the National Committee for the IYM was engaged in various poverty reduction activities to provide food and training and help in long-term planning.
- In Peru, the National Working Group on Mountain Ecosystems (GNTEM) was created by the government as a co-ordination mechanism between the state and civil society. GNTEM is now being decentralized and regional sub-groupings representing different mountain zones of Peru are appearing.
- In the Philippines, the Negros Committee for the IYM was a key participant in the on-going advocacy campaign to prosecute illegal loggers in forests of the region.
- The national committee in Ukraine supported a WWF-funded project to protect endangered brown bears in the Carpathians.

All of these events and initiatives involved large numbers of people and raised awareness in various ways. However, to provide a long-term basis for the sustainable development of mountain regions, appropriate legislation and policies are required. New mountain laws were passed in Kyrgyzstan and drafted in Morocco and Romania; and in Korea, the Korea Forest Service, which took the lead for the IYM, prepared a Forest Management Law which was passed on 30 December 2002. National mountain strategies and plans were developed in Madagascar, Spain and Turkey. It is anticipated that IYM national committees, or their successors, will be at the forefront of similar initiatives in other countries.

2.3.3 Regional initiatives

Many mountain ranges are shared by numerous countries, and many national frontiers run along mountain ridges. Equally, all of the world's largest rivers rise in mountain areas, and most of their basins are shared by more than one country (Liniger & Weingartner, 1998). Hence, regional co-operation on mountain issues is vital, complementing national initiatives. The first regional instrument addressing an entire mountain range was the Alpine Convention, signed in 1991; 11 years later, during the IYM, the vital decision on the location of its Secretariat was made. Learning from the experience of the Alpine Convention, the IYM saw an acceleration of the process towards a Carpathian Convention (Angelini *et al.*, 2002), leading to its signature in May 2003.

While such legal initiatives have not been developed outside Europe, mountains received regional-scale attention in other parts of the world. In Africa, the ninth session of the African Ministerial Conference on the Environment in Uganda in July 2002 produced the Kampala Declaration on the Environment for Development. This complemented the various joint projects and programmes underway to counter land degradation in African highlands and conserve them for water resources. In Central Asia, Azerbaijan, Kazakhstan, Kyrgyzstan and Turkey initiated discussing a Regional Watershed Management Training project to build capacity to manage watersheds effectively. These new initiatives complemented many existing ones, which are particularly well-developed in the Andes and Himalaya-Hindu Kush (Price, 1999).

2.3.4 Global meetings and processes

In addition to the many local, national, and regional meetings, there were eight major global meetings associated specifically with the IYM (Table 2.1). Four of these, in India, Bhutan, Peru and Ecuador, specifically addressed the needs and interests of mountain people: respectively, children, women, indigenous people, and mountain populations. Two, both in Switzerland, addressed various aspects of development, particularly with regard to communities and agriculture; the latter linking Chapter 13 of 'Agenda 21' with Chapter 14 on sustainable agriculture and rural development. The 'High Summit' was a truly global event, with simultaneous events on four continents bringing together mountain people, scientists and representatives of NGOs, UN agencies and the media through internet and videoconference technology.

All of these meetings produced final documents (all of which can be found on the Mountain Forum website) which fed in to the final global event of the IYM, the Bishkek Global Mountain Summit, held in Kyrgyzstan, which produced the Bishkek Mountain Platform (BMP) which also draws on a series of background papers prepared by global experts and further revised in the light of an electronic consultation through the Mountain Forum (Price *et al.*, 2004). The BMP formulates recommendations for concrete action towards sustainable mountain development, providing guidance to governments and others on how to improve the livelihoods of mountain people, protect mountain ecosystems, and use mountain resources more wisely. The BMP was circulated at the 57th session of the UN General Assembly later in 2002, leading to the adoption of a resolution which, *inter alia*, designated 11 December as International Mountain Day, and encourages the international community to organize on this day events at all levels to highlight the importance of sustainable mountain development.

Not only were there many meetings specifically considering mountains during 2002; they were also an important focus at other meetings. During the World Food Summit: Five Years Later, held at the FAO in Rome in June, attention was drawn to the almost 392 million people living in the Andes, the Hindu Kush Himalaya, and the mountains of East Asia and Oceania who are chronically vulnerable to food shortages and malnutrition. At this meeting, the IYM Focus Group, which brings together Permanent Representatives of concerned governments at UN Headquarters in New York, underlined their support for an International Partnership for Sustainable Development in Mountain Regions. The outline of this partnership had been developed by the Swiss Government, FAO, and UNEP during the fourth Preparatory Meeting for the World Summit on Sustainable Development (WSSD) in Bali, earlier in June. The Partnership was launched on 2 September at the WSSD in Johannesburg (see below); as at UNCED ten years before, the meeting's final document specifically refers to mountains – this time, in paragraph 42 of the Plan of Implementation.

Table 2.1 Global meetings associated with IYM 2002.

Title	Dates, Location	Participants	Organisers	Outcome
World Mountain Symposium 2001: Community Development between Subsidy, Subsidiarity and Sustainability	30 September–4 October 2001, Interlaken, Switzerland	150 participants from 56 countries	Swiss Agency for Development and Cooperation, Centre for Development and Environment, University of Bern	Proceedings, CD
High Summit 2002: International Conference around the Continents' Highest Mountains	6-10 May 2002, Mendoza, Argentina; Nairobi, Kenya; Kathmandu, Nepal; Milan and Trento, Italy	Mountain people, scientists, representatives of NGOs, UN agencies and the media	Italian National Committee for the IYM	Recommendations for action on five cornerstones of mountain development: water, culture, economy, risk and policy
International Conference of Mountain Children	15-23 May 2002, Uttaranchal, India	Children from 13 to 18 years of age from over 50 countries	Research Advocacy and Communication in Himalayan Areas	Recommendations for the Bishkek Mountain Platform, Internet-based Mountain Children's Forum
2nd International Meeting of Mountain Ecosystems, "Peru, country of mountains towards 2020: water, life and production"	12-14 June 2002, Huaraz, Peru	300 participants from 16 countries, especially indigenous people from Peru, Ecuador, and the Himalayas	National Committee of Peru for the IYM	Huaraz Declaration
International Conference on Sustainable Agriculture and Rural Development in Mountain Regions	16-20 June 2002, Adelboden, Switzerland	200 people from 50 countries	Swiss Federal Office for Agriculture	Adelboden Declaration
Second World Meeting of Mountain Populations	17–22 September 2002, Quito, Ecuador	Representatives of 115 countries	World Mountain Peoples Association, El Centro de Investigación de los Movimientos Sociales del Ecuador	Quito Declaration: Draft Charter for World Mountain People
Celebrating Mountain Women	1-4 October 2002, Thimphu, Bhutan	250 participants from 35 countries: civil society, NGOs, media, academia, development agencies, donors	International Centre for Integrated Mountain Development and Mountain Forum	Thimphu Declaration
Bishkek Global Mountain Summit	28 October-1 November 2002, Bishkek, Kyrgyzstan	Over 600 people from 60 countries	Government of Kyrgyzstan, with assistance from UN Environment Programme	Bishkek Mountain Platform

In addition, the IYM influenced two major global processes: the Millennium Ecosystem Assessment and the Convention on Biological Diversity. The Millennium Ecosystem Assessment (MA) is a global exercise with the objective to provide decision-makers, civil society and the private sector with the latest scientific knowledge about the relationships between ecosystem change and human well-being (Millennium Ecosystem Assessment, 2003). The MA will present options for addressing those changes and will specify opportunities and consequences. Beyond tackling today's need for reliable information, the MA is building a framework for comparing the state of ecosystems through time and across scales. Within the block of 'conditions and trends' of the MA, Chapter 24 is specifically devoted to mountain ecosystems. A strong team of mountain scholars has been working hard to finalise the contents for this chapter. The MA will present its results in 2005.

Within the Convention on Biological Diversity (CBD), a specific work programme on mountain biological diversity has been elaborated (Anon., 2004). Its overall purpose is the significant reduction of mountain biological diversity loss by 2010 at global, regional and national levels; its implementation aims to make a significant contribution to poverty alleviation in mountain ecosystems, and in the lowlands dependent on the goods and services of mountain ecosystems. This work programme was largely developed and negotiated in 2003, during two sessions of the Subsidiary Body on Scientific, Technical and Technological Advice of the CBD, as well as a session of an *ad hoc* technical expert group. The work programme was discussed and then endorsed by the 7th Conference of the Parties (COP7) of the CBD in February 2004 in Kuala Lumpur, and countries are now asked to implement the work programme at the national level. It is a very comprehensive document which is structured into three broad, and complementary, programme elements with regard to the conservation, sustainable use, and sharing of the benefits of mountain biodiversity: direct actions, means of implementation, and supporting actions.

2.4 The Mountain Partnership: looking ahead

The Mountain Partnership, originally called the International Partnership for Sustainable Development in Mountain Regions, was launched at the WSSD in Johannesburg in 2002. In its vision statement, which is very similar to the second of the two goals of Chapter 13 of 'Agenda 21', the Partnership envisages the improved well-being, livelihoods and opportunities of mountain people and the protection and stewardship of mountain environments around the world. The Partnership is a voluntary alliance of interested parties committed to working together, with the common goal of achieving sustainable mountain development around the world. The Partnership addresses the challenges of mountain regions by tapping the wealth and diversity of resources, knowledge, information and expertise, from and through its members, in order to stimulate concrete initiatives at all levels that will ensure improved quality of life and environments in the world's mountain regions.

The secretariat of the Mountain Partnership is hosted by FAO and is jointly funded by the governments of Switzerland and Italy. UNEP and the Mountain Forum are key partners in implementing the secretariat functions. During 2003 and 2004 the organization, governance and membership of the Partnership were defined. Two global meetings of members, one in Merano (Italy) in 2003 and one in Cusco (Peru) in 2004, were milestone events in this collaborative process. By the end of 2004, 43 countries, 14 intergovernmental organisations and 55 major groups had joined the Mountain Partnership.

The dynamic core of the Mountain Partnership is action on the ground through specific thematic or regional initiatives. To date, members have identified and are actively engaged in developing seven thematic initiatives (Education, Gender, Policy and Law, Research,

Sustainable Agriculture and Rural Development in Mountain Regions or SARD-M, Sustainable Livelihoods, Watershed Management) as well as six regional initiatives (Andes, Central America and the Caribbean, Central Asia, East Africa, Europe, Hindu Kush Himalaya). As the Mountain Partnership evolves and its members exchange further information, experiences and best practices, other initiatives will develop.

Many of these Partnership Initiatives build on events, processes and concrete activities that took place or were initiated within the framework of the IYM: for instance, the SARD-M Initiative builds on the global conference held in Adelboden, Switzerland, from 16-20 June 2002; and the Gender Initiative on the Global Meeting 'Celebrating Mountain Women', held in Thimpu, Bhutan, from 1-4 October 2002. The Partnership Initiative on Watershed Management emerges from an extended watershed management review process which was carried out between 2002 and 2003 by FAO in collaboration with various partners worldwide. The Partnership Initiative on Mountain Research is of particular importance: it recognizes that projects, policies and laws and other activities which support sustainable development in mountain areas have to be based on sound information and knowledge (Royal Swedish Academy of Sciences, 2002), and that this is typically achieved by cooperation amongst partners. In addition, this Partnership Initiative builds on a number of scientific programmes which were initiated or intensified by the IYM. Of particular interest in this regard is the Mountain Research Initiative, which brings together three global research programmes (Becker & Bugmann, 2001; www.mri.scnatweb.ch). Similarly, the Global Mountain Biodiversity Assessment (www.unibas.ch/gmba/) published its first major report during the IYM (Körner & Spehn, 2002). Another important activity was the preparation and publication of 'Mountain Watch', the first systematic global assessment of mountain ecosystems, which was launched by the UNEP World Conservation Monitoring Centre before the Bishkek Global Mountain Summit (Blyth *et al.*, 2002).

While such human and institutional investments in mountain areas are vital, so are financial investments based on informed decisions, recognising not only the relative disadvantages of these areas, but also that mountain people, and the ecosystems they manage, provide vital services for a significant proportion of humankind. Such investments have been increasing in recent years. For instance, by 2002, the Global Environment Facility (GEF) had committed over $620 million and leveraged additional funding of about $1.4 billion in support of 107 mountain-related projects in 64 countries (Walsh, 2002). In addition, the World Bank had contributed $1.3 billion to projects in mountain areas which focus on integrated ecosystems management, partnerships, and innovative funding mechanisms such as payment for environmental services, debt for nature swaps, and environmental trust funds (MacKinnon, 2002). The World Bank, UNEP, and UNDP are the implementing agencies of the GEF; all of these, as well as other UN agencies and the Asian Development Bank, are members of the Mountain Partnership, as is PlaNet Finance, which supports the development of sustainable microfinance institutions worldwide.

2.5 Conclusion

In retrospect, the Mountain Partnership is possibly the most visible global outcome of the International Year of Mountains in 2002, a year which created awareness and further commitment to mountain regions, providing unique opportunities to launch and realise joint initiatives for mountain regions, their communities, and their resources. Cooperation is one of the distinguishing characteristics of mountain societies; it has long been recognised that sharing and pooling resources and working together is essential for long-term survival in these environments. However, the effectiveness of such co-operative structures is often diminished when mountain areas are integrated into regional and global economies where profit-orientated interests can become prevalent. Therefore, in these special regions, the uncertainties of the modern world – of which the most profound worldwide manifestations are the globalization of economies and climate change – are magnified, with increasingly unpredictable effects for both mountain communities and those living downstream. A key indicator of the long-term success of the International Year of Mountains will be the number of effective structures developed to avoid conflicts and to increase cooperation both between mountain people and between them and other stakeholders concerned with the long-term security of mountain environments and the billions who depend on them.

References

Angelini, P., Egerer, H. & Tommasini, D. (Ed) (2002). *Sharing the Experience: Mountain Sustainable Development in the Carpathians and the Alps.* Accademia Europea Bolzano, Bolzano.

Anonymous (2004). *Programme of Work on Mountain Biodiversity.* Secretariat of the Convention on Biological Diversity, Montreal.

Becker, A. & Bugmann, H. (2001). *Global Change and Mountain Regions: The Mountain Research Initiative.* IGBP Secretariat, Stockholm.

Blyth, S., Groombridge, B., Lysenko, I., Miles, L. & Newton, A. (2002). *Mountain Watch: Environmental Change and Sustainable Development in Mountains.* UNEP World Conservation Monitoring Centre, Cambridge.

Food and Agriculture Organization of the United Nations (FAO) (2000). *International Year of Mountains: Concept paper.* FAO, Rome.

Huddleston, B., Ataman, E., da Salvo, P., Zanetti, M., Bloise, M., Bel, J., Franceschini, G. & Fè d'Ostiani, L. (2003). *Towards a GIS-Based Analysis of Mountain Environments and Populations.* FAO, Rome.

Kapos, V., Rhind, J., Edwards, M., Price, M.F. & Ravilious, C. (2000). Developing a map of the world's mountain forests. In *Forests in Sustainable Mountain Development: A State-of-Knowledge Report for 2000,* ed. by M.F. Price & N. Butt. CAB International, Wallingford. pp. 4-9.

Körner, C. & Spehn, E.M. (Ed) (2002). *Mountain Biodiversity: A Global Assessment. Parthenon,* New York.

Liniger, H. & Weingartner, R. (1998). Mountains and freshwater supply. *Unasylva,* **49**, 39-46.

MacKinnon, K. (2002). *Conservation of Biodiversity in Mountain Ecosystems – At a Glance.* World Bank, Washington DC.

Messerli, B. & Ives, J.D. (Ed) (1997). *Mountains of the World: A Global Priority.* Parthenon, Carnforth.

Meybeck, M., Green, P. & Vorosmarty, C. (2001). A new typology for mountains and other relief classes: an application to global continental water resources and population distribution. *Mountain Research and Development,* **21**, 34-45.

Millennium Ecosystem Assessment (2003). *Ecosystems and Human Well-being: A Framework for Assessment.* Island Press, Washington DC.

Price, M.F. (1998). Mountains: globally important ecosystems. *Unasylva,* **195**, 49, 3-12.

Price, M.F. (1999). *Chapter 13 in Action 1992-97 - A Task Manager's Report.* FAO, Rome.

Price, M.F., Jansky, L. & Iatsenia, A.A. (Eds) (2004). *Key Issues for Mountain Areas.* United Nations University Press, Tokyo.

Price, M.F. & Kim, E.G. (1999). Priorities for sustainable mountain development in Europe. *International Journal of Sustainable Development and World Ecology,* **6**, 203-219.

Royal Swedish Academy of Sciences (2002). *The Abisko Document: Research for Mountain Area Development. Ambio Special Report Number 11.* Royal Swedish Academy of Sciences, Stockholm.

Stone, P.B. (2002). The fight for mountain environments. *Alpine Journal,* **107**, 117-131.

Taylor, D.A. (2000). Mountains on the move. *Américas,* **52**, 36-43.

Villeneuve, A., Castelein, A. & Mekouar, M.A. (2002). *Mountains and the Law: Emerging Trends.* FAO, Rome.

Walsh, S. (Ed.) (2002). *High Priorities: GEF's Contribution to Preserving and Sustaining Mountain Ecosystems.* Global Environmental Facility, Washington DC.

3 Managing mountain ecosystems to meet human needs: a case for applying the ecosystem approach

Stephen R. Edwards

Summary

1. Mountains are essential for human survival. Whether we live in or near mountains, we are dependent on them in some way to meet our daily needs – even if it is not immediately obvious. By the end of the last century virtually every inch of the planet was chronicled, explored, and photographed. We went from walking the land to walking in space. To fuel this evolution of technological capacity we have steadily eroded the natural resources on which we depend – often at our peril. Another consequence as our numbers have grown is that we have occupied ever more marginal landscapes – sometimes with cleverness, but sometimes without a hint of intelligence.

2. This chapter explores the different roles mountains play in our lives; how they contribute to our subsistence and pleasure; our security and survival. Mountains are examined in the context of broader landscapes and in that context how they affect other elements in these landscapes. The services provided by mountains on which we depend are highlighted. The effect degraded mountain ecosystems have on humans is, noted and the mountains' capacities to sustain the array of services on which humans depend are examined in light of the increasing pressures of population growth and consumption.

3. The inescapable conclusion of this analysis is that we must promote an evolution of technological capacities to manage (and in some cases restore) mountain ecosystems if human populations are to obtain the ecosystem goods and services they require for survival in most parts of the world.

3.1 Introduction

Mountains cover about one quarter of the Earth's land surface; only 46 countries have no mountains (People & the Planet, 2004). About 1.48 billion people live in or beside mountain regions, representing about a quarter of the world's population. However, more than half of the world's population is "… directly or indirectly dependent on mountain resources and services, the foremost being water for drinking and home use, irrigation, hydro power, industry and transportation" (Price, 2001a).

Edwards, S.R. (2005). Managing mountain ecosystems to meet human needs: a case for applying the ecosystem approach. In *Mountains of Northern Europe: Conservation, Management, People and Nature*, ed. by D.B.A. Thompson, M.F. Price & C.A. Galbraith. TSO Scotland, Edinburgh. pp. 23-38.

Of those people who live in mountains, 80% are poor and dependent on agriculture and wild harvests to meet their daily requirements. Globally, a large number of minority ethnic groups are associated with mountains. Whether by their physical or cultural isolation, mountain peoples tend to be marginalised and have relatively little influence over political processes and decisions.

In order to meet our needs for ecosystem goods and services, we can no longer focus management on a single biome – like a wetland, a mountain stream, a forest or a farmer's land. What happens in the forest surrounding the wetland affects our capacity to manage the wetland, whether on a mountain or in the lowlands. Ecosystem processes and how we influence them on a mountain slope can, and do, have a direct impact on the availability and quality of ecosystem goods and services delivered in the lowlands hundreds or even thousands of kilometres away.

When considering the ecosystem services provided by mountains, we most often think about those products that we consume that originate in mountains, such as water, food, firewood, construction materials, fibres and minerals. We also recognize the role that mountains play in meeting cultural or spiritual needs and the recreational opportunities we enjoy in mountains. Mountain ecosystems also provide a number of services that are not obvious or directly linked to economic or societal benefits, such as shaping climate across broad regions, controlling erosion, building soils and providing habitat for species. When faced with the challenge of managing these ecosystems we must consider that all of these services are delivered simultaneously in a spatial context that often spans several governmental jurisdictions and a temporal scale that can span several millennia.

An excellent review of different mountain ecosystem services can be found at www.peopleandplanet.net/.

In this chapter, I provide a context for examining management needs by

- briefly reviewing the status and trends of some characteristic ecosystem services;
- identifying scale issues that must be considered to achieve effective and sustainable management;
- considering how the principles of the Ecosystem Approach might provide a framework to promote more effective management practices; and
- examining the role scenario planning could play as a tool to foster more creative problem solving to achieve our management objectives.

3.2 Mountain ecosystem goods and services

An ecosystem is defined under the Convention on Biological Diversity as

"A dynamic complex of plant, animal and micro-organism communities and their non-living environment interacting as a functional unit." (Convention on Biological Diversity, 2000).

In defining 'ecosystem' in the context of the Ecosystem Approach, the Parties noted that an ecosystem can "… refer to any functioning unit at any scale … [where] … the scale of analysis and action should be determined by the problem being addressed" (Convention on Biological Diversity, 2000). Thus, ecosystems have no definitive borders, but are defined in terms of the question or issue being addressed, which can focus on a single bromeliad or the entire planet. Because we are striving to manage mountain ecosystems for the express purpose of delivering goods and services on which humans depend – both in the mountains and the lowlands, which can be attributed to mountains – the ecosystem will be defined at the scale of a watershed. Its boundaries therefore will be determined by the mountain structure that supports the ecosystem we are striving to manage.

The following ecosystem services highlight several benefits that humans derive from them. The consequence of emphasising one service over another (e.g. economic benefits from timber harvests over maintenance of forests to sustain water quality) are noted. Conflicts amongst competing interests for different ecosystem services are highlighted.

3.2.1　Water

All of the world's major river systems originate in mountains. In Europe, Switzerland's Alps are the origin of four major river systems serving a large part of the continent: Rhine, Rhone, Danube and Po Rivers. Depending on where you live, between 30% and 95% of your fresh water originates in mountains (Liniger *et al.*, 1998).

Mountains are critical to the capture and storage of water in glaciers and snow pack at higher elevations. With rising temperatures through the spring and summer months this 'stored' water is released to feed the streams and rivers that flow to the lowlands, where it is used to irrigate crops, and meet an array of other needs. Where weather conditions are right along montane slopes, cloud forests 'capture' water from the clouds. In some areas this has been shown to represent a significant proportion of the water that enters streams and eventually makes its way to the lowlands.

Today, however, mountain glaciers are melting in all parts of the world as the global temperature continues to rise. The result is that a significant reservoir of water is being depleted. Where people are dependent on glacial water to irrigate their crops their long-term security is also disappearing. The consequence will be more conflicts, between those in the highlands and the lowlands, and between nations.

Based on these trends, there is no doubt that water must be considered the highest priority ecosystem 'good' which we need to manage.

3.2.2　Forests

Based on area of coverage in km², 64% of the world's forests are found at elevations over 1,000 m, and nearly half of those occur in the 500 m band between 1,000 and 1,500 m (Kapos *et al.*, 2000; UNEP - World Conservation Monitoring Centre website, 2005). This is also the zone in mountain landscapes where most people reside, grow their crops, and manage their livestock. It is not surprising, therefore, that the status of mountain forests and agricultural uses of mountain landscapes are closely linked. In developed countries, where the human population is stable or declining, there appears to be an increase in mountain forest cover. In Switzerland the forest area has increased by 60% since the mid-1800s (Price, 2001b).

In developing countries in tropical Southeast Asia, however, as lowland human populations increase, mountain forests are being cut down to provide land for agricultural uses. Another trend is seen in tropical mountain ecosystems in Africa and Latin America where highland peoples are moving down mountain slopes in search of new resources as the resources at higher elevations are exhausted (United Nations Environment Programme, 2002).

Deforestation, for timber harvests or to meet increasing demands for agricultural production (more often limited to meeting the subsistence needs of local mountain people), is having a substantial impact on the stability of mountain slopes, the security of people downstream and the supply and quality of water to the lowlands. Management for the better good in this situation would clearly favour the needs of the lowlanders. How do we reconcile this need with the fundamental rights of the people living in the mountains to meet their food and other subsistence needs; or the right of the landowners to sell the timber from their land?

3.2.3 Biodiversity

Mountains support high levels of biodiversity of special interest, in part because they have not been easily accessible. In some instances, such as the wild rice strains found in the mountains of Nepal or the wild corn strains in the mountains of southern Mexico, these species represent invaluable genetic assets.

As long as people have lived in mountains they have benefited from the harvest of plants and animals, for food, fibres, medicines and construction materials. In Peru, 25,000 vascular plant species have been catalogued from the mountains. Of these, people use 3,140 for a variety of purposes (Price, 2001c). Where the value of these species resources becomes established through external demand, increasing harvests to meet that demand can have a major impact on their status. In these cases it is rare that the local harvesters (or managers) receive anywhere near an equitable share of the value realized in the end market. For example, of the 724 species of medicinal plants harvested in Nepal's mountains, 100 are harvested for commercial use. It is estimated that these harvests represent 15,000-42,000 million metric tons per year worth between US$8.6 and 26.7 million. Very little of this value is realized by the local people who harvest the plants. In another example, some 7 million ha of bamboo grow in China's mountains. In 1997, commercial bamboo harvests for furniture, paper, ply-bamboo, food, medicine, and handicrafts generated US$2.2 billion, including exports worth over US$320.0 million, which represented 25% of China's total forest exports (Price, 2001b).

The deteriorating status of key species and their habitats and the high degree of endemism of montane species are often cited as the rationale for the establishment of protected areas. Another factor that has led to the formation of protected areas in mountains is simply that it was land that was not valued for other purposes, and the government could gain substantial political value in designating such areas as protected areas. Globally 32% of designated protected areas occur in mountains (Blyth *et al.*, 2002).

Mountains are providing the most direct (and dramatic) evidence that the world is experiencing significant warming that is affecting the distribution and survival of biodiversity (Thompson *et al.,* this volume; Fowler & Batterbee, this volume). Temperature-sensitive species at the highest elevations are being lost. Species adapted to slightly higher temperatures at lower elevations are gradually moving higher up the mountains. In the European Alps, the eight most common species are moving up the mountains at the rate of 4 m per decade. In Australia's Snowy Mountains, alpine ecosystems could disappear in 70 years (Price, 2001d). In these mountains, sub-alpine trees are growing 40 m higher then they did 25 years ago. Today, only 100 m separate the tree-line from the top of some mountains in the range. As a result the 250+ species of alpine plants which grow in this shrinking zone of habitat are at risk of extinction.

Sustainability of some uses can be directly controlled or affected by the people who are using the resource. In other cases sustainability is dependent on actors well removed from the site of the harvest or use. In the extreme, where a resource is traded internationally, the end market, which determines the demand, and hence harvest levels, can be several thousands of kilometres away, well beyond the policies and laws of the nation in which the harvest occurs. Even more complicated to address in a management plan is the impact of global warming, where decisions taken today will not likely benefit present-day stakeholders, yet may be crucial to meeting basic service needs of people several generations into the future.

3.2.4 Recreation

Mountains are important tourist destinations. More than 500 million people visit mountains annually (UNEP, 2002). In 2001 it was estimated that between 15 and 20% of the annual tourism industry, totalling US$70 to 90 billion, was linked to mountain destinations (Price, 2001e). With increasing wealth and leisure time, mountain sport-based tourism is beginning to be promoted in non-traditional sites, such as the Caucasus, Central Asia and the Andes. Non-winter uses of mountains have also expanded, ranging from adventure tourism to hiking, nature viewing, and 'opportunities for contemplation and meditation' (UNEP, 2002).

With increased tourism have come greater opportunities for economic development in mountain communities, which range from the sale of handicrafts to the construction of hotels and villas to service-based operations such as restaurants, guide services and room rentals. The impact of these combined uses is now taking its toll. In regards to tourism, the UNEP International Year of Mountains website (www.mountains2002.org/i-tourism.html), notes "… that mountain tourism can have a range of damaging effects. It can degrade and stress fragile mountain ecosystems, destroying the qualities that make these environments so alluring. Mountains are among the world's most important repositories of biodiversity, yet construction, pollution and noise all threaten this precious asset."

Here again, a picture emerges that illustrates the prospect of conflict between those seeking short-term economic gains, which could be substantial, and the impact of development that can, over time, lead to deterioration of ecosystem capacity. In such cases the cost for restoring even minimal capacity will likely require large investments, which are rarely factored into the cost/benefit analysis of the project at its inception.

3.3 The Ecosystem Approach – framing the issues

In this section I look at how the Ecosystem Approach (Convention on Biological Diversity, 2000) can be used to promote greater integration of technical capacities to manage mountain ecosystems to conserve biodiversity and deliver the array of goods and services on which people depend.

3.3.1 Defining ecosystem goods and services

The Millennium Ecosystem Assessment (MA) has adopted an ecosystem service classification scheme (Alcamo et al., 2005) that is useful to identify the full spectrum of services that will need to be considered when framing the issues and designing management approaches to sustain delivery of mountain ecosystem goods and services. These categories are as follows.

- *Provisioning services*, which include the array of products and services that are harvested from, or passively provided by, ecosystems. In the context of mountains, these would include forest products (including timber and non-timber products harvested for food and medicines), agricultural production, grass necessary for livestock pasturage, water, minerals that are extracted, and even the high value genetic resources, such as the maize and rice genes noted above.
- *Cultural services*, which generally include the 'non-material' benefits from ecosystems. From mountain ecosystems these would include the spiritual and cultural benefits we realize; the unique educational/knowledge systems associated with mountain ecosystems, most often ascribed to as traditional or local knowledge. Also included in this category would be the diversity of cultures, languages, and understandings that are promoted by the physical partitioning of people in mountain landscapes. And, with the increase of expendable wealth, the rapidly increasing dependence on mountains to fulfil recreation and ecotourism demands.
- *Regulating services*, which regulate overall environmental conditions on Earth. In mountain ecosystems these services would include maintenance of air and water quality, erosion control, and storm protection, e.g. forests enhance stability of soil and decrease risk of landslides.
- *Supporting services*, which are the "services that maintain the conditions of life on earth …" according to Alcamo *et al.* (2005). Mountains, when viewed across a time scale, taking into account the highly diverse ecosystem components found in mountains (e.g. wetlands, forests, streams, paramo and peatlands), are central deliverers of supporting services like pollination for plant reproduction, soil retention, production of oxygen and capture of carbon, and nutrient cycling.

3.3.2 Balancing supply against demand

When considering mountain ecosystems, management should strive to balance the supply of all of these different services against the array of demands humans have for them. *A priori* some of these 'services' are of critical importance for humans to survive on the planet. Others, primarily related to the provisioning and cultural service categories, tend to have strong influence over our thinking. We need them now – and in many cases, humans are not less dependent on their sustained delivery. A smaller number of these services, such as mineral extraction or the construction of a ski resort, are a matter of societal choice, but in terms of emotion, may rank very high.

'Agenda 21' (United Nations Conference on Environment and Development, 1992) provided compelling arguments for the need to address the human condition as a concomitant requirement, if not prerequisite, to achieving conservation goals. In regards to mountains, Chapter 13 of 'Agenda 21' articulates the need to address development and conservation in mountains as a priority. This priority was to be addressed through two 'Programme Areas', one dealing with building "… knowledge about ecology and sustainable development of mountain ecosystems; …", the other focusing on "… integrated watershed development and alternative livelihood opportunities." According to Sene & McGuire (1997) these two areas provide a "framework of an integrated ecosystems approach to sustainable mountain development." The implication here is that the traditional ecologist or rural community development specialist was not sufficient to address the problems alone, or even together. Other specialists are needed, with skills that go beyond those that were applied previously. At the same time they note that, subsequent to Rio de Janeiro, new emphases are being placed on "… culture, sacred values and landscape diversity." Spirituality and the special relations that mountain peoples hold with their land are now important considerations in conservation and development planning. More attention is being given to the economic condition of the people who live in mountains. Initiatives are beginning to focus on the consequences of growing human populations who, with increased expendable income, are using mountain areas for recreation and development in rapidly increasing numbers.

In *Mountains of the World: Water Towers for the 21st Century* (Liniger *et al.*, 1998), six concrete recommendations are presented on a) managing water; b) conserving biodiversity; c) links between mountains and lowlands; d) water assessment; e) economic investment; and f) conflicts. The paper concludes that "Integrated basin management encompassing both the mountains and the lowlands is the key to meeting … challenges" in relation to the sectors/issues that were targeted. This perspective introduces the guiding principle that we must work at a large geographic scale in order to promote more effective linkages across inter-linked parts of landscapes. The paper also notes that "Planning and management tools at the local, national and international levels should be further developed, together with better cooperation between decision makers, researchers, planners, and users at all levels" which scopes some needs on an institutional scale.

In September 2002, at the World Summit on Sustainable Development, two multi-sectoral Partnership initiatives related to mountains were announced (Sustainable Agriculture and Rural Development (SARD) Initiative and the International Partnership for Sustainable Development in Mountain Regions (Price & Hofer, this volume)). These are particularly important to our analysis of management because they emphasize needs for development in mountains, including agriculture, as means to address poverty and livelihood issues in these ecosystems. Yet, to achieve these goals it is crucial that they be pursued in the context of policies that promote the sustained delivery of key ecosystem services, including the sustainable use of natural resources on which the development will be based. Indiscriminate harvest of species resources or abuse of habitats by ecotourists will soon undermine the development, with the opposite conditions prevailing – increasing poverty and unsustainable livelihoods.

The Parties to the Convention on Biological Diversity (CBD) have endorsed the Ecosystem Approach (Convention on Biological Diversity, 2000), comprising 12 principles and five operational guidance points (Box 3.1) that were crafted to promote a balanced approach to management of ecosystems that took account of ecological and social factors. Shepherd (2002), in an analysis of the principles of the Ecosystem Approach in relation to forest management, noted that they account for scale, diversity of uses in landscapes, the importance of local involvement in management, and "encourage outsiders … to build on these capacities rather than ignoring or destroying them." She also pointed out that the understanding of what an 'ecosystem' is varies substantially, depending on the user of the ecosystem. Local people tend to consider the ecosystem to comprise the land that they use for agriculture or their livestock and the forest or natural habitat from which they harvest firewood or medicinal plants. However, sub-national government officials may consider the forest or protected area they are managing to be the ecosystem. Governments, because of their

Box 3.1 Ecosystem Approach. Source: Convention on Biological Diversity (2000)

Principles

1. The objectives of management of land, water and living resources are a matter of societal choice.
2. Management should be decentralised to the lowest appropriate level.
3. Ecosystem managers should consider the effects (actual or potential) of their activities on adjacent and other ecosystems.
4. Recognising potential gains from management, there is usually a need to understand and manage the ecosystem in an economic context. Any such ecosystem-management programme should:
   - reduce those market distortions that adversely affect biological diversity;
   - align incentives to promote biodiversity conservation and sustainable use;
   - internalise costs and benefits in the given ecosystem to the extent feasible.
5. Conservation of ecosystem structure and functioning, in order to maintain ecosystem services, should be a priority target of the ecosystem approach.
6. Ecosystems must be managed within the limits of their functioning.
7. The ecosystem approach should be undertaken at the appropriate spatial and temporal scales.
8. Recognising the varying temporal scales and lag-effects that characterise ecosystem processes, objectives for ecosystem management should be set for the long term.
9. Management must recognise that change is inevitable.
10. The ecosystem approach should seek the appropriate balance between, and integration of, conservation and use of biological diversity.
11. The ecosystem approach should consider all forms of relevant information, including scientific and indigenous and local knowledge, innovations and practices.
12. The ecosystem approach should involve all relevant sectors of society and scientific disciplines.

Operational guidance

a. Focus on the functional relationships and processes within ecosystems.
b. Enhance benefit-sharing.
c. Use adaptive management practices.
d. Carry out management actions at the scale appropriate for the issue being addressed, with decentralisation to lowest level, as appropriate.
e. Ensure intersectoral cooperation.

responsibility of the 'body politic' tend to focus on ecosystems defined at the level of biome or watershed within their jurisdiction. International bodies often consider an entire river basin as an ecosystem, which generally involves several national jurisdictions, governmental systems and policies. This variance between views along the institutional scale, all of which could relate to the same landscape, requires that we take account of different 'stakeholder perspectives' when considering needs and approaches to mountain ecosystem management.

It follows that balancing the ecosystem services equation will require tradeoffs between different sectors or recipients. It also means that this will be possible only within the inherent biological and ecological limitations of the components of the ecosystems with which we are working. Finally, such tradeoffs can only be implemented where the supporting policy framework exists and institutions at all levels have the capacity to act effectively.

3.3.3 Interpreting the Principles of the Ecosystem Approach

The Principles of the Ecosystem Approach, as they were adopted (see Box 3.1), provided little guidance in how to interpret or apply them to address ecosystem management problems. To guide ecosystem management planning and actions in mountains I have organized them in Table 3.1 in relation to three categories of components that reflect broad capacity

Table 3.1 Principles of the Ecosystem Approach organized according to their relevance to ecological/biological, social and policy aspects of a management regime (bracketed numbers refer to the numbering of the principles as adopted by the Parties to the CBD; see Box 3.1).

Ecological/Biological Principles	Social Principles	Policy Principles
[Ecosystem managers should] Consider the effects (actual or potential) of management activities on adjacent and other ecosystems (3).	Management must recognise that change is inevitable (9).	[Government policies should ensure that] Objectives of management of land, water and living resources [can] be established as a matter of choice (1).
Conservation of ecosystem structure and functions should be a priority to maintain ecosystem services (5).	Management should be decentralised to the lowest appropriate level (2).	Policies, related to management of ecosystems should be structured to: a) Reduce market distortions that adversely affect biological diversity; b) Align incentives to promote biodiversity conservation and sustainable use; and c) Internalise costs and benefits in the given ecosystem to the extent feasible (4).
Ecosystems must be managed within their functioning limits (6).	The ecosystem approach should involve all relevant sectors of society and scientific disciplines (12).	Management should be undertaken at the appropriate spatial and temporal scales (7).
Management objectives should be set for the long-term to account for lag-effects that characterise ecosystem processes (8).	Management should take into account all forms of relevant information, including scientific, indigenous and local knowledge, and technological innovations and practices (11).	Policies promoting the principles of the ecosystem approach should seek the appropriate balance between, and integration of, conservation and use of biological diversity (10).

requirements necessary to achieve effective integrated ecosystem management: ecological/biological principles, social principles, and policy principles. In organizing the principles in relation to these three categories I have reordered them and in a few instances taken the liberty to rephrase the text to clarify what I interpret to be the intent of the principle. Each of the five guidance points is also organized according to these three categories of principles in Table 3.2. The number in brackets at the end of each principle or guidance point refers to the order in which the principles were presented at the time they were adopted.

Table 3.2 Guidance to the Ecosystem Approach organized according to their relevance to ecological/biological, social and policy components of a management regime (bracketed letters refer to the letter reference for the guidance points, as adopted by the Parties to the CBD; see Box 3.1).

Ecological/Biological Components	Social Components	Policy Components
Focus on the functional relationships and processes within ecosystems (a).	Enhance benefit-sharing (b).	Carry out management actions at the scale appropriate for the issue being addressed, with decentralisation to lowest level, as appropriate (d).
	Use adaptive management practices (c).	
	Ensure inter-sectoral co-operation (e).	

These relationships are illustrated graphically in Figure 3.1 as elements being balanced in a set of scales. The left side of the scale 'holds' the principles related to the ecological and biological components; the right the social principles. In the centre beam of the scale are the policy-related principles.

The ecological principles explain the capacity of ecosystems to sustain delivery of the array of goods and services defined above. Put another way, these principles describe the limits within which management is constrained if the goods and services are to be sustained. The social principles describe the human-based aspects of ecosystem management that have generally been found to enhance the sustainability of management systems. The details of how each of these principles is applied are going to vary according to the local culture and existing governments' policies. The policy principles describe a framework in which the social principles can be structured to ensure that management remains within the capacity of ecosystems to sustain delivery of the goods and services on which people and biodiversity depend. The five guidance points of the Ecosystem Approach, which provide more detailed explanations of the principles, are arrayed across the top of the scale in relation to those principles with which they are associated.

At the 7th Meeting of the Conference of the Parties to the CBD (February 2004), Parties agreed that facilitating adoption of the ecosystem approach as "… the primary framework for addressing the three objectives of the Convention in a balanced way …." (Convention on

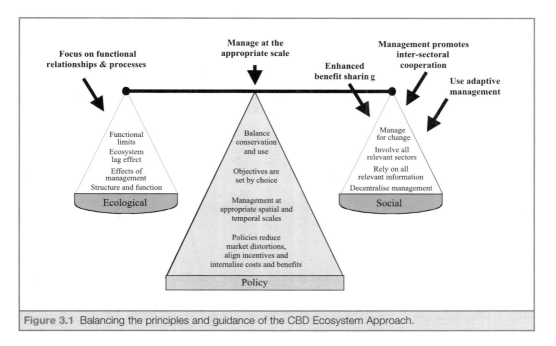

Figure 3.1 Balancing the principles and guidance of the CBD Ecosystem Approach.

Biological Diversity, 2004). In so doing, they acknowledged and welcomed the use of Guidance provided to help the Parties apply the principles appropriately which can be seen at www.biodiv.org/decisions/default.aspx?m=COP-07&id=7748&lg=0.

3.4 Dealing with scale

In order to manage uses of mountain ecosystems to sustain delivery of goods and services for all stakeholders, governments and managers are required to consider how the principles of the ecosystem approach will be applied in relation to spatial, institutional and temporal scales (see Figure 3.2). The spatial scale, noted at the top of the figure, relates to the size of the ecosystem to be managed, from a single pond at the bottom to a multinational landscape at the highest level.

The institutional scale, provided on the left in the figure, accommodates the scope of institutions needed for decision-making and governance to manage ecosystems at the different spatial scales. In this model they range from a local landowner to international conventions or commissions. By implication, this scale draws attention to the policy and capacity needs that must be met across the range of institutions. In the case of the policies that apply to each level, there would be need for them to be congruent, from the local to multinational level, for example in granting authority and providing means for the different actors to be accountable.

The temporal scale, along the right of the figure, provides a frame for accounting for different time-dependent needs that apply at different levels on the ecological or institutional scale. For example, there could be time-based constraints between an action and an effect in an ecosystem, which should be factored into the management. In the context of mountains the temporal scale could range from the life cycle of an insect, to an annual cycle of migratory species, to the generation time of a species, or the decades to centuries required to re-establish

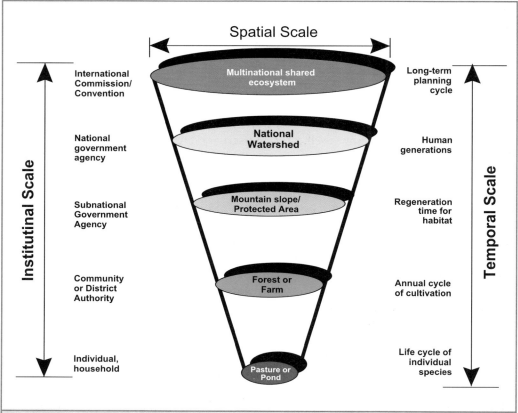

Figure 3.2 Graphic representation of three scales influencing the application of the principles of the Ecosystem Approach.

key components of an ecosystem. While implied in the model, there is not necessarily a direct relationship between particular levels on the three scales.

Balancing needs at the appropriate levels across these three scales is crucial to successful management. For example, to balance the services goal, because of the diversity of the services (and human behaviours) we are endeavouring to manage, will require that management be addressed at a large spatial scale on the order of watersheds. However, specific actions designed to enhance delivery of an ecosystem service may take place at a local site, such as the reforestation of a slope or the restoration of a wetland ecosystem.

Management activities must accommodate needs that would be determined on the temporal scale in terms ranging from the lifecycle of key species (which could be measured in days or weeks for a pollinator) to human generations to millennia into the future.

Finally, on the institutional scale, consideration must range from the requirements (policies, institutional roles and capacities) at the local, subnational, national, regional and even global levels. At the local level, the institutional components might be a business, household, community, or government agency. At the subnational, national and regional and global levels, most often the key actors will be government agencies and non-governmental organizations. In considering these different elements on the institutional scale, there is a fundamental necessity that managers have the authority to act and be accountable for those actions at each level within the context of the ecosystem being managed. At the highest level, this will require trans-national institutions to be created if the management regime spans more than one national governance jurisdiction.

3.5 Synthesis

The emerging field of scenario planning may prove to be an excellent tool to deal with the complexities which we are striving to address. A comprehensive overview paper on the approach can be found at www.well.com/~mb/scenario/, a web page provided by Martin Börgesson or, for non-profit organizations, a website provided by the Global Business Network at www.gbn.com/ArticleDisplayServlet.srv?aid=32655.

Scenario planning accepts that there are many different possible solutions to any problem and that there are many possible futures. Constructing alternative scenarios of what would be desired as a future is one way to elucidate obstacles and possible means for countering them. Of course, the effectiveness of this approach is dependent on our ability to go beyond our fixed perceptions of our world – to be creative, to think outside of the box – which is likely to be the most difficult part of the process.

Those who have developed and apply this approach underscore the need for flexibility, recognizing that there are no fixed answers and managers will need to constantly adjust their decisions and actions to respond to every changing condition that will influence the sustainability of the management. This conditional approach to planning and action is by its very nature anathema to governments, which rely on 'simple messages' designed to embrace an 'average' constituent and which are replicable across their constituency.

To get beyond this conundrum, government policies must target higher level concepts, which allow – and even encourage through incentives – the degrees of creativity and flexibility required by local managers pursuing their objectives.

To ensure that alternative scenarios meet our conditions, there will be a need to identify and understand the scope of the services provided in a management regime in terms of the obvious provisioning and cultural categories of service, as well as the less obvious regulatory and supporting services. Once the array of services that are provided or needed is understood by the range of stakeholders, the challenge will be to construct alternative scenarios, each of which leads to a 'future' in which critical ecosystem functions are assured, and the remainder are balanced equitably across the different stakeholders. Where the capacity of the ecosystem to deliver needed goods and services has been lost or severely eroded, there will be a need to consider alternative means to restore or recover that capacity in the scenario.

Some problems may be readily recognizable, e.g. there is an immediate need to clean up the water in a river or steps must be taken to stabilize the slopes above a village. Others may not be so readily identifiable. For example, those problems related to delivery of services that will not be realized for some time – like ensuring a sustained capacity for cloud forests to 'capture' water or the role mountain forests play in sequestering carbon.

Mechanisms must be put in place for the spectrum of stakeholders related to the diverse ecosystem goods and services we are striving to sustain, at least through representation, to contribute to the construction and engage in the analyses associated with the alternative scenarios. Because of the high degree of investment in one potential outcome over another (e.g. one that favours one service over another), it is crucial that the process be facilitated and that a condition of participation would be acceptance of an approach that would examine several alternative scenarios and a commitment to remain engaged in the process.

Construction of alternative scenarios will require iterative, collective 'brainstorming' involving the full array of stakeholders from within a mountain ecosystem and living downstream. Promoting dialogues amongst the different stakeholders, ensuring that local 'visionaries' contribute to the process, and building in mechanisms to ensure that all ideas and visions are considered and not thrown out because they do not meet present-day perceptions are important considerations in the building of the scenarios.

At the same time, multi-sectoral, interdisciplinary action is very much easier to say than to actually do. In part because we have access to more information, have had more varied experiences, and because more diverse actors are engaging in conservation and development issues today, we all recognize that our own disciplinary skills are not adequate to meet the challenges of the problems we are addressing today. Further, it is not as simple as establishing a smorgasbord of specialists and calling each as needed. Rather, the challenge is to have the key disciplines represented from the beginning so that an added-value synergy can be realized from the interactions of the different specialists. Again, this is a concept that is far easier to visualize than to implement.

Across the board, specialists are locked into their own paradigms of language, process, hierarchy, incentives and rewards. Specialists who elect to work outside of their field of specialization generally find great resistances amongst their peers, and in the extreme are penalized for being so adventurous. To address these restraints will require conscious effort to counteract these forces, which reinforce isolation rather than integration. Policies will necessarily have to provide clear incentives to promote the degree of interdisciplinarity that future decisions will require.

Each scenario will likely incorporate decisions at a number of key junctions, based on the best available information at the time. However, over time, management progressing along a particular path, or in the context of a given scenario, may generate information that favours a change in the guiding scenario that had not been considered previously. Therefore, policies must be flexible enough to alter the structure on the basis of new information as management proceeds.

Finally, when crafting alternative scenarios, consideration should be given to how the sustained delivery of the different priority services will be financed. By and large, the more obvious provisioning services, which are based on harvests or extractions have established markets, have established economic values that are set on the basis of the supply and demand for the particular good or service. The challenge will be to develop viable estimates of the costs to maintain the full spectrum of priority goods and services that are identified, many of which will be less 'tangible' (e.g. the role forests play in maintaining water quality). Only with this information available will it be possible for scenario planners to develop alternative 'future patterns of use' of mountain ecosystems, and associated government policies and incentives, which optimise the delivery of those services determined by all stakeholders to be of highest priority.

In conclusion, the complexity of the problems we are called on to address across the spatial, institutional, and temporal scales that frame our actions are so diverse that no one individual, discipline, or institution has the capacity to achieve the joint conservation and development goals. We must find ways that promote the 'interdisciplinarity' that is often referred to as being essential – but rarely is practiced. I believe that scenario planning holds promise as a technique that will bring the different stakeholders together, along with the spectrum of skill needed to resolve the problems. But, like the village management programmes I have helped design over my career, it will only be effective if all stakeholders 'own' the process, are able to grasp the tradeoffs and consequences from following one management/policy course over another, and are a party to decisions at each stage in the process.

References

Alcamo, J., Ash, N.J., Butler, C.D., Callicott, J.B., Capistrano, D., Carpenter, S.R., Castilla, J.C., Chambers, R,, Chopra, K., Cropper, A., Daily, G.C., Dasgupta, P., de Groot, R., Dietz, T., Duraiappah, A.K., Gadgil, M., Hamilton, K., Hassan, R., Lambin, E.F., Lebel, L., Leemans, R., Jiyuan, L., Malingreau, J.-P., May, R.M., McCalla, A.F., McMichael, A.J., Moldan, B., Mooney, H., Naeem, S., Nelson, G.C., Wen-Yuan, N., Noble, I., Zhiyun, O., Pagiola, S., Pauly, D., Percy, S., Pingali, P., Prescott-Allen, R., Reid, W.V., Ricketts, T.H., Samper, C., Scholes, R., Simons, H., Toth, F.L., Turpie, J.K., Watson, R.T., Wilbanks, T.J., Williams, M., Wood, S., Shidong, Z. & Zurek, M.B. (2005). Ecosystems and their services. In *Ecosystems and Human Well-being: a Framework for Assessment,* ed. by J. Alcamo, N.J. Ash, C.D. Butler *et al.* Island Press, Washington, DC. pp. 49-70.

Blyth, S., Groombridge, B., Lysenko, I., Miles, L. & Newton, A. (2002). *Mountain Watch: Enviromental Change and Sustainable Development in Mountains.* UNEP World Conservation Monitoring Centre, Cambridge.

Convention on Biological Diversity (2000). Decision V/6: Ecosystem Approach. 5th Conference of the Parties to the Convention on Biological Diversity. Nairobi, 15-26 May 2000. www.biodiv.org/decisions/default.asp?lg=0&m=cop-05&d=06.

Kapos, V., Rhind, J., Edwards, M., Price, M.F. & Ravilious, C. (2002). Developing a map of the world's mountain forests. In *Forests in Sustainable Mountain Development: A State-of-Knowledge Report for 2000,* ed. by M.F. Price & N. Butt. CAB International, Wallingford. pp. 4-9.

Convention on Biological Diversity (2004). Decision VII/11: Ecosystem Approach. 7th Conference of the Parties to the Convention on Biological Diversity. Kuala Lumpur, 9-20 February 2004. www.biodiv.org/decisions/default.aspx?m=COP-07&id=7748&lg=0.

Liniger, H.P., Weingartner, R., Grosjean, M., Kull, C., MacMillan, L., Messerli, B., Bisaz, A. & Lutz, U. (Eds) (1998). *Mountains of the World, Water Towers for the 21st Century - A Contribution to Global Freshwater Management.* Mountain Agenda, Paul Haupt, Bern.

People & the Planet (2004). Mountains: vital for human survival. www.peopleandplanet.net/doc.php?id=966&section=11.

Price, M. (compiler) (2001a). Water towers for humanity. People & the Planet. www.peopleandplanet.net/pdoc.php?id=968.

Price, M. (compiler) (2001b). Mountain forests. People & the Planet. www.peopleandplanet.net/pdoc.php?id=1032.

Price, M. (compiler) (2001c). Wildlife hotspots. People & the Planet. www.peopleandplanet.net/doc.php?id=1031.

Price, M. (compiler) (2001d). Mountains in a changing climate. People & the Planet. www.peopleandplanet.net/pdoc.php?id=1036.

Price, M. (compiler) (2001e). Mountain tourism. People & the Planet. www.peopleandplanet.net/pdoc.php?id=1034.

Sene, E.H & McGuire, D. (1997). Mountain Agenda for the 21st century. In *Mountains of the World - A Global Priority*, ed. by B. Messerrli & J.D. Ives. Parthenon Publishing Group, Ltd., Carnforth. pp 447-453.

Shepherd, G. (2002). Managing forest ecosystems for sustainable livelihoods. Proceedings of a Workshop entitled 'Managing forest ecosystems for sustainable livelihoods'. 16th Global Biodiversity Forum, The Hague, Netherlands. www.iucn.org/themes/cem/library/reports/gbf_articles/managing_forest_ecosystems_gill_shepherd.doc.

United Nations Conference on Environment and Development (1992). *Agenda 21*. Division for Sustainable Development, UN Department of Social & Economic Affairs, United Nations, New York. www.un.org/esa/sustdev/documents/agenda21/english/agenda21toc.htm.

United Nations Environment Programme (2002). International Year of Mountain website: www.mountains2002.org/i-tourism.html.

UNEP – World Conservation Monitoring Centre (2002). Mountain Watch. (See 'mountains' under 'habitats' at www.unep-wcmc.org/). UNEP – World Conservation Monitoring Centre, Cambridge.

PART 2:
Mountain Environments: Perspectives

Tryfan and Llyn Idwal, Glyderau, Wales, UK (Photo: Martin Price).

In their introduction to the nature of mountains, Thompson *et al.* (Chapter 4) outline the many interactions between natural and human influences in shaping mountain landscapes and nature. They provide a detailed map of mountain areas, and outline the principal global and European policies governing the conservation of these areas in northern Europe.

In Chapter 5, Jonasson *et al.* explore the links between geodiversity and biodiversity in European mountains, and provide three case studies from Sweden, Scotland and the Czech Republic. The authors provide a vivid account of the dynamic nature of mountain environments, and compare the relative effects of natural geomorphological processes and human activities on mountain landscapes. Like other authors, they emphasise the need to manage mountains at the ecosystem level, rather than focusing on individual species or habitats. The authors stress geodiversity and biodiversity cannot be separated, and that geomorphological processes, both in the past and at present, are fundamental in influencing biodiversity and landscape sensitivity. For instance, in parts of Scandinavia, slush torrents are responsible for much of the relocation of significant amounts of sediment on steeper slopes of mountains, but may be active in a given location for only a few minutes every 10 years, equivalent to only one and a half days of activity, if extrapolated over the entire Holocene. Yet, each time there is a slush torrent, the landscape changes dramatically (see Figure 5.4). Much novel research is reported here, which looks set to continue through collaboration between the authors.

There are major concerns about the potential impacts of climate change and pollution across mountain environments. Fowler & Battarbee provide a substantial overview of the issues (Chapter 6), and in particular explore reasons for mountain areas being so sensitive to change. The authors emphasise the value of mountain ecosystems for detecting early signs of developing environmental problems. Indeed, they conclude that 'mountains are one of the best places to look for potential problems'. Perhaps what is most surprising is the high levels of pollution recorded in mountain areas; many people think of these as clean and pure environments.

Building on this review, the next four chapters report on more specialised studies. Kernan *et al.* (Chapter 7) have studied the history of pollutant levels in Scottish mountain lochs. They report on the distribution of lead, persistent organic pollutants and spheroidal carbonaceous particles in the surface sediments of upland lochs in Scotland. Interestingly, the authors find that the distribution maps for each pollutant is different, because the major emission sources and sediment accumulation rates vary. At Lochnagar, their work raises concerns about the potential effects of pollutants on fish and higher predators.

In Chapter 8, Ellis & Good explore the effects of climate change on mountain ecosystems in Scotland and Wales. They begin their chapter with the statistic that, between the years 1861 and 2000, the global mean annual temperature rose by 0.6°C, with the northern hemisphere warming at a rate of about twice as great as the southern hemisphere. The authors go on to detail graphically the implications of climate change for wildlife. Essentially, this is a summary of ongoing work, and there is reference to major reports produced in the last five years.

Coll *et al.* (Chapter 9) describe their research on climate modelling in the Scottish Highlands. This work is based on a new approach to obtaining local representations of future climates. One of the challenges here has been to cope with the substantial spatial variability in the climate in the Scottish Highlands, with much of this influenced by mountain landforms. This research is, again, work in progress, but clearly the authors have developed an approach which should be applicable in other parts of Northern Europe.

In developing our understanding of the effects of pollutants and climate on mountain ecosystems, researchers are carrying out more detailed studies of mountain vegetation. Britton *et al.* (Chapter 10) provide a concise account of their studies of the effects of nitrogen deposition on heather (*Calluna vulgaris*) and woolly hair moss (*Racomitrium lanuginosum*). They have found some complex impacts of nitrogen deposition, but with these two species responding differently. There remains much debate about whether or not nitrogen deposition, grazing pressure or a combination of both are causing deterioration in the condition of *Racomitrium lanuginosum*-dominated heaths in some of the more southern parts of N. Europe.

This part of the book closes with a review of recreation in mountain environments. John Mackay reviews the history of recreational enjoyment of mountains, and analyses data collected through a number of surveys in the Scottish Highlands. He presents detailed information on the characteristics of visitors to mountain areas that raise all sorts of questions about the participation of different age classes and people from different backgrounds. Mackay closes Chapter 11 by considering the evolving debate on access to mountain areas. He concludes that 'places of home and work are the landscapes of daily life, and for those who visit the mountains, often from urban settings, these are landscapes of escape and refreshment.'

4 The nature of mountains: an introduction

D.B.A. Thompson, L. Nagy, S.M. Johnson & P. Robertson

Summary

1. Almost a quarter of the world's land surface is classified as mountains, and a quarter of this is found in Europe.
2. Mountain areas are characterised by: (a) mostly shallow soils and low productivity above the tree line; (b) a high degree of endemism; and (c) anthropogenic changes in natural habitats, principally through deforestation below the tree line, and grazing, burning and pollution.
3. The changing nature of mountains is the product of historical interactions between environmental and human-related factors. At any location or period in history, the landscape and its biota are the product of this interaction. This has important implications for the range of policies and practices deployed to manage mountains.
4. This introduction points to the need to integrate work involving research, policy development and conservation management in order to sustain mountain areas for people and nature.

4.1 Introduction

Covering around a quarter of the earth's surface and embracing a quarter of the world's human population, mountains are important globally (e.g. Huddleston *et al*, 2003; Price & Hofer, this volume; M. Edwards, this volume). Mountains influence at least half of the world's population, by providing water, agriculture, energy, food, wood and enjoyment (Messerli & Ives, 1997). Added to this are the physical, biological, social and cultural facets of mountain areas, contributing immensely to the world's biological, economic and societal diversity. Mountains are now at the forefront of the world's greatest contemporary environmental challenge – global warming. The changes in species distribution and richness – as well as the rapid retreat of glaciers – in the highest mountains portend much broader changes across the globe, and the potential rapidity of change means that society has to consider new courses of action in managing mountain areas.

Thomson, D.B.A., Nagy, L., Johnson, S.M. & Robertson, P. (2005). The nature of mountains: an introduction. In *Mountains of Northern Europe: Conservation, Management, People and Nature*, ed. by D.B.A. Thompson, M.F. Price & C.A. Galbraith. TSO Scotland, Edinburgh. pp. 43-56.

4.2 What are mountains?

Many definitions and classifications apply to mountains, with most authorities agreeing that they have one thing in common – a summit. Elevation, relief, steepness, bulk and many other characteristics have all been used to classify mountain areas (e.g. Gerrard, 1990). The use of the term 'upland' by Ratcliffe (1977) and Ratcliffe & Thompson (1988), referring to areas above the upper limits of enclosed farmland, is near-synonymous with the definition of mountains by the European Commission Council Regulation 1257/99 (Article 18) which states:

> *"Mountain areas shall be those characterised by a considerable limitation of the possibilities for using the land and an appreciable increase in the cost of working it, due to: (a) the existence, because of the altitude, of very difficult climatic conditions the effect of which is substantially to shorten the growing season; (b) at a lower altitude, the presence over the greater part of the area in question of slopes too steep for the use of machinery or requiring the use of very expensive special equipment; or (c) a combination of these two factors, where the handicap resulting from each taken separately is less acute, but the combination of the two gives rise to an equivalent handicap."*

Recently, Kapos *et al.* (2000) developed a global map of the world's mountains, published as part of the work of the UNEP - World Conservation Monitoring Centre. That global map indicated that approximately one-quarter of the world's land surface is covered by mountains; these are predominantly areas which have more than 300 m elevation within a 10 km radius of a target cell. The digital database from which this map was developed has subsequently been used by UNESCO (2002) to create a global database of its mountain activities, by FAO to assess the proportion of the global population living in mountain areas (Huddleston *et al.*, 2003), and by the European Commission (Price *et al.*, 2004) to determine the extent of Europe's mountain areas.

While useful in assessing the extent of mountains by arbitary elevation classes, the map developed by Kapos *et al.* (2000) does not allow for bioclimatological factors, which essentially represent elevation belts or zones. Figure 4.1 demonstrates the considerable altitudinal and latitudinal variation in the occurrence of bioclimatological zones or belts in mountain areas across the world. Towards the north and especially towards the north-west of Europe, habitat 'life' zones are compressed, and occur at lower altitudes. For example, the limit of 2,500 m elevation used by Kapos *et al.* (2000) (below which human physiology is not affected by altitude) falls in the montane forest zone in the tropics, whilst in Scandinavia nival conditions prevail towards that altitude (the highest Scandinavian peak is 2,469 m). Equally, at the lower boundary of the alpine zone, the tree line shows considerable variation with latitude and distance from the Atlantic Ocean in north-west Europe alone (Figure 4.2).

At the European scale, Nagy *et al.* (2003a) provided an overview of the ecology and biodiversity of the alpine zone – the areas above the tree line – within the principal mountain regions of Europe. Three characteristics link the north European alpine areas together: (a) shallow soils and low productivity, largely as a result of geological change; (b) changes in natural habitats, principally through long-term grazing, burning and more recently pollution; and (c) high biological richness relative to the size of the area occupied, and a high degree of endemism.

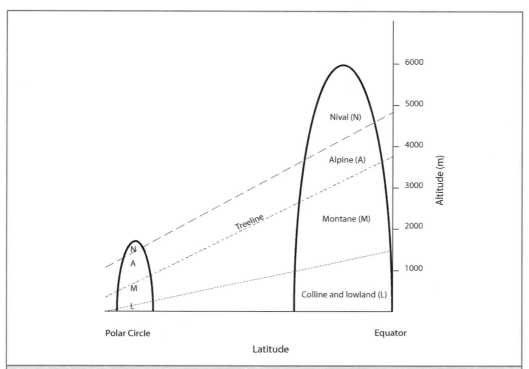

Figure 4.1 Schematic representation of latitudinal and altitudinal variation in of the bioclimatological altitude zones or belts in mountain areas. The principal zones shown are: lowland and colline (L), montane (M), alpine (A) and nival (N); Nagy *et al.* (2003a) provide full definitions of these zones.

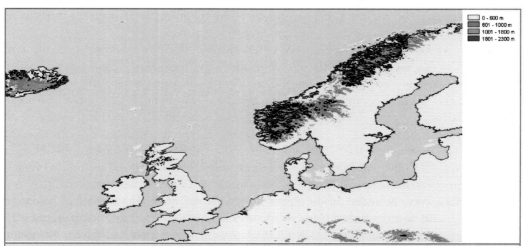

Figure 4.2 The extent of mountain/alpine areas across Northern Europe. The altitude classes are adapted from Nagy *et al.* (2003b). The approximate altitude of the tree line is outlined in black and largely follows Ozenda (1994): 450 m for Iceland, 750 m for Scotland, 1,100 m for the southern Scandes and 700 m for the northern Scandes. The map was compiled using public data provided by the US National Geophysical Data Center (Global Ecosystems Database and The Global Land One-kilometer Base Elevation (GLOBE) Digital Elevation Model). Source: these data are distributed by the Land Processes Distributed Active Archive Center (LP DAAC), located at the US Geological Survey's EROS Data Center (http://LPDAAC.usgs.gov).

The mountain landscapes of today are a product of an ever-changing interaction between natural and human influences (Table 4.1). Reviewing the nature of change in the British uplands, Ratcliffe & Thompson (1988) identified six principal elements in this interaction: extensive forest clearance; extensive use as grazing range by domestic livestock; land improvement for agriculture and hunting/field sports; persecution of wildlife (especially predators) relating to livestock and game management; industrial acidification; and extensive conifer afforestation. Looking more widely across Northern Europe, we can add the growing expansion of tourism and related recreation, and the more recent expansion of renewable energy developments, notably hydroelectric schemes and wind farms.

Table 4.1 Some of the principal characteristics of the mountain areas of Northern Europe. Sources: McVean & Ratcliffe (1962); Pearsall (1971); Sissons (1976); Barry (1981); Ratcliffe & Thompson (1988); Thompson & Brown (1992); Whittow (1992); Ozenda (1994); Vaisänen (1998); Firbank et al., (2000); Thompson et al. (2001); Hallanaro & Pylvänäinen (2002); Ozenda (2002); Nagy et al. (2003a); Averis et al. (2004); Ellis & Good (this volume); Fowler & Battarbee (this volume); Simmons (this volume); M. Edwards (this volume).

Features	Characteristics
Orogenesis	Mainly Palaeozoic (Caledonian) folding (Britain, N Europe) and pre-Cambrian shields (S Scandinavia); west – steep, dissected glacial geomorphology; east – less steep and less dissected (Scandes, Scottish Highlands); volcanic and glacial (Iceland).
Climate	Atlantic (most of British Isles). Boreo-Atlantic (N Scotland, SW coast of Norway), low latitudinal lapse rates; steep altitudinal lapse rate in NW Highlands of Scotland, and SW Norway, life zones descend north-westwards in the British Isles and Scandinavia, and towards the centre in Iceland. Boreal (Iceland, rest of Scandinavia).
Geology	Crystalline (gneiss, granite, mica schist); sandstone, occasional limestone.
Soils	Leached, infertile acid soils throughout NW Europe; local pockets of less acid, richer soils.
Taxonomic diversity	Low species richness, variable degree of endemism; local richness associated with edaphic conditions; high cryptogam richness in western fringes. Genetic diversity reflects microhabitat diversity.
History of land use	Forest clearance, heavy grazing, trampling and fires throughout N Europe, though relative extent of deforestation greatest in Britain. Episodes of pollution.
Pattern of contemporary land use	'Pockets' of human activity – agriculture, forestry, tourism, hunting/field sports and renewable energy. Agriculture predominant land use, followed by tourism and forestry. Management for field sports prevalent in Britain (principally for red grouse (*Lagopus lagopus scoticus*) and red deer (*Cervus elaphus*)), though hunting widespread throughout N European mountains. Intensive reindeer (*Rangifer tarandus*) husbandry in northern Norway and northern Sweden; sheep (*Ovis aries*) grazing in Iceland and Britain.
Pollution	High wind speed and precipitation contribute to high rates of deposition of pollutants. High exposure to acidification, eutrophication, heavy metal deposition and persistent organic pollutants.

No part of Europe's mountains is truly natural, though substantial areas are near-natural or semi-natural (with a full complement of native species). As a general rule, the hillfoot regions, and those close to principal areas of agricultural activity are the least natural. The impact of human activity has produced some unique 'cultural' ecosystems, such as heather-dominated moorland in Britain and other Atlantic regions of Europe (e.g. Behre, 1988; Ratcliffe & Thompson, 1988; Thompson *et al.*, 1995; Scottish Natural Heritage, 2002), and the species-rich montane meadows and pastures of continental Europe (with the alpine pastures also bearing the impact of millennia-long summer grazing by herds of domestic ungulates, e.g. Grabherr, this volume). As a result of socio-economic changes and fiscal policies, the cultural landscapes have been changing in two different directions. Some areas have been abandoned and are shifting to less species-rich grassland formations and those in the montane forest zone are becoming re-colonised by secondary forest and scrub (e.g. Andre, 1998, for Massif Central, France). Other areas have been affected by intensification of animal husbandry, resulting in the conversion of hay meadows to managed grasslands (e.g. Garcia, 1992, for Spain), or moorland to rough grassland and re-seeded pastures (e.g. Thompson *et al.*, 1995, for the UK).

4.3 Conservation management issues – some policy drivers and research needs

The mountain areas of Northern Europe are affected by a wide range of global, national and local policy and management issues (e.g. Baldascini *et al.*, 2002; Thompson, 2002; Price *et al.*, 2002). Table 4.2 gives examples of some of the main global and European agreements, laws and strategies that drive policy development in relation to the conservation of nature across Europe's mountains; the listing is by no means exhaustive. Many of the global conventions are focused on the conservation and sustainable use of ecosystems and biodiversity. With the growing number of protected areas, there are growing obligations to monitor change. For instance, at the European Union level there are Special Protection Areas and Special Areas of Conservation and, at the international level, RAMSAR sites and Biosphere Reserves (designated by UNESCO's Man and Biosphere Programme to be sites of excellence linking conservation and sustainable development at the regional level; Price, 2002). Several international co-operative programmes have been developed to undertake assessments of biodiversity (e.g. DIVERSITAS Global Mountain Biodiversity Assessment, Körner & Spehn, 2002), and the potential impacts of climate change, globally, on alpine plants and vegetation (GLORIA; see Pauli *et al.*, 2004).

At the research level, three important areas for development are evident. First, there is a need to develop our understanding of geo-biological and socio-economic processes to be able to understand landscape and biological change, and the factors influencing different elements of this (see Jonasson *et al.*, this volume). Second, there remains a considerable amount of work to be done on managing mountain ecosystems, habitats and species. There are considerable problems of scale, where research at the experimental plot level has revealed a very fine and precise understanding of, for instance, vegetation change, but not at the local landholdings or larger topographic scale. Indeed, there is considerable work to be done on the socio-economic forces which have influenced, and continue to influence, the nature of change at the landscape level. Here, there would appear to be considerable potential gain from international collaboration to share experience of land management practices.

Table 4.2 Examples of global and European conventions and legislation influencing the conservation management of Northern Europe's mountains. Information sources are provided for the conventions. Adapted from the European Centre for Nature Conservation (2005).

A. GLOBAL CONVENTIONS, STRATEGIES AND AGREEMENTS		
Name	**Purpose**	**Information source**
UNESCO's Man and Biosphere (MAB) Programme (1971)	Develops the basis, within the natural and the social sciences, for the sustainable use and conservation of biological diversity, and for improvements in the relationship between people and their environment, globally.	World Network of Biosphere Reserves UNESCO, Division of Ecological Science, 1, rue Miollis, 75732 Paris Cédex 15, France. www.unesco.org/mab/
Convention concerning the Protection of the World Cultural and Natural Heritage (WHC or World Heritage Convention, 1972)	To encourage the identification, protection and preservation of the cultural and natural heritage, around the world which is considered to be of outstanding value to humanity. This includes the listing of World Heritage Sites.	The World Heritage Centre UNESCO, 7, place de Fontenoy, 75352 Paris 07 SP, France. www.unesco.org/whc/
World Conservation Strategy (1980)	Developed and launched by IUCN/UNEP /WWF to help advance the concept of sustainable development through the conservation of living resources. The Strategy: explains the contribution of living resource conservation to human survival and to sustainable development; identifies the priority conservation issues and the main requirements for dealing with these; and proposes effective ways for achieving the Strategy's aim.	Strategy by IUCN/UNEP/WWF
World Charter for Nature (1982) United Nations General Assembly Resolution 37/7	Sets out 24 principles of conservation by which all human conduct affecting nature is to be guided and judged.	Charter by IUCN/UNEP/WWF
Convention on Wetlands of International Importance Especially as Waterfowl Habitat (Ramsar Convention, 1971)	To ensure the conservation of wetlands, especially those of international importance, by fostering wise use, international cooperation and reserve creation referred to as (Ramsar sites)	IUCN – Ramsar Convention Bureau Rue Mauverney 28, 1196 Gland, Switzerland. www.ramsar.org
Convention on the Conservation of Migratory Species of Wild Animals (Bonn Convention or CMS, 1979)	To provide a framework for the conservation of migratory species and their habitats by means of, as appropriate, strict protection and the conclusion of international agreements. The Convention seeks to ensure strict protection for species listed in appendices, notably migratory species in danger of extinction throughout all or a significant portion of their range.	UNEP/CMS Secretariat Martin-Luther-King Straße 8, 53175 Bonn, Germany. www.wcmc.org.uk:80/cms/

A. GLOBAL CONVENTIONS, STRATEGIES AND AGREEMENTS (cont.)

Name	Purpose	Information source
AEWA - African-Eurasian Migratory Waterbird Agreement (comes under the Bonn Convention)	The conservation of African-Eurasian migratory waterbirds through coordinated measures to restore species to a favourable conservation status or to maintain them in such a status.	UNEP/CMS Bonn, Germany. www.wcmc.org.uk/aewa/home.html
Convention on International Trade in Endangered Species (CITES or Washington Convention, 1973)	To regulate the trade in endangered species of wild fauna and flora.	CITES Secretariat 15, chemin des Anémones, 1219 Châtelaine-Genéve, Switzerland. www.cites.org
Global Biodiversity Strategy (1992)	The Strategy includes 85 specific proposals for action to conserve biodiversity at the national, international and local levels. It is intended to stimulate fundamental changes in how individuals, nations, and organisations perceive, manage, and use the earth's biological wealth.	Report by World Resources Institute/IUCN/UNEP
UN Convention on Biological Diversity (CBD or Rio Convention, 1992)	The conservation of biological diversity, the sustainable use of its components and the fair and equitable sharing of the benefits arising out of the utilization of genetic resources. The Convention provides a framework for conserving biodiversity. Most of its articles set out policy guidelines which Parties can follow, rather than establishing precise obligations or targets.	Secretariat for the Convention on Biological Diversity World Trade Centre, 393 St. Jacques Street, Office 300, Montréal, Québec H2Y 1N9, Canada. www.biodiv.org
UN Framework Convention on Climate Change (UNFCCC or Climate Change Convention, 1992)	Sets an overall framework for intergovernmental efforts to tackle the challenge posed by climate change. It recognizes that the climate system is a shared resource whose stability can be affected by industrial and other emissions of CO_2 and other greenhouse gases. Under the Convention, governments: gather and share information on greenhouse gas emissions, national policies and best practices; launch national strategies for addressing greenhouse emissions and adapting to expected impacts, including the provision of financial and technological support to developing countries; and co-operate in preparing for adaptation to the impacts of climate change. The Kyoto Protocol, adopted in December 1997, strengthens the commitment of governments to achieving the aims of the Convention.	UNFCCC Secretariat PO Box 260124, 53153 Bonn, Germany. www.unfccc.int

B. EU NATURE CONSERVATION POLICIES: PROGRAMMES, DIRECTIVES, REGULATIONS, DECISIONS, DESIGNATIONS AND STRATEGIES

Name	Purpose
Sixth Environmental Action Programme 2001-2010 – *Environment 2010: Our Future, Our Choice*	The Programme takes a wide-ranging approach and gives a strategic direction to the Commission's environmental policy over the next decade. There are four priority areas for urgent action: (i) Climate Change; (ii) Nature and Biodiversity; (iii) Environment and Health and Quality of life; and (iv) Natural Resources and Waste. The programme sets out the following main avenues for action to be explored: • effective implementation and enforcement of environmental legislation (necessary to set a common baseline for all EU countries); • integration of environmental concerns (environmental problems have to be tackled where their source is, and this is frequently in other policies); • use of a blend of instruments (all types of instruments have to be considered, the essential criterion for choice being that it has to offer best possible efficiency and effectiveness); and • stimulation of participation and action by all 'actors', from business to citizens, NGOs and social partners, through provision of better and more accessible information on the environment and joint work on solutions.
DIRECTIVES	
Directive on the conservation of natural and semi-natural habitats and of wild fauna and flora (92/43/EEC) (Habitats Directive, 1992)	To conserve fauna, flora and natural habitats of EU importance. The fundamental purpose of this Directive is to establish a network of protected areas throughout the Community designed to maintain both the distribution and the abundance of threatened species and habitats, both terrestrial and marine. The network of Special Areas of Conservation (SAC) is called Natura 2000, and includes SPAs classified under the Birds Directive. Criteria for selection include priority habitats and species, as identified in the Annexes.
Directive on the conservation of wild birds (79/409/EEC) (Birds Directive, 1979)	This Directive imposes strict legal obligations on European Union Member States to maintain populations of naturally occurring wild birds at levels corresponding to ecological requirements, to regulate trade in birds, to limit hunting to species able to sustain exploitation, and to prohibit certain methods of capture and killing. Article 1 applies to the conservation of birds and also to their eggs, nests and habitats. Article 4 requires Member States to take special measures to conserve the habitats of certain listed threatened species through the designation of Special Protection Areas (SPA).
Directive on the quality of fresh waters needing protection or improvement in order to support fish life (78/659/EEC) (1978)	To establish quality requirements for waters intended to support fish life. Regulates sampling frequency, measuring methods and measurement requirements, as well as conditions in which the quality requirements are to be achieved.

B. EU NATURE CONSERVATION POLICIES: PROGRAMMES, DIRECTIVES, REGULATIONS, DECISIONS, DESIGNATIONS AND STRATEGIES (cont.)

Name	Purpose
Directive establishing a framework for Community action in the field of water policy (2000/60/EC) (Water Framework Directive, 23 October 2000)	Although not explicitly aimed at the protection of particular species, communities, biotopes or habitats, biodiversity is the central indicator used by this Directive to define what constitutes good ecological status.
Directive on the assessment of the effects of certain public and private projects on the environment (85/337/EEC) (EIA Directive, 1985)	Requires that projects are made subject to an assessment regarding their effects which, according to type, size and location, may have considerable effects on the environment.
	The content of the EIA is outlined in Article 3, which defines the term 'environment'. According to this article, the EIA has to identify, describe and assess the direct and indirect effects of a project on human beings, fauna and flora, soil, water, air, climate and the landscape as well as on the interaction of these factors and on material assets and the cultural heritage. Due importance is given to articles under the Habitats and Birds Directives.
REGULATIONS	
Regulation on the protection of the Community's forests against atmospheric pollutants (3528/86/EEC) (1986)	To establish a scheme for the production of periodic forest health reports, and financing for field experiments and pilot projects aimed at maintaining and restoring damaged forests.
Regulation on the establishment of the European Environment Agency and the European Environment Information and Observation Network (1210/90/EEC) (1990)	To establish a European Environment Agency (EEA), to provide the EU and the Member States with objective, reliable and comparable information at the European level, so that requisite measures are taken to protect the environment, to assess the results of these measures, and to inform the public about the state of the environment. The Agency publishes or makes available to the public environmental data supplied to it or emanating from it, subject to Community and Member States rules on the confidentiality of information.
Regulation on protection of the Community's forests against fire (2158/92/EEC) (1992)	To establish a Community financing scheme comprising measures to identify the causes of forest fires and the means to combat them, as well as measures to set up or improve systems of prevention, such as the launching of protective infrastructure.
DECISIONS	
Decision on the conclusion of the Convention on the conservation of European wildlife and natural habitats (82/72/EEC) (1982); and Convention on the conservation of migratory species of wild animals (82/461EEC) (1982)	Approval of the Bern Convention, to enable closer co-operation between countries in activities to protect wild flora and fauna in their natural habitat, and to enable concerted action to preserve and manage endangered migratory species.

B. EU NATURE CONSERVATION POLICIES: PROGRAMMES, DIRECTIVES, REGULATIONS, DECISIONS, DESIGNATIONS AND STRATEGIES (cont.)	
Name	**Purpose**
DESIGNATIONS	
Natura 2000	The network of protected areas set up under the Birds and Habitats Directives. Under the Habitats Directive, Natura 2000 is defined in Article 3(1) as a coherent European ecological network of special areas of conservation. This network, composed of sites hosting the natural habitat types listed in Annex I, and habitats of species listed in Annex II, would enable the natural habitat types and the species' habitats concerned, to be maintained or, where appropriate, restored at a favourable conservation status in their natural range.
	Article 10 indicates that European Union Member States should endeavour, in their land-use planning and development policies, to encourage and manage features of the wider landscape which are of importance for wild fauna and flora. Linear features, such as rivers and hedgerows, and isolated elements, such as lakes and ponds, are essential for migration, dispersal and genetic exchange of wild species.
	Based on the Birds and Habitats Directives, the network sets the minimum standard for biodiversity conservation in the Member States, encompassing a wide range of issues and containing a number of concrete obligations. This concept is strengthened by the Maastricht Treaty, according to which all Community policies and instruments must comply with the Community's environmental statutes, including the Habitats and Birds Directives.
STRATEGIES	
European Community Biodiversity Strategy (1998)	This strategy aims to anticipate, prevent and tackle the causes of significant reduction or loss of biodiversity at the source. This will help both to reverse present trends in biodiversity reduction or losses and to place species and ecosystems, including agro-ecosystems, at a satisfactory conservation status, both within and beyond the territory of the European Union.

Third, much remains to be done on interpreting long-term data and collecting quality data for ongoing and future monitoring, and therefore understanding the nature of change.

4.4 Some pointers to the future

The chapters in this book draw together research, policy and management-related activities across Northern Europe. To date, there has been something of a dislocation in the links between policy, science and conservation management activities. For instance, the Common Agricultural Policy, comprising some 50% of the European Union budget, was, until the late 1990s, only marginally influenced by a growing knowledge of environmental responses to grazing and agricultural intensification. Policy development, in terms of agriculture, forestry, tourism and, more recently, renewable energy, has had sectoral histories, with little attention focused on integration. We can characterise the main policy, science and conservation management activities under just a few broad headings (Figure 4.3); the linkages between these three domains of activity need to be strong.

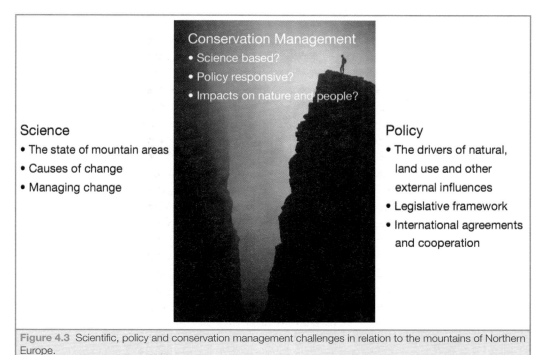

Figure 4.3 Scientific, policy and conservation management challenges in relation to the mountains of Northern Europe.

The challenge ahead is to integrate research which is able to address questions about consequences of socio-economic and environmental change and their interactions. Researchers will need tools and techniques to help cope rapidly with the impacts of interacting land uses, and the windows of opportunity presented in policy development, touching on agriculture, forestry, tourism, renewable energy and recreation. The chapters in this book demonstrate the complexity of human and natural influences on mountain ecosystems and societies.

However, the different disciplines of research appear now to be more aware of the need to integrate approaches and interests in order to develop a more holistic understanding of what is happening across Europe's mountains, and what needs to be done to improve the conservation and management of nature and other interests. Perhaps the greatest challenge facing land users, researchers and policy makers is the large-scale and rapid nature of environmental and societal changes. This will place a heavy burden on the development of research sufficiently robust to predict these changes, and action needed to improve the conservation and management of mountain ecosystems.

Acknowledgements

The following are thanked for their contributions to this introductory paper: Ian Bainbridge, Rick Battarbee, Stephen Edwards, John Gordon, Georg Grabherr, Bill Heal, John Mackay, Martin Price and an anonymous referee.

References

Andre, M.F. (1998). Depopulation, land-use change and transformation in the French Massif Central. *Ambio*, **27**, 351-353.

Averis, A.B., Averis, A.M., Birks, H.J.B., Horsfield, D., Thompson, D.B.A. & Yeo, M. (2004). *An Illustrated Guide to British Upland Vegetation.* Joint Nature Conservation Committee, Peterborough.

Baldascini, A., Perlis, A. & Romeo, R.L. (2002). *International Year of Mountains: concept paper.* Food and Agriculture Organization of the United Nations, Rome.

Barry, R.G. (1981). *Mountain Weather and Climate.* Methuen, London.

Behre, K.-E. (1988). The role of man in European vegetation history. In *Vegetation History,* ed. by B. Huntley & T. Webb. Kluwer, Dordrecht. pp. 633-672.

EC Regulation 1257/1999. Support for rural development from the European Agricultural Guidance and Guarantee Fund (EAGGF). European Commission, Brussels.

European Centre for Nature Conservation (2005). EU Biodiversity and policy instruments. www.ecnc.nl/EuBiodiversityAndPol/Index_442.html.

Firbank, L.G., Smart, S.M., van de Poll, H.M., Bunce, R.G.H., Hill, M.O., Howard, D.C., Watkins, J.W. & Stark, G.J. (2000). Causes of change in British vegetation. ECOFACT vol. 3. DETR, HMSO, Norwich.

Garcia, A. (1992). Conserving the species-rich meadows of Europe. *Agriculture Ecosystems & Environment*, **40**, 219-232.

Gerrard, A.J. (1990). *Mountain Environments: An Examination of the Physical Geography of Mountains.* Belhaven, London.

Hallanaro, E.-L. & Pylvänäinen, M. (Ed.) (2002). *Nature in Northern Europe - Biodiversity in a Changing Environment.* Nord 2001: 13. Nordic Council of Ministers, Helsinki.

Huddleston, B., Ataman, E., da Salvo, P., Zanetti, M., Bloise, M., Bel, J., Franceshini, G. & Fè d'Ostiani, L. (2003). *Towards a GIS-Based Analysis of Mountain Environment and Population.* Food and Agriculture Organisation and the United Nations, Rome.

Kapos, V., Rhind, J., Edwards, M., Price, M.F. & Ravilious, C. (2000). Developing a map of the world's mountain forests. In *Forests in Sustainable Mountain Development: a state-of-knowledge report for 2000,* ed. by M.F. Price & N. Butt. CAB International, Wallingford. pp. 4-9.

Körner, C. & Sphen, E.M. (Ed.) (2002). *Mountain Biodiversity: A Global Assessment.* Parthenon, London.

McVean, D.N. & Ratcliffe, D.A. (1962). *Plant communities of the Scottish Highlands.* HMSO, Edinburgh.

Messerli, B. & Ives, J.D. (Ed.) (1997). *Mountains of the World: a Global Priority.* Parthenon, Carnforth.

Nagy, L., Grabherr, G., Körner, C.H. & Thompson, D.B.A. (Eds) (2003a). *Alpine Biodiversity in Europe.* Springer-Verlag, Berlin.

Nagy, L., Thompson, D.B.A., Grabherr, G. & Körner, C. (2003b) *Alpine Biodiversity in Europe: an Introduction.* Joint Nature Conservation Committee, Peterborough.

Ozenda, P. (1994). *La Végétation du Continent Européen.* Delachaux et Niestle, Lausanne.

Ozenda, P. (2002). *Perspectives pour une Géobiologie des Montagnes.* Presses Polytechniques et Universitaires Romandes, Lausanne.

Pauli, H., Gottfried, M., Hohenwallner, D., Reiter, K., Casale, R. & Grabherr, G. (2004). *The GLORIA Field Manual – Multi-Summit Approach.* DG Research, EUR 21213, Official Publications of the European Communities, European Commission, Luxembourg.

Pearsall, W.H. (1971). *Mountains and Moorlands.* The New Naturalist, London, Collins.

Price, M.F. (2002). The Periodic Review of Biosphere Reserves: A mechanism to foster sites of excellence for conservation and sustainable development. *Environmental Science and Policy*, **5**, 13-19.

Price, M.F., Dixon, B.J., Warren, C.R. & Macpherson, A.R. (2002). *Scotland's Mountains: Key Issues for their Future Management.* Scottish Natural Heritage, Perth.

Price, M.F., Lysenko, I. & Gloersen, E. (2004). La delimitation des montagnes européennes/Delineating Europe's mountains. *Revue de Geographie Alpine/Journal of Alpine Research,* **92**, 61-86.

Ratcliffe, D.A. (Ed.) (1977). *A Nature Conservation Review.* 2 vols. Cambridge University Press, Cambridge.

Ratcliffe, D.A. & Thompson, D.B.A. (1988). The British Uplands: their ecological character and international significance. In *Ecological Change in the Uplands,* ed. by M.B. Usher & D.B.A. Thompson. Blackwell, Oxford. pp. 9-36.

Scottish Natural Heritage (2002). *Scotland's moorland: unique and important. A Statement of Intent* (marking the establishment of Scotland's Moorland Forum). Scottish Natural Heritage, Perth,

Sissons, J.B. (1976). *The Geomorphology of the British Isles: Scotland.* Methuen, London.

Thompson, D.B.A. (2002). The importance of nature conservation in the British Uplands: nature conservation and land use changes. In *The British Uplands: Dynamics of Change,* ed. by T.P. Burt, D.B.A. Thompson & J. Warburton. Joint Nature Conservation Committee, Peterborough. pp. 36-40.

Thompson, D.B.A. & Brown, A. (1992). Biodiversity in montane Britain: habitat variation, vegetation diversity and some objectives for conservation. *Biodiversity and Conservation,* **1**, 179-208.

Thompson, D.B.A., Gordon, J.E. & Horsfield, D. (2001). Montane landscapes in Scotland: are these natural, artefacts, or complex relics? In *Earth Science and the Natural Heritage: Interaction and Integrated Management,* ed. by J.E. Gordon & K.F. Leys. The Stationery Office, Edinburgh. pp. 105-119.

Thompson, D.B.A., MacDonald, J.A. & Hudson, P.J. (1995). Upland moors and heaths. In *Managing Habitats for Conservation,* ed. by W.J. Sutherland & D.A. Hill. Cambridge University Press, Cambridge. pp. 292-362.

UNESCO (2002). *UNESCO in the Mountains of the World.* UNESCO, Paris.

Väisänen, R. (1998). Current research trends in mountain biodiversity in NW Europe. *Pirineos,* **151-152**, 131-156.

Watson, A. (1996). Introduction. In *Scotland's Mountains: an Agenda for Sustainable Development,* ed. by A. Wightman. Scottish Wildlife and Countryside Link, Perth.

Whittow, J. (1992). *Geology and Scenery in Britain.* Chapman Hall, London.

5 Links between geodiversity and biodiversity in European mountains: case studies from Sweden, Scotland and the Czech Republic

Christer Jonasson, John E. Gordon, Milena Kociánová, Melanie Josefsson, Igor J. Dvořák & Des B.A. Thompson

Summary

1. Spatial patterning in mountain habitats and species is a function of geological and climatic factors, migration routes, geomorphological changes and human impacts all operating over different timescales in the past and present. This applies at the landscape as well as the local scale.
2. Past geological and geomorphological events continue to influence present functional links between geomorphology and biodiversity.
3. Mountain environments are episodically dynamic in response to geomorphological processes, which operate with different magnitudes and frequencies and are in many cases based on short-term climatic events.
4. The nature, rate and location of geomorphological processes reflect complex responses to antecedent conditions, trigger events and readjustment times. Dynamic equilibrium can be disturbed by human stresses (trampling, recreation activities, pollution, deforestation), grazing pressure by herbivores and long-term climate change. There is potential for irreversible changes on human timescales if geomorphological and ecosystem thresholds are crossed.
5. Integrated management of mountain areas and ecosystems needs to recognise and understand the functional links between climate, geomorphological process dynamics, terrain sensitivity, ecological processes and biodiversity.

5.1 Introduction

Compared with biodiversity, geodiversity is a relatively new concept, although it is one that is receiving increasing attention both in its own right and through the links to biodiversity, landscape and natural heritage management (e.g. Johansson, 2000; Gordon & Leys, 2001; Gray, 2003). Geodiversity is the complex variety of rocks, sediments, landforms, soils and geomorphological processes that form landscapes. The geodiversity of natural landscapes is a function of changes in climate and geomorphological processes through time. It also reflects the cumulative effects of these changes, which operate over different timescales (e.g. from glaciations over thousands of years to seasonal freezing of the soil). Geodiversity also

Jonasson, C., Gordon, J.E., Kociánová, M., Josefsson, M., Dvořák, I.J. & Thompson, D. B.A. (2005). Links between geodiversity and biodiversity in European mountains: case studies from Sweden, Scotland and the Czech Republic. In *Mountains of Northern Europe: Conservation, Management, People and Nature,* ed. by D.B.A. Thompson, M.F. Price & C.A. Galbraith. TSO Scotland, Edinburgh. pp. 57-70.

involves the triggering mechanisms that initiate episodic landscape forming processes, such as soil erosion, debris flows, snow avalanches and slush avalanches. This is particularly important in terms of climate change scenarios, where mountain areas might experience quite different conditions and responses compared with what we see today, for example, as a result of the melting of permafrost (Rapp *et al.*, 1997). In many areas, geodiversity is closely linked with human activities. Therefore, from a management perspective, it is important to know how geomorphological processes are acting, especially since these might vary considerably over time. It follows that artificial constructions and installations in the mountains need to be planned to take account of changes in the magnitude and frequency of extreme events (Lundkvist, 2001).

There is also growing recognition by ecologists and geomorphologists of the fundamental links and dependencies between geodiversity and biodiversity (e.g. Swanson *et al.*, 1988; Köhler *et al.*, 1994; Palacios & Sánchez-Colomer, 1997; Nichols *et al.*, 1998; Gordon *et al.*, 2001; Thompson *et al.*, 2001; Kozłowska & Rązkowska, 2002; Barrett *et al.*, 2004) and allied to this, the value of maintaining dynamic natural processes and change as part of a more functional approach to conservation of biodiversity and ecosystem management (Fiedler *et al.*, 1997; Poff *et al.*, 1997; Poiani *et al.*, 2000; Nagy *et al.*, 2003). In mountain areas, these links are particularly strong, reflecting steeper slopes, more extreme climatic conditions and more intense geomorphological activity. They are often expressed directly through interactions with soils, topography, hydrology and snow cover and also indirectly through interactions with climate (e.g. Grabherr, 1997; Palacios & Sánchez-Colomer, 1997; Gordon *et al.*, 2001; Kozłowska & Rązkowska, 2002). Using examples from three mountain areas in Europe – the Abisko Mountains (Sweden), the Giant Mountains (Czech Republic) and the Cairngorms (Scotland) – we examine some of these dependencies, links over different timescales and across different spatial scales, and their implications for biodiversity conservation frameworks. These areas are located in different climate zones but display many similarities in terms of past and present geomorphology and habitats (Gordon *et al.*, 2002).

5.2 Landform inheritance and habitat diversity at different spatial scales

5.2.1 Large-scale landform patterns

Mountain landscapes situated in northern or mid-Europe generally comprise geological and geomorphological features of different age, size and origin. For example, in the Abisko area in northern Sweden, it is possible to detect a number of landforms of varying age and origin (Figure 5.1). The relatively flat surface seen in the background is an old peneplain that has existed for several tens of millions of years. The large valleys, like those occupied by the lake, are cut down into the erosion surface. They mainly result from processes of weathering and erosion that have been active over many millions of years.

It is only during the last two million years that the climate deteriorated and ice sheets and glaciers expanded across Northern Europe and in the mountainous parts of mid-Europe. The Abisko area was heavily glaciated during long periods of the Quaternary. Several major landforms were created by glacial activity – for example, glaciated valleys and ice-scoured bedrock. However, even if many of the landforms in the Abisko area are of glacial origin, most of the valleys and bigger landforms originated before the Quaternary

Figure 5.1 The Abisko area and Torneträsk viewed from the west. The broad outlines of the landscape are essentially pre-glacial in origin and have been only locally modified by selective glacial erosion. (Photo: L. Bäck).

period (Stroeven *et al.*, 2002). The reason that these landforms still exist is that the glacier ice that covered them was cold-based and was 'selective' in its landscape modification (Kleman, 1992). Hence the main elements of the ancient landscape are more or less preserved.

Similar patterns are evident in the Cairngorms and the Giant Mountains, where the broad outlines of the landscape reflect the geological character of the massifs and their long-term evolution through a range of geomorphological processes during warmer pre-glacial climatic conditions. Here, too, the glacial signature is superimposed on these older landscape elements (Sekyra, 1964; Brazier *et al.*, 1996; Migoń, 1999). In the Cairngorms, the broad outlines of the landscape reflect the form of the granite pluton which was unroofed during the Devonian; weathering and erosion then formed a series of etchplains and selectively exploited linear zones of altered granite to form the main pre-glacial valleys, which in turn were selectively enhanced by glacial erosion (Thomas *et al.*, 2004). Over a wider area in Scotland, the variations in geological history and landscape evolution have produced notable contrasts in present-day mountain landforms (e.g. between the heavily dissected mountains of the North-West Highlands and the plateau surfaces of the Cairngorms). This, combined with wind exposure, precipitation gradients and other climatic factors, has produced a remarkable assemblage of arctic, temperate and oceanic plant communities within a relatively small geographical area (Thompson *et al.*, 2001).

The Giant Mountains originated through uplift during the Cadomian (Late Proterozoic) and Variscan (Palaeozoic) orogenies. The large-scale landforms comprise two mostly southward – draining peneplains (etchplains) formed by tectonic uplift during the Mesozoic and Tertiary, bordered by hard rock (granite and contact metamorphic rock) summits and W-E directed fault lines in the northern part in Poland. This gives rise to an asymmetrical topography, with steeper slopes, locally covered by blockfields, in the north and more gently inclined slopes to the south and east with more developed drainage patterns (Sekyra, 1964). The northern Polish part is also characterised by glacial corries in the alpine zone and locally by short glaciated valleys. The Czech part to the south has long windward–leeward glaciated valleys and corries, the floors of which lie in the forested montane zone.

The implication in all three areas is that the broad-scale diversity of habitats, from the valley bottoms to the mountain tops, is linked to large-scale landform patterns that developed cumulatively under similar climate conditions and geomorphological processes, but different from those of the present day, and over millions to tens of thousands of years.

5.2.2 Intermediate-scale landform patterns

Habitat diversity also displays close dependencies on intermediate-scale landform patterns. In the Abisko Mountains, glacial deposits cover most of the valley bottoms and there is a strong zonal pattern linked to elevation, landform and soils (Rapp, 1960; Darmody *et al.*, 2004). Along the slopes facing north towards the Torneträsk valley, there are many flat areas of sandy material. These are deltas that were formed when meltwater collected between the ice remaining in the valley bottom and the mountain side. From a drainage perspective, these areas differ from their surroundings, and show different plant distributions, with a tendency for birch forest on drier glaciofluvial deposits, and bogs and palsas on wetter valley floors (Melander, 1977).

In the Cairngorms, the glacial legacy has given rise to a diversity in landform assemblages, each with its own mosaic of habitats – arctic-like plateau surfaces, corries with cliffed headwalls and complex patterns of glacial deposits and meltwater channels on the lower slopes. This, combined with wind exposure, precipitation gradients and other climatic factors, has produced a remarkable assemblage of arctic, temperate and oceanic plant communities (Thompson *et al.*, 1993).

In the Giant Mountains, the present distribution of habitats is closely linked to interactions between the intermediate-scale landforms (mainly the glacially shaped topography) and the prevailing climate (Figure 5.2) (Jeník, 1997). The Giant Mountains are exceptional among the middle mountains of mid-Europe for their relict periglacial features, including tors, blockfields, different types of patterned ground, solifluction lobes and nivation hollows (Sekyra, 1964; Soukupová *et al.*, 1995). The relict periglacial landforms and contemporary interactions between snow, frost and wind and vegetation provide the basis for what has been termed 'arctic-alpine tundra' on the exposed surfaces. Three aspects of this arctic-alpine tundra are of particular significance (Štursa, 1998):

- lichen tundra developed on the periglacial blockfields of the higher cryoplains;
- alpine grassland with vegetated patterned ground, scattered stands of dwarf pine (*Pinus mugo*) and subarctic-subalpine mires developed on intermediate plateau surfaces; and
- plant and animal communities on the leeward slopes of former glacial corries and nivation niches related to short- to long-lying snowbeds and snow avalanche debris pathways.

These landform-vegetation links clearly demonstrate the dependencies of the biodiversity on the geodiversity at this scale.

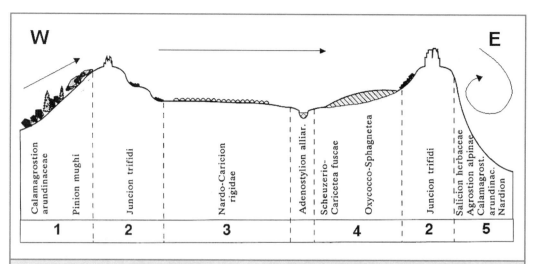

Figure 5.2 Schematic profile from west to east across the Giant Mountains, showing the 'anemo-orographic system' and the close links between geodiversity and biodiversity (after Jeník, 1961, 1997; Kociánová & Sekyra, 1995). The present distribution of habitats is closely linked to the interactions between the glacially shaped topography and the prevailing climate. 1. On the exposed upper windward slopes, krummholz pine and *Vaccinium* heath give way to 'arctic-alpine tundra' on the exposed plateau surfaces. 2. On the summit tors and blockfields, lichen predominate, with scattered *Juncus* communities on the gravelly granite regolith and grasses and *Carex* on the finer material in relict sorted circles. 3. Alpine grassland (*Nardo-Caricion rigidae*) with vegetated patterned ground (non-sorted stripes), scattered stands of dwarf pine (*Pinus mugo*) and subarctic-subalpine mires (*Oxycocco-Sphagnetea* and *Scheuchzerio-Caricetea fusci*) are developed on intermediate plateau surfaces. 4. Short- to long-lying snowbeds in sheltered hollows support *Nardo-Callunetea*. 5. Steep slopes of corrie and glaciated valley headwalls with snow avalanche-related plant communities (*Salicion herbaceae*, *Agrostion alpinae* and *Calamagrostion arundinaceae*) on the leeward slopes. Arrows indicate prevailing wind direction. Vegetation classed according to Braun-Blanquet (1928).

5.2.3 Small-scale landform patterns

At a small scale, the detailed mosaics of habitats and plant communities in all three study areas display close functional links with regolith and soil properties, small-scale landform patterns, exposure, snow-lie patterns and microclimate. In the Cairngorms, there are close dependencies between individual plant communities and the local geomorphology, soils and climate (Gordon *et al.*, 1998; Haynes *et al.*, 1998). In the Giant Mountains, there are broadly three types of frost-sorted soils (stony, partly overgrown, completely overgrown), each with a different vegetation cover (varying dominance of lichens, mosses, *Nardus stricta–Deschampsia flexuosa*, *Calluna vulgaris*, *Vacciniaceae* and invasive plant species), dependent mainly on soil properties and microclimate (Soukupová & Kociánová, 1995). The mosaic of montane habitats there is linked to topography, slope, snow-lie and drainage (Figure 5.2).

In the Abisko Mountains, Kozłowska & Rązkowska, (2002) demonstrated similar broad links between geomorphological processes, landform types and vegetation associations at a local scale, although a precise characterisation of landforms by phytoindicators was complicated by several factors: different geomorphological processes may produce similar habitat conditions; variations in process intensity may influence vegetation development; some processes such as solifluction may occur over a relatively broad spatial and altitudinal area; and variations in bedrock geology may affect soil chemistry. This illustrates the importance of scale and interactions of other factors such as microclimate, hydrology, topographic position and exposure and the geological substrate in mediating links between geomorphological diversity and biodiversity.

5.3 Species distributions and past events

Biodiversity reflects changes in climate and geomorphological processes over different time scales. For example, patterns of Quaternary glaciation and distributions of refuges have played an important role in present biodiversity. In the Giant Mountains, corries and major valley heads acted as sources for valley glaciers during the Quaternary glaciations, but the higher summits and corrie headwalls are believed to have been unglaciated (nunataks) and acted as refuges. The maximum limits of the Scandinavian ice sheet lie just to the north of the mountains. Botanically and zoologically, therefore, the Giant Mountains occupy a biogeographical 'crossroads' in relation to the Alps to the south and the Scandinavian mountains to north, with a merging of arctic and alpine communities on the ice-free slopes during Quaternary glaciations (Soukupová *et al.*, 1995).

The Giant Mountains have been described as 'the arctic island in the middle of Europe' (Štursa, 1995). Botanically and zoologically, glacial relics include, for example, cloudberry (*Rubus chamaemorus*), arctic saxifrage (*Saxifraga nivalis*), Sudetic lousewort (*Pedicularis sudetica*) and the mountain dragonflies (*Somatochlora alpestris* and *Aeschna caerulea*) and Gyllenhal carabid beetle (*Nebria gyllenhali*). There are also numerous endemic plant and animal species. Today the Giant Mountains are unique among the Hercynian ranges in rising above the timberline, thereby allowing the preservation of a remarkable biodiversity which has its origins during the Quaternary glaciations.

5.4 Biodiversity and active geomorphological processes

Active geomorphological processes, including debris flows, avalanches and slush avalanches/torrents, are important in maintaining dynamic environments for habitat development and as transporters of seeds along mountain slopes. The location and distribution of these processes is dependent on factors such as topography, debris supply, snow accumulation patterns and climate. In the Giant Mountains, for example, the form of the relict topography, comprising extensive plateau surfaces and steep headwalls of glacial valley heads, combined with the prevailing climatic conditions and active geomorphological processes (Jeník, 1997), play an important role in habitat and species diversity. Snow accumulations, both on the plateaux and on lee slopes, are significant (1.5-3.0 m average; up to 17.0 m maximum) and avalanches are frequent in winter and spring. These avalanches help to maintain the biodiversity of the mountain slopes by creating clear tracks through the forest and giving rise to the remarkable floristic diversity and the so-called 'natural gardens' on the leeward slopes (Figure 5.3).

In the Cairngorms, the main montane vegetation zones are related to altitude, exposure and snow-lie duration (Watt & Jones, 1948): *Calluna* heath is replaced at higher levels by *Vaccinium*-lichen heath and on the highest slopes by alpine and boreal grasslands including *Juncus-trifidus-Racomitrium lanuginosum* rush-heath and *Carex bigeolowii-Racomitrium lanuginosum* moss-heath. Within each zone, the mosaic of communities varies with the main geomorphological processes affecting the soil and regolith – wind, frost heave, solifluction, runoff from snowmelt and rainstorms, and debris flows. A combination of these processes,

Figure 5.3 In the Giant Mountains, snow avalanches are a major influence on the pattern of vegetation zones on the lee slopes of east facing corries and glacial valley heads. They maintain clear tracks through the forest, seen here in Úpská jáma corrie, and give rise to a remarkable floristic diversity and the so-called 'natural gardens' on the leeward slopes. (Photo: Jan Vanek)

the wet and windy climate, the well-drained, gravelly soils and the local variations in topography, means that soil disturbance is common on the higher and more exposed slopes and that conditions are hostile for plant growth (Gordon *et al.*, 1998).

5.5 Role of extreme episodic events

Several landforms of postglacial origin are caused by distinct processes such as physical weathering, or soaked wet soil moving downslope during the snow-melt season. Until some decades ago it was unknown how fast this kind of landscape evolved. It was previously thought that the impact of slides and debris flows was greatest immediately after deglaciation, but it has been demonstrated that these kinds of processes have been active during the entire Holocene (Jonasson, 1991; Sletten, 2002). In the 1950s, scientific development within geomorphology was largely concerned with quantification of present-day geomorphological processes (e.g. Jäckli, 1957; Rapp, 1960). During this decade, geomorphologists realised the importance of extreme events. The increased knowledge in the field of extreme events has shown that the development of several types of landform has been highly episodic. For example, field studies in northern Sweden have shown that landform development of valley bottoms might be a result of rare events.

The valley bottom seen in Figure 5.4a was previously believed to be largely of fluvial origin with some avalanche material added on top (Melander, 1977). However, studies on the very coarse-grained sediments, as well as observations after the snowmelt period, have indicated that other kinds of processes have been responsible (Lundkvist, 1997). In particular, studies by a Swiss-German team in 1995 revealed the geomorphological significance of slush torrents which release tremendous amounts of water and sediments in a short time (Figure 5.4b) (Gude & Scherer, 1995). The frequency of occurrence of the slush torrents responsible for much of the sediment relocation on steeper slopes might be once in 10 years. Documentation of these slush torrents shows that they are active only for a few minutes. The effects of such episodic processes is potentially of major significance; for example, 2 minutes of activity every 10 years is equivalent to only one and a half days (33 hours) when extrapolated over the entire Holocene.

As another example, during a rainstorm at Abisko in 1998, river discharge and the transportation of suspended sediment increased substantially. Investigations showed that the transport of sediments during 2 hours of this event was equivalent to what was normally transported during 3-5 years (Jonasson & Nyberg, 1999). Lake sediments provide an important record of such events and are allowing the reconstruction of the magnitude and frequency of different landforming events (Jonasson, 1991; Rubensdotter, 2001; Sletten, 2002).

Debris flows, together with water erosion, are the most important active geomorphological processes on the valley sides of the Giant Mountains. During the last 100 years or so, debris flows have displaced about 200,000–300,000 m³ of weathered material and remodelled more than 80 ha of slopes. The bigger slides often occur in connection with localised heavy rain (minimum of 20 mm/1-2 hours), with major events documented in 1882, 1887, 1964, 1974 and 1997 (Pilous, 1973, 1975, 1977).

Figure 5.4 Slope modification by extreme events in Kärkevagge, near Abisko: a) alluvial/slush torrent fan (Photo: C. Jonasson); b) slush torrent event in 1995. (Photo: M. Gude/D. Scherer).

Snow avalanches also play an important role in the Giant Mountains. Monitoring of 51 avalanche sites in the Czech part between 1961/1962 and 1999/2000 has documented 780 avalanches. On 60% of the sites, a high frequency of avalanches has permanently destroyed the forest, producing areas of open habitat supporting diverse shrub vegetation, rich alpine tall-herb meadows and numerous endemic species. On 40% of the sites, episodic large avalanches penetrate deep into the forest and over a period of 16 years or more, the upper tree-line has fluctuated over a distance of 200-300 m along these avalanche paths (Jeník, 1961; Kociánová & Spusta, 2000). In April 1999 and May 2002, snow avalanches along the ground repeatedly displaced large amounts of material, comprising about 287 m^3 of rock and 1,408 m^3 of soil, sods and vegetation.

It is difficult to determine to what extent the different extreme events represented in such records reflect climatic trends or provide proxy climatic information, or indeed if they only reflect chaotic, occasional events. From monitoring activity at Abisko Scientific Research Station, as well as from preliminary results from ongoing research in the Giant Mountains and in Scotland, it seems as if many of the recorded extreme events have been a result of meteorological coincidences. However, such events have occurred repeatedly during the Holocene (Jonasson, 1991; Ballantyne, 2002). The significance of such events for biodiversity is threefold. First, periods of habitat stability or linear change may be punctuated by short episodes of instability and non-linear or episodic change. Second, such events contribute to geomorphological heterogeneity and hence habitat diversity over different timescales, so that geomorphological disturbance may provide new opportunities for biodiversity, from small patch to landscape scales (Burnett *et al.*, 1998; Nichols *et al.*, 1998). Third, the effects of disturbance may persist in the landscape for a long time and, depending on the magnitude and frequency of the events, ecological equilibrium may not be re-established or may be established in a different form.

5.6 Geomorphological sensitivity, landscape stability and montane ecosystems

Geomorphological systems may be described as robust or sensitive in terms of their responses to external perturbations (Brunsden, 2001; Werritty & Leys, 2001). Robust systems are able to absorb the impact or self-correct through feedback mechanisms. Such systems operate within limiting thresholds and the original landscape character is re-established after a triggering event such as an intense rainstorm. An example would be the recovery of a talus slope after a debris flow event; another would be stone stripes reforming the following winter after disturbance by trampling. Conversely, in sensitive or responsive systems, a fundamental change in the nature and rate of the landforming process occurs as an extrinsic threshold is crossed. The system moves from one process state to another where it may attain a new equilibrium. An example of a sensitive system would be deflation of a soil cover, and hence habitat loss, following disturbance of the surface vegetation. A new process environment is formed (a deflation surface) accompanied by vegetation and habitat change or loss. In ecology, ecosystem stability has been assessed in terms of resistance (ability to withstand displacement) and resilience (ability to recover after disturbance) (Holling, 1973; Hill, 1987).

The alpine/montane landscape is an intricate mosaic reflecting the interactions between endogenous factors (type of bedrock and its characteristics) and exogenous processes (such as weathering, deflation, frost activity in the soil, snow-lie patterns, hydrology, topography and vegetation patterns). Biodiversity is linked to the geological and geomorphological heterogeneity of the landscape, as well as to climate and migration routes. In the Cairngorms, some areas are inherently stable and are underlain by well-developed alpine podzols, indicating stability over thousands of years (Grieve, 2000); others are inherently dynamic. In many cases, there is a dynamic equilibrium that can be easily disturbed by human stresses (trampling, pollution, overgrazing, recreation activities, deforestation) and long-term climate change. For example, in some areas, a threshold has been crossed, leading to accelerated soil erosion and habitat loss (Haynes *et al.*, 1998).

5.7 Looking ahead

Under conditions of future climate change, how will the geomorphological process systems respond? Will there be more or less stability in the landscape? Warmer and wetter conditions predicted for some areas (e.g. for the tundra areas of Northern Europe, including Abisko, and for Scotland) may lead to more extensive grass- and sedge-dominated vegetation cover and changes in bottom-layer and canopy-layer species. This is supported by International Tundra Experiment (ITEX) studies in the Abisko Mountains (Molau & Alatalo, 1998; Arft *et al.*, 1999) and by the contemporary natural state of vegetation cover in both the Cairngorms and especially the Giant Mountains, which, due to their more southern geographical position, developed under warmer climatic conditions. However, it is not known if such changes are likely to mean potentially greater slope stability, or conversely lead to a higher probability of slope failure in saturated soils.

Human stresses – trampling, grazing and pollution – also have the potential to push sensitive systems closer to thresholds for irreversible change. We need to be able to identify which areas and habitats are most sensitive to change. In previous work, we have provisionally identified high summit plateaux, summit ridges, exposed spurs, cols and crests of corrie headwalls as likely to be particularly sensitive (Thompson *et al.*, 2001; Gordon *et al.*, 2002). However, we need better empirical data to allow the development of terrain sensitivity classifications and to improve the prediction of other areas most at risk. Current investigations of soil properties and root mat strength of the vegetation in the uplands of Scotland will provide much needed empirical data (Morrocco, this volume) and should help to inform erosion risk modelling (Lilly *et al.*, 2002, this volume). By using similar, combined approaches in areas of different geographical and climatic position, as in the sub-arctic Abisko mountains and the mid-Europe Giant Mountains, we should be better placed to understand the sensitivity of the geodiversity and biodiversity of mountain systems.

Several lessons emerge from this discussion in the light of current developments in geomorphology and ecological theory and a growing focus in conservation management on the functional integrity of ecosystems and landscapes. We need to move away from the narrow focus on preservation of biodiversity or individual species (e.g. Christensen *et al.*, 1996; Fiedler *et al.*, 1997; Woodwell, 2002; Nagy *et al.*, 2003). Ecologists now recognise that ecosystems are dynamic entities, that episodic change is a fundamental characteristic of ecosystems, and that ecosystems can switch rapidly (and sometimes irreversibly) from one state to another (Holling *et al.*, 1995). Mountain ecosystems are dynamic in space and time, in terms of both ecological and geomorphological processes. It follows that land management activities need to focus on maintaining natural processes and ecosystem functioning, rather than trying to address the individual landforms or species as if they were static and single entities frozen in time (Christensen *et al.*, 1996). We propose that this requires an integrated approach to understanding and managing the geo- and bio-components of ecosystems as part of a whole system, geo-ecological approach.

Human activity in the mountains may lead to a gradual loss of robustness/resilience and hence greater susceptibility to break down suddenly under pressure from disturbances that could previously be absorbed (Holling, 1986). This is potentially an important complication, because historical landscapes may have been very different from those studied today (e.g. in terms of pollution impacts). The problem is to know how close individual montane systems are to their critical thresholds for irreversible change, and the likelihood of human activity tipping the balance (Gordon *et al.*, 2001), potentially from stasis to drastic change. We propose that addressing this problem is fundamental to our understanding of the management needs of mountain ecosystems.

References

Arft, A.M. & 28 others (1999). Responses of tundra plants to experimental warming: meta-analysis of the International Tundra Experiment. *Ecological Monographs*, **69**, 491-511.

Ballantyne, C.K. (2002). Debris flow activity in the Scottish Highlands: temporal trends and wider implications for dating. *Studia Geomorphologica Carpatho-Balcanica*, **36**, 7-27.

Barrett, J.E., Virginia, R.A., Wall, D.H., Parsons, A.N., Powers, L.E. & Burkin, M.B. (2004). Variation in biogeochemistry and soil biodiversity across spatial scales in a polar desert ecosystem. *Ecology*, **85**, 3105-3118.

Braun-Blanquet, J. (1928). *Pflanzensoziologie. Grundzüge der Vegetationskunde*. Springer Verlag, Berlin.

Brazier, V., Gordon, J.E., Hubbard, A. & Sugden, D.E. (1996). The geomorphological evolution of a dynamic landscape: the Cairngorm Mountains, Scotland. *Botanical Journal of Scotland*, **48**, 13-30.

Brunsden, D. (2001). A critical assessment of the sensitivity concept in geomorphology. *Catena*, **42**, 99-123.

Burnett, M.R., August, P.V., Brown Jr., J.H. & Killingbeck, K.T. (1998). The influence of geomorphological heterogeneity on biodiversity. II. A patch-scale perspective. *Conservation Biology*, **12**, 363-370.

Christensen, N.L. & 12 others. (1996). The report of the Ecological Society of America Committee on the scientific basis for ecosystem management. *Ecological Applications*, **6**, 665-691.

Darmody, R.G., Thorn, C.E., Schlyter, P. & Dixon, J.C. (2004). Relationship of vegetation distribution to soil properties in Kärkevagge, Swedish Lapland. *Arctic, Antarctic, and Alpine Research*, **36**, 21-32.

Fiedler, P.L., White, P.S. & Leidy, R.A. (1997). The paradigm shift in ecology and its implications for conservation. In *The Ecological Basis of Conservation. Heterogeneity, Ecosystems, and Biodiversity*, ed. by S.T.A. Pickett, R.S. Ostfield, M. Shachak & G.E. Likens. Chapman & Hall, London. pp. 83-92.

Gordon, J.E. & Leys, K.F. (eds) (2001). *Earth Science and the Natural Heritage: Interactions and Integrated Management*. The Stationery Office, Edinburgh.

Gordon, J.E., Thompson, D.B.A., Haynes, V.M., MacDonald, R. & Brazier, V. (1998). Environmental sensitivity and conservation management in the Cairngorm Mountains, Scotland. *Ambio*, **27**, 335-344.

Gordon, J.E., Brazier, V., Thompson, D.B.A. & Horsfield, D. (2001). Geo-ecology and the conservation management of sensitive upland landscapes in Scotland. *Catena*, **42**, 323-332.

Gordon, J.E., Dvořák, I.J., Jonasson, C., Josefsson, M., Kociánová, M. & Thompson, D.B.A. (2002). Geo-ecology and management of sensitive montane landscapes. *Geografiska Annaler*, **84A**, 193-203.

Grabherr, G.G. (1997). The high-mountain ecosystems of the Alps. In *Ecosystems of the World 3. Polar and Alpine Tundra*, ed. by F.E. Wielgolaski. Elsevier, Amsterdam. pp. 97-121.

Gray, J.M. (2003). *Geodiversity: Valuing and Conserving Abiotic Nature*. John Wiley & Sons Ltd, Chichester.

Grieve, I.C. (2000). Effects of human disturbance and cryoturbation on soil iron and organic matter distributions and on carbon storage at high elevations in the Cairngorm Mountains, Scotland. *Geoderma*, **95**, 1-14.

Gude, M. & Scherer, D. (1995). Snowmelt and slush torrents – preliminary report from a field campaign in Kärkevagge, Swedish Lappland. *Geografiska Annaler*, **77A**, 199-206.

Haynes, V.M., Grieve, I.C., Price-Thomas, P. & Salt, K. (1998). The geomorphological sensitivity of the Cairngorm high plateaux. Scottish Natural Heritage Research, Survey and Monitoring Report, No. 66.

Hill, A.R. (1987). Ecosystem stability: some recent perspectives. *Progress in Physical Geography*, **11**, 315-333.

Holling, C.S. (1973). Resilience and stability of ecological systems. *Annual Review of Ecology and Systematics*, **4**, 1-23.

Holling, C.S. (1986). The resilience of ecosystems; local surprise and global change. In *Sustainable Development and the Biosphere*, ed. by W.C. Clark & R.E Munn. Cambridge University Press, Cambridge. pp. 292-317.

Holling, C.S., Schindler, D.W., Walker, B.W. & Roughgarden, J. (1995). Biodiversity in the functioning of ecosystems: an ecological synthesis. In *Biodiversity Loss: Economic and Ecological Issues*, ed. by C. Perrings, K.-G. Maler, C. Folke, C.S. Holling & B.-O. Jansson. Cambridge University Press, Cambridge. pp. 44-83.

Jeník, J. (1961). Alpinská Vegetace Krkonoš, Králického Sněžníku a Hrubého Jeseníku. *Teorie Anemo-Orografickỳch Systémů*. Academia, Prague.

Jeník, J. (1997). Anemo-orographic systems in the Hercynian Mts and their effects on biodiversity. *Acta Universitatis Wratislaviensis, No. 1950*. Prace Instytutu Geograficznego, Seria C. Meteorologia i Klimatologia, 4, 9-21.

Jäckli, H. (1957). Gegenwartsgeologie des Bündnerischen Rheingebietes. Beiträge zur Geologie der Schweiz, Geotechn. Serie 36. Kummerli and Frey, Bern.

Johansson, C.E. (2000). *Geodiversitet i Nordisk Naturvård*. Nordisk Ministerråd, Copenhagen.

Jonasson, C. (1991). Holocene slope processes of periglacial mountain areas in Scandinavia and Poland. Uppsala Universitet Naturgeografiska Institutionen, UNGI Rapport 79.

Jonasson C. & Nyberg, R. (1999). The rainstorm of August 1998 in the Abisko area, northern Sweden: preliminary report on observations of erosion and sediment transport. *Geografiska Annaler*, **81A**, 387-390.

Kleman, J. (1992). The palimpsest glacial landscape in northwestern Sweden – Late Weichselian landforms and traces of older west-centered ice sheets. *Geografiska Annaler*, **74A**, 305-325.

Kociánová, M. & Sekyra, J. (1995). Distribution of vegetated patterned grounds. In Arctic-alpine tundra in the Krkonose, the Sudetes, ed. by L. Soukupová, M. Kociánová, J. Jeník & J. Sekyra. *Opera Corcontica*, **32**, 54-69.

Kociánová M. & Spusta V. (2000). Influence of avalanche activity on the fluctuation of tree-line in the Giant mountains. *Opera Corcontica*, **36**, 473-480.

Köhler, B. Löffler, J. & Wundram, D. (1994). Probleme der kleinräumigen Geoökovarianz im mittelnorwegischen Gebirge (Problems of local geoecovariance in the central Norwegian mountains). *Norsk Geografisk Tidsskrift*, **48**, 99-111. Kozłowska & Rązkowska, Z. (2002). Vegetation as a tool in the characterisation of geomorphological forms and processes: an example from the Abisko Mountains. *Geografiska Annaler*, **84A**, 233-244.

Kozłowska, A. & Rązkowska, Z. (2002). Vegetation as a tool in the characterisation of geomorphological forms and processes: an example from the Abisko Mountains. *Geografiska Annaler*, **84A**, 233-244.

Lilly, A., Hudson, G., Birnie, R.V. & Horne, P.L. (2002). The inherent geomorphological risk of soil erosion by overland flow in Scotland. Scottish Natural Heritage Review No. 183.

Lundkvist, M. (1997). Slushflows and other rapid slope processes in Northern Scandinavia – attributes and hazards. Unpublished undergraduate thesis, Uppsala University, Institute of Earth Sciences, Physical Geography.

Lundkvist, M. (2001). Natural and anthropogenic dynamics in applied geomorphic environmental risk assessment – A case study from the Lake Torne Area Biosphere Reserve, northern Sweden. Licentiate thesis, Uppsala University, Uppsala.

Melander, O. (1977). Geomorphological map 30H Riksgränsen (east), 30I Abisko, 31H Reurivare and 31I Vadvetjåkka. Description and assessment of areas of geomorphological importance. Statens Naturvårdsverk, SNV PM 857.

Migoń, P. (1999). The role of 'preglacial' relief in the development of mountain glaciation in the Sudetes, with the special reference to the Karkonosze Mountains. *Zeitschrift für Geomorphologie, Supplement Band,* **113**, 33-44.

Molau, U. & Alatalo, J.M. (1998). Responses of subarctic-alpine plant communities to simulated environmental change: biodiversity of bryophytes, lichens, and vascular plants. *Ambio,* **27**, 322-329.

Nagy, L., Grabherr, G., Körner, C. & Thompson, D.B.A. (eds) (2003). *Alpine Biodiversity in Europe.* Springer, Berlin.

Nichols, W.F., Killingbeck, K.T. & August, P.V. (1998). The influence of geomorphological heterogeneity on biodiversity. II. A landscape perspective. *Conservation Biology,* **12**, 371-379.

Palacios, D. & Sánchez-Colomer, M.G. (1997). The distribution of high mountain vegetation in relation to snow cover: Peñalara, Spain. *Catena,* **30**, 1-40.

Pilous, V. (1973). Strukturní mury v Krkonoších I. *Opera Corcontica,* **10**, 15-69.

Pilous, V. (1975). Strukturní mury v Krkonoších II. *Opera Corcontica,* **12**, 7-50.

Pilous, V. (1977). Strukturní mury v Krkonoších III. *Opera Corcontica,* **14**, 7-94.

Poff, N.L., Allan, J.D., Bain, M.B., Karr, J.R., Prestegaard, K.L., Richter, B.D., Sparks, R.E. & Stromberg, J.C. (1997). The natural flow regime. A paradigm for river conservation and restoration. *BioScience,* **47**, 769-784.

Poiani, K.A., Richter, B.D., Anderson, M.G. & Richter, H.E. (2000). Biodiversity conservation at multiple scales: functional sites, landscapes, and networks. *BioScience,* **50**, 133-146.

Rapp, A. (1960). Recent development of mountain slopes in Kärkevagge and surroundings, Northern Scandinavia. *Geografiska Annaler,* **42**, 65-200.

Rapp, A., Jonasson, C. & Nyberg, R. (1997). Extrema sommarregn med översvämningar och jordskred. Ett utslag av klimatändring i Norden efter 1950-talet? *Svensk Geografisk Årsbok* 1997, 67-79.

Rubensdotter, L. (2001). Geomorfologiska processer avspeglade i sjösediment och deras koppling till Holocena miljöförändringar i norra Lappland. Ph.lic. thesis, Department of Physical Geography and Quaternary Geology, Stockholm University.

Sekyra, J. (1964). Kvartérně geologické a geomorfologické problémy krkonošského krystalinika. Opera Corcontica, **1**, 7-24.

Sletten K. (2002). Holocene mass movement processes in Norway, and the development of a moraine complex on Svalbard. PhD thesis, Bergen University.

Soukupová, L. & Kociánová, M. (1995). Plant communities of patterned grounds. In Arctic-alpine tundra in the Krkonose, the Sudetes, ed. by L. Soukupová, M. Kociánová, J. Jeník & J. Sekyra. *Opera Corcontica,* **32**, 48-55.

Soukupová, L., Kociánová, M., Jeník, J. & Sekyra, J. (eds) (1995). Arctic-alpine tundra in the Krkonose, the Sudetes. *Opera Corcontica*, **32**, 5-88.

Stroeven, A.P., Fabel, D., Harbor, J., Hättestrand, C. & Kleman, J. (2002). Quantifying the erosional impact of the Fennoscandian ice sheet in Torneträsk-Narvik corridor, northern Sweden, based on cosmogenic radionuclide data. *Geografiska Annaler*, **84A**, 275-287.

Štursa J. (1995). *Das Riesengebirge – eine Insel der arktis inmitten Europas Riesengebirge.* Deutsche Sonderausgabe der Monatzeitschrieft Krkonoše, Vrchlabí.

Štursa, J. (1998). Research and management of the Giant Mountains' arctic-alpine tundra (Czech Republic). *Ambio*, **27**, 358-360.

Swanson, F.J., Dratz, T.K., Caine, N. & Woodmansee, R.G. (1988). Landform effects on ecosystem patterns and processes. *BioScience*, **38**, 92-98.

Thomas, C.W., Gillespie, M.R., Jordan, C. & Hall, A.M. (2004). Geological structure and landscape of the Cairngorm Mountains. Scottish Natural Heritage Commissioned Report No. 64 (ROAME No. F00AC103).

Thompson, D.B.A., Horsfield, D., Gordon, J.E. & Brown, A. (1993). The environmental importance of the Cairngorm massif. In *The Cairngorms - Planning Ahead.* Proceedings of Conference, ed. by A. Watson & J. Conroy. Kincardine and Deeside District Council, Ballater. pp. 15-23.

Thompson, D.B.A., Gordon, J.E. & Horsfield, D. (2001). Montane landscapes in Scotland: are these natural, artefacts or complex relics? In *Earth Science and the Natural Heritage: Interactions and Integrated Management*, ed. by J.E. Gordon & K.F. Leys. The Stationery Office, Edinburgh. pp. 105-119.

Watt, A.S. & Jones, E.W. (1948). The ecology of the Cairngorms. Part 1. The environment and altitudinal zonation of the vegetation. *Journal of Ecology*, **36**, 283-304.

Werritty, A. & Leys, K.F. (2001). The sensitivity of Scottish rivers and upland valley floors to recent environmental change. *Catena*, **42**, 251-273.

Woodwell, G.M. (2002). On purpose in science, conservation and government. *Ambio*, **31**, 432-436.

6 Climate change and pollution in the mountains: the nature of change

David Fowler & Rick Battarbee

Summary

1. The physical processes leading to higher wind speeds and precipitation, and lower temperatures in the uplands than on low ground, are well known and understood. These processes also cause rates of deposition of pollutants to be larger, as a consequence of the increased turbulence and precipitation.

2. Whether the increased deposition rates lead to a larger deposition is mainly controlled by the nature of the pollutant. Concentrations of primary pollutants such as SO_2 are reduced by the enhanced turbulence, as there are few sources in the uplands. However, for secondary pollutants including nitric acid (HNO_3), ozone (O_3) or secondary particulate matter, the flux to the surface is increased.

3. The interactions between mountains and the secondary pollutants, present as aerosols, therefore lead to enhanced exposure of terrestrial surfaces to a wide range of inorganic and organic compounds, including those responsible for acidification and eutrophication, heavy metal deposition as well as persistent organic pollutants.

4. The lack of physical disturbance of upland lakes, together with their relatively large inputs of pollutants, make these ecosystems ideal locations to study the history of chemical and biological change in the uplands, using the diatom, heavy metal trace organic and particulate pollution record in lake sediments. Thus uplands provide opportunities to monitor, assess and provide early warning of the ecological consequences of anthropogenic modification of the chemical climate of the country.

6.1 Introduction

Mountains, where anthropogenic influences are minimal, are widely regarded as refuges within the landscape for the natural environment. The beauty of these upland landscapes is enhanced by the perception that they remain largely unaffected by development. This view stems in part from the climatic conditions in the mountains, which are generally hostile to normal human activity, but also because most mountainous regions are distant from the major urban centres. Mountains are also areas in which direct land use changes by human activities are generally smaller than in the lowland landscape, and hence the flora

Fowler, D. & Battarbee, R. (2005). Climate change and pollution in the mountains: the nature of change. In *Mountains of Northern Europe: Conservation, Management, People and Nature*, ed. by D.B.A. Thompson, M.F. Price & C.A. Galbraith. TSO Scotland, Edinburgh. pp. 71-88.

and fauna may appear less at risk. However, human activities extend well beyond urban conurbations, and contaminants resulting from industrial and transport activities are found throughout the global atmosphere and oceans. Thus the upland areas in Europe, North America, and increasingly the uplands throughout other continents, are exposed to a wide range of contaminants. The extent to which terrestrial surfaces are influenced by atmospheric contaminants clearly depends on the chemical properties and scale of the input relative to the sensitivity of the soil, fresh water and biota to chemical or biological change.

An examination of the processes responsible for the major features of mountain climates shows that the unique climate of upland areas generates conditions which greatly increase the exposure of the uplands to the effects of all the reactive and soluble pollutants. Thus mountain climates strongly influence their chemical climate. Furthermore, the mountainous regions in northern Europe comprise some of the most sensitive mineralogy to the effects of acidity and a flora which is very sensitive to nitrogen inputs. Some of the most beautiful and remote upland landscapes of Europe therefore combine large exposure to reactive pollutants with ecosystems which are among the most sensitive to inputs of acidity and nitrogen from the atmosphere.

The modification of soils, vegetation and freshwaters of the uplands therefore occurs through the chemical effects of pollutants deposited from the atmosphere, rather than from the direct effects of physical developments of infrastructure or farming in the landscape. These areas are unique within the terrestrial landscape in being affected indirectly by anthropogenic activity, often by contaminants emitted many hundreds of kilometres upwind. An important consequence of this process of modification is that the chemical and biological signatures of pollutant deposition remain undisturbed in lake sediments that act as natural environmental archives and provide a record of the chemical, physical and biological changes over decade and century timescales (Battarbee, 1984).

In recent years, as a result of a series of major EU-funded projects on mountain lakes (Battarbee *et al.*, 2002), our understanding of the status of these regions with respect especially to air pollution and climate change has been transformed. Mountain lakes and their sediment records can be used as sensors of impacts on both lakes themselves and on other mountain ecosystems subject to the same pollution and climatic conditions. Lakes situated above (altitude) or beyond (latitude) the local natural timberline that have catchments undisturbed by human activity are the most useful and in Europe, because of the spatial distribution of mountain systems and strong pollution and climate gradients, study sites can be arranged along environmental gradients, principally air pollution and climate gradients, from north to south and west to east.

In this chapter we describe the processes which predispose mountains to the effects of major pollutants emitted by human activity, and show direct evidence of the chemical changes in the uplands from sediment records in lakes. Examples are taken from the extensive programmes of research in the uplands of the United Kingdom on acid deposition and eutrophication. These problems remain current threats to the ecology of extensive tracts of the semi-natural habitats of upland Britain despite the extensive control measures taken to reduce pollutant emissions. The combination of very effective deposition processes and great sensitivity of the ecosystems in upland Britain to the deposited pollutants is the primary cause. The processes leading to this combination of factors provides a logical pathway between human activities and widespread effects in remote areas of the country.

But current understanding, however simple, was not predicted, or accepted readily. The final section of the chapter therefore considers whether mountains should be considered more sensitive than lowlands to air pollutants in general, regardless of the chemical properties of the pollutant and the environmental target in the uplands.

6.2 Mountains and the atmosphere

The sheer bulk of the upland topography provides an obstacle to airflow over the landscape, increasing the drag of terrestrial surfaces on the wind, and forces air to rise and cool as it ascends. As air descends from high ground it warms and dries, providing relatively dry climates in the lee of high ground. The physical drag of mountains on the movement of air results in increased turbulence and much larger wind speeds in the uplands. Generalising the effect is complicated by the three-dimensional structure of the topography and its interaction with airflow. However, the tops of hills in the UK at elevations in the range 600 m to 800 m experience average winds a factor of between 2 and 4 larger than lowlands in the same region (Chandler & Gregory, 1976). As turbulence is responsible for the vertical exchange of gases and particles throughout the atmosphere, terrestrial surfaces in the uplands are more effectively coupled to the atmosphere than lowland surfaces. The consequence for exposure of the uplands to pollutants is that the chemical species which are deposited rapidly, due either to their chemical reactivity or particle size, are deposited at much larger rates.

6.3 Orographic enhancement of wet deposition

One of the most important effects of mountains on physical climate is the orographic effect on precipitation. As air is forced to rise, it cools and relative humidity increases. With increasingly moist air, a point is reached in mildly supersaturated air where water droplets form. The droplets form on aerosols through a process known as activation and, as air continues to ascend, the droplets grow with further condensation, increasing their liquid water content. This process leads to the formation of hill or orographic cloud in the uplands of northern Europe. The washout of this low-level cloud by precipitation falling from higher-level cloud by seeder-feeder scavenging (the seeder droplets originate in the higher-level cloud and these washout the hill cloud (feeder) droplets) is the primary cause of the very high rainfall in the uplands close to the western coastline (Figure 6.1).

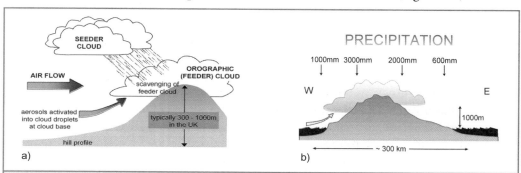

Figure 6.1 (a) Illustrating the seeder-feeder process of orographic enhancement of precipitation. (b) A west to east cross section of upland UK illustrating effects of orographic enhancement on annual precipitation with the rate of precipitation increase with altitude in four upland areas (mm per 100 m altitude): N. Wales 460; C. Wales 230; Cairngorms 250; Pennines 200.

The major reactive pollutants emitted during the combustion of fossil fuels include sulphur dioxide and nitric oxide, which are oxidised in the atmosphere to sulphate (SO_4^{2-}) and nitrate (NO_3^-) respectively. These oxidation products of the emitted pollutants are present mainly as particles in the size range 0.1 to 1.0 µm (dia) which are too small to deposit at significant rates and rely on precipitation scavenging for their removal from the atmosphere (Fowler, 1984).

The seeder-feeder scavenging of pollutants therefore enhances both the precipitation and pollutant deposition. The sulphur- and nitrogen- containing aerosols form effective condensation nuclei and are readily activated into cloud droplets in orographic cloud. The hill cloud, so commonly associated with the uplands, may therefore be contaminated with significant concentrations of the oxidation products of the major pollutants. The concentrations of ionic species in orographic cloud are largest close to cloud base and decline as the droplets are forced to rise further up the hill slope, as illustrated in Figure 6.2.

6.4 Cloud droplet deposition (occult deposition)

The orographic processes lead to the enhancement of wet deposition in the uplands, but a further important effect is also widespread in the uplands, the direct deposition of cloud droplets onto vegetation, a process also known as occult deposition. The aerosols in low-level air are present as small, sub-micron particles, and these small particles deposit onto vegetation at small rates, generally about an order of magnitude slower than the precursor pollutants (SO_2 and NH_3). The slow removal by dry deposition is the primary cause of long-range transport of pollutants as aerosols, the form in which the majority of the sulphur and nitrogen pollutants are exchanged between countries (Hjellbrekke & Tarrason, 2001). However, following activation into cloud droplets, particle size increases from sub-micron to several µm in the first 100 m above cloud base (Figure 6.2).

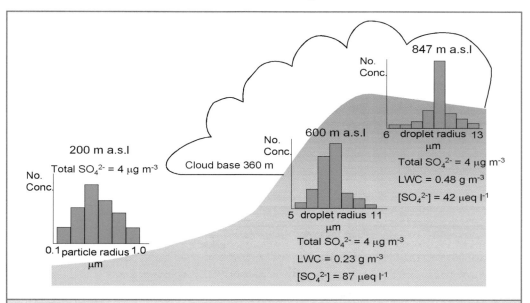

Figure 6.2 The activation of aerosols, using the sulphate (SO_4^{2-}) content to illustrate the process, into cloud droplets as moist air is forced over hills and the change in droplet size and liquid water content (LWC) of the orographic cloud with height. 'No Conc.' on the diagram means 'Droplet number concentration'.

The cloud droplets deposit rapidly to vegetation, at rates typically an order of magnitude faster than the sub-micron aerosols from which they were formed. The process of cloud droplet capture by vegetation is seldom observed except in the presence of wind-driven sea fog captured by isolated trees wetting the ground beneath them (Hori, 1953). The process has been termed occult deposition as it is not observed by standard precipitation collectors, and is thus 'hidden'. Such effects are less commonly observed in the uplands, except when super-cooled cloud droplets are captured by trees and freeze as the droplet meets terrestrial surfaces, resulting in characteristic rime accumulation on the exposed shoots (Figure 6.3). Cloud droplet deposition is only important as a deposition process in the uplands at elevations in excess of 500 m in northern and western Europe. However, at some locations this deposition process may deposit large quantities of the major pollutants and, as illustrated in Figure 6.3, the concentrations may be very large relative to those in precipitation. As concentrations decline with elevation above cloud base, the elevation at which the greatest effect on vegetation occurs is close to cloud base. This may be the reason for the very distinct height-dependent damage to some of the upland conifer forests in the polluted regions of Europe during the 1980s.

Figure 6.3 A photograph of rime, the result of several hours collection of super-cooled cloud droplets on spruce at an elevation of 600 m in the Scottish Borders. 12 hours exposure; -4°C; SO_4^{2-} 800 meq l^{-1}.

Having described some of the important physical processes in the uplands which are responsible for the chemical climate of mountains, we now consider examples of environmental issues to illustrate the effects in mountain ecosystems.

6.5 Acid deposition

The combustion of coal and oil gives rise to the emission into the atmosphere of sulphur and nitrogen oxides as gases and fly ash particles. Sulphur emissions are mainly derived from coal and oil combustion in power stations. Nitrogen emissions include both oxidised (NO_x) and reduced nitrogen species (NH_3). Oxidised compounds are derived from both power stations and vehicle exhausts while reduced compounds come principally from agricultural sources.

Evidence for the contamination of mountain lakes by these products can be demonstrated from the large concentration of non-marine sulphate and nitrate in the water column and the presence of fly ash particles, especially spheroidal carbonaceous particles (SCP) in lake sediments (Rose *et al.*, 2002). The spatial variation in concentration of these substances is in good agreement with the distribution of industrial regions within Europe, and the temporal variation in the concentration of SCPs measured in sediment cores reflects the progressive industrialisation of Europe from the beginning of the 19th century.

Whilst many mountain lakes in Europe remain un-acidified either because they have adequate natural alkalinity to neutralise the acidity or because they occur in regions of low acid deposition, lakes in high acid deposition regions that also have low natural alkalinity have been acidified (Battarbee, 1990). This is indicated most clearly by changes in the composition of diatom assemblages preserved in recent lake sediments (Jones *et al.*, 1993) (Figure 6.4). The data from the Acidification of Mountain Lakes: Palaeolimnology and Ecology (AL:PE) project show that the most severe acidification has taken place in Central and Western Europe (Wathne *et al.*, 1997). In these regions, macro-invertebrate populations are impoverished and lack species sensitive to acidification such as *Baetis rhodani* (Raddum & Fjellheim, 2002), and fish populations (mainly brown trout (*Salmo trutta*) or arctic charr (*Salvelinus alpinus*)) show signs of acid stress (Rognerud *et al.*, 2002). Whereas sulphur and nitrogen deposition together are responsible for surface water acidification, nitrogen deposition additionally causes problems of eutrophication of soils where N is often the limiting nutrient for vegetation.

Following international agreements on the reduction of acidifying gases in Europe (1st Sulphur Protocol/"30% club" in 1985, 2nd Sulphur protocol/Oslo Protocol in 1994 and Gothenburg Protocol in 1999, as well as the EU-5th Environmental Action Programme in 1993 and EU - National Emissions Ceilings Directive in 1999), acid deposition in mountain regions has begun to decline; the pH and alkalinity of some, but not all sites, has begun to recover (NEGTAP, 2001; Monteith & Evans, in press) and, although slight at this stage, there is evidence of biological recovery beginning to take place across Europe (Monteith & Evans, in press).

6.6 Toxic trace metals and persistent organic pollutants

In addition to acidity and fly ash contamination, mountain lakes are also contaminated by toxic metals and organic compounds. The increases that have taken place over the last century or so can be clearly observed from analyses of sediment cores (Yang *et al.*, 2002).

Changes in the concentration of trace metals such as Pb, Hg and Cd and in polycyclic aromatic hydrocarbons (PAHs) correspond closely to the patterns observed for SCPs suggesting a common fossil-fuel combustion source (Fernandez *et al.*, 2002). The long-range transport of these metals occurs as aerosols, predominantly in the size range 0.1 to 1.0 µm. The accumulation of metals in upland ecosystems occurs as a consequence of three quite different processes. First, the wet deposition process provides an efficient mechanism for the removal of small particles in upland regions, as shown in Figure 6.3. Second, the dry

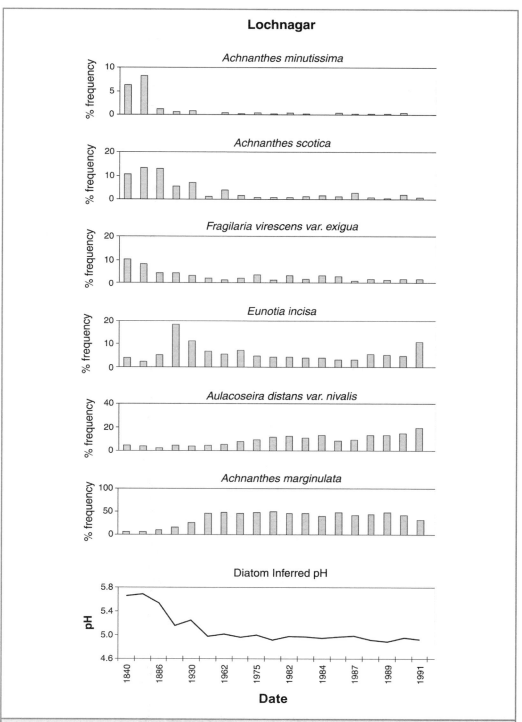

Figure 6.4 Changes in the composition of diatoms and pH reconstruction from 1840 to 1990 based on the analysis of a radiometrically-dated sediment core from Lochnagar, Scotland.

deposition process, which for most of the lowland landscape is an inefficient removal mechanism for small particles (Nemitz *et al.*, 2002), becomes a much more efficient deposition mechanism in the uplands due to the much larger wind speeds, and the increase in particle size as the particles are lifted to higher and cooler elevations. Lastly, the organic soils in the uplands accumulate metals, especially the heavy metals, which bind to components of the organic matter (Fowler *et al.*, 1998).

The largest concentration of Pb, Cd and PAH in the atmosphere occurs in central parts of Europe (Fernandez *et al.*, 1999) and is reflected clearly in the relative concentration of trace metals in fish tissue in high mountain lakes in this region (Figure 6.5). Schwarzee ob Solden, the lake with the highest concentration of Pb and Cd, is located in the Austrian Alps (Rosseland *et al.*, 1997).

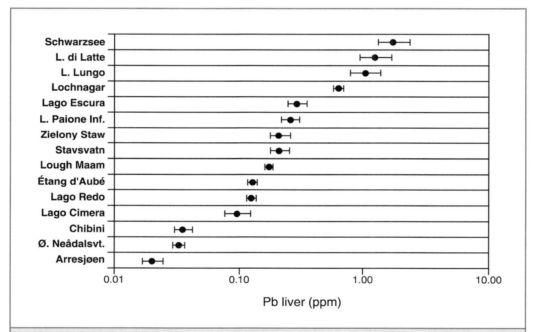

Figure 6.5 Concentrations of Pb in fish liver for a series of mountain lakes along a gradient of atmospheric inputs across Europe: Schwarzsee (Austria), L. di Latte, L. Lungo, L. Paione Inf. (Italy), Lochnagar (Scotland), Lago Escura (Portugal), Zielony Staw (Poland), Stavsvatn, Ø. Neådalsvatn, (Norway), Lough Maam (Ireland), Étang d'Aubé (France), Lago Redo, Lago Cimera (Spain), Chibini (Russia), Arresjoen (Svalbard).

6.7 Persistent Organic Pollutants (POPs) in the uplands

The presence of some persistent organic compounds, such as toxaphene, not used in Western Europe but found in European mountain lakes, indicates long-range transport from North America (Rose *et al.*, 2001). For some POPs there is evidence that concentrations increase with altitude (Figure 6.6), as these substances become progressively redistributed to colder and more remote regions by volatilization and cold trapping processes (Grimalt *et al.*, 2001), contaminating lakes that are distant (>1,000 km) from the production and use of the compounds. The accumulation pattern of these compounds depends on local climatic conditions, whereas their atmospheric fallout is quite uniform and

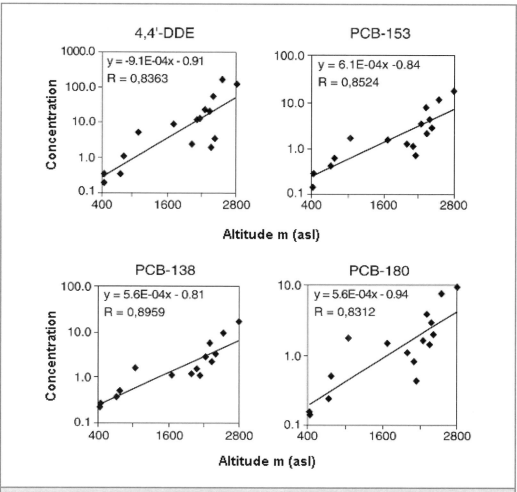

Figure 6.6 Variation with altitude in concentrations of persistent organochlorine compounds in fish tissue in mountain lakes (re-drawn from Grimalt *et al.*, 2001).

seasonally dependent (Carrera *et al.*, 2002). Temperature is also the main factor controlling the deposition fluxes of lower molecular weight PAHs, but particle deposition and wet precipitation determine the atmospheric deposition fluxes of the higher molecular weight PAH homologues. Some of the largest concentrations in both metals and POPs are found in Svalbard, one of the most remote regions of the world, where levels of Hg and a number of PCB congeners are double those at other sites as a result of food chain biomagnification. Rognerud *et al.* (2002) have shown that these higher values are the result of a progressive shift to cannibalism in the diet of arctic charr between the ages of 11 and 20.

6.8 Climate change

Until recently it has been assumed that climate, although variable, imposed a relatively constant influence on the ecology of mountain lakes. It is now becoming clear, however, that this is not the case, and that climate change is imposing an additional stress on upland ecosystems. The climatic impact on mountain lakes is driven mainly through changes in temperature, precipitation and wind regimes that affect snow and ice cover, catchment hydrology, and water column stratification and mixing. These, in turn, control many chemical and biological processes such as primary production, nutrient cycling, hypolimnetic O_2 consumption, alkalinity generation and water column pH, and have a strong influence directly on habitat characteristics and distribution, and on biological life cycles. Data from European and various national projects show that the mountain lakes studied have clearly experienced significant variability in climatic conditions over the last 200 years. Instrumental temperature reconstructions by Agustí-Panareda & Thompson (2002) show decadal-scale fluctuations in mean annual temperature with amplitude of up to 2°C. These are sufficient to cause ecologically-important changes in lake heat balance and ice-cover. Palaeolimnological studies covering this time period suggest that climate warming has indeed caused ecological change to have taken place, specifically through its influence on lake acidity and lake productivity. For acidity, Psenner & Schmidt (1992) have shown from diatom analysis of recent sediments in the Austrian Alps that inter-annual and decadal fluctuations in pH are closely related to changes in mean annual air temperature. They argue that temperature plays an important role in driving the generation of alkalinity in lakes and lake catchments, and that for naturally low alkalinity lakes this can cause significant shifts in lake water pH. At other sites, e.g. in the western highlands of Scotland, increased winter rainfall may lead to a reduction in alkalinity as a result of hydrochemically-driven dilution of stream base-flow (NEGTAP, 2001).

At other sites, recent warming appears to be having mainly a productivity effect. This can be seen from both changes in diatom plankton at sites in the Pyrenees (Catalan *et al.*, 2002a), Finland (Sovari & Korhola, 1998) and Austria (Koinig *et al.*, 2002) and from recent increases in the amount of organic matter accumulating in sediments at almost all sites (Battarbee *et al.*, 2002).

These changes appear to be the result of changes in lake productivity linked to changes in nutrient, especially phosphorus, loading either through water temperature or water column mixing driving internal nutrient recycling. Alternatively, changes to the catchment, where an increase in the delivery of nutrients to the lake may be caused by reduced snow cover, enhanced soil development, and increased soil erosion could also be responsible (Catalan *et al.*, 2002b). The specific diatom responses observed may be a combination of increased nutrient loading and re-cycling coupled with changes to lake stratification and mixing patterns that selectively favour different taxa. An example from Lake Redo in the Pyrenees is the increasing abundance of *Cyclotella pseudostelligera*, an autumn blooming planktonic diatom, as October temperatures have gradually risen over the last few decades (Figure 6.7).

Recent climate change as a consequence of the atmospheric accumulation of greenhouse gases, which trap long wavelength thermal radiation, is indicated by instrumental records of air temperature as shown in the report of the Intergovernmental Panel for Climate Change (Houghton *et al.*, 2002). The same authoritative report catalogues changes in precipitation, clouds and, importantly, the very large regional variability in expected trends with time. In particular, the changes in temperature are expected to be largest in polar regions and smallest in the tropics. The uplands will be especially sensitive to changes in clouds and precipitation since the uplands amplify the precipitation amounts and wet deposition by the

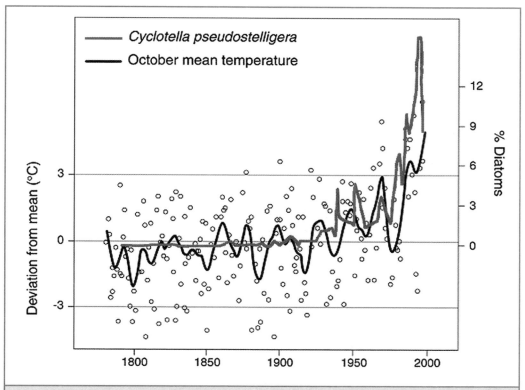

Figure 6.7 *Cyclotella pseudostelligera* abundance in a lake sediment core from Lake Redo, Spanish Pyrenees, compared with October mean temperatures (from Catalan *et al.*, 2002a)

orographic processes described above. If, as seems probable, the uplands of western Britain become wetter and warmer through the coming decades, the increased deposition of a wide range of chemicals, from acidifying and enriching substances to toxic metals and organics, may offset ecological recovery expected from reductions in pollutant emissions. There is some evidence of these effects already with trends in cloud-water concentrations of major ions suggestive of an increase in the duration of orographic cloud during the last decade. As measurements of cloud composition in the uplands of the UK are relatively new, beginning in the late 1980s, it is too early to be confident of the longer term. However, it is clear that monitoring in the uplands is important as these regions experience amplified pollution and climate influences and are coupled with ecosystems that are very sensitive to changes in their physical and chemical climates.

6.9 Ozone

Ozone is a reactive oxidant, formed both by natural processes and also from photochemical processing of pollutants in the atmosphere (PORG, 1997). In the troposphere, a background concentration results from the downward diffusion of ozone from the stratosphere, but the majority of ozone within the troposphere is formed from the oxidation of hydrocarbons in the presence of concentrations of NO_2 in excess of 50 ppt. Current average concentrations of ozone in the UK are in the range 20 ppbV to 30 ppbV, with the largest average concentrations in the uplands and the smallest in urban areas and close to roads. Why should the uplands be exposed to the largest concentrations of this potentially damaging pollutant? The photochemical production of ozone requires short wavelength solar radiation, and formation is therefore confined to daylight hours and ambient concentrations at the surface represent the balance between photochemical production and losses at the surface by dry deposition. At night, terrestrial surfaces continue to remove ozone from surface air while the calmer conditions, caused by cooling at the surface and stratification of surface air, reduce vertical transport of ozone from higher levels. In these conditions concentrations at the surface decline during the hours of darkness in the lowlands. However, in the uplands, the windy conditions continue to provide an ozone supply from higher levels in the troposphere during the night, preventing the surface concentration declining as much during the night as is commonly observed at lowland locations. Ozone concentrations in urban areas are further reduced by gas phase titration with nitric oxide (NO), the form of most of the emitted oxidized nitrogen. The effect is observed close to all major sources of NO, but in the rural environment away from sources of NO, dry deposition to vegetation or soil is the main sink for ozone (Figure 6.8).

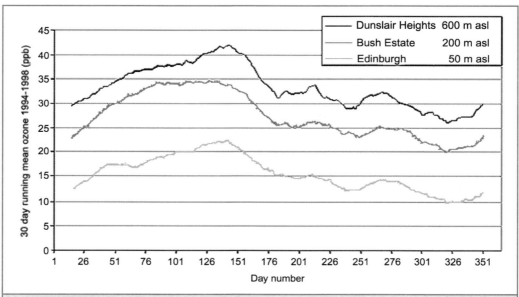

Figure 6.8 Effects of elevation and urban areas on surface ozone concentrations from data collected at three sites in Midlothian, Scotland between 1994 and 1998. Average 30 day-running mean ozone cycles at three sites: Edinburgh (urban); Bush Estate 15 km from Edinburgh centre; Dunslair Heights 600 m asl, approx 30 km from central Edinburgh (remote rural hill top).

The effects of altitude and of the associated exposure on surface ozone concentrations are illustrated by the data in Figure 6.8 from three monitoring stations in southern Scotland, along a 30 km transect from Edinburgh due south to the summit of the Moorfoot Hills at Dunslair Heights. These measurements show the largest concentrations at the hill top site at Dunslair Heights. The effect of the uplands results from increased vertical mixing, preventing deposition at the surface, reducing ambient concentration.

The competition between deposition at the surface and vertical mixing is best seen in the relationship between wind speed and ozone concentration (Figure 6.9). The further effect of local sources of NO on ozone concentration, which is clear in the urban measurements, is also absent at these upland sites.

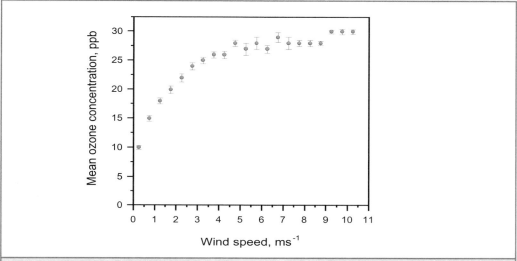

Figure 6.9 The effect of wind speed on surface ozone concentrations from data collected at Bush Estate, Scotland, November 1991 to January 1992.

Mountains and the flora and fauna they contain are therefore exposed to larger average ozone concentration than the lowlands. Furthermore, surface ozone concentrations are now substantially larger than at the turn of the 20th century, probably by a factor of two (Volz & Kley, 1988) and there is strong evidence that increasing global emissions of the ozone precursors (NO_x and volatile organic compounds) will drive average surface ozone concentrations above thresholds for effects on vegetation and human health during this century (Prather *et al.*, 2003). As mountains experience the largest average ozone concentrations, these will be the areas in which the effects of increasing ozone concentration will probably be observed first, and where the effects may be largest. In this sense the uplands will, as in the case of acidification and eutrophication, be the target within the landscape for the largest terrestrial effects.

Temperature interactions between ozone and climate have been recognised in both ozone production and loss processes. The emission of ozone precursor gases (volatile organic compounds (VOCs)) from vegetation increase rapidly with temperature (Guenther *et al.*, 1995) and air temperatures and the energy partitioning at terrestrial surfaces influence the rate of ozone deposition to vegetation. In general, these effects lead to increased ambient ozone concentration with increasing temperatures.

6.10 Conclusions

The chemical climate of the mountains has been shown here to exhibit extremes, which are analogous to those of their physical climate. The exposure of vegetation to large concentrations of the major ions present in orographic cloud and aerosols may result in direct foliar injury, as in the case of red spruce in the Appalachians (Sheppard *et al.*, 1993). The pollutants are present over the low ground, but it is the effect of the mountains in transforming the physical form of the pollutants that enhances the terrestrial inputs in these regions. The very large exposure of mountains to ozone concentrations provides a further example of the effect of the enhanced coupling of the atmosphere to terrestrial surfaces in the uplands. The increased wet deposition in uplands provides a disproportionate fraction of the annual input to the upland regions, relative to low ground.

Some species of both flora and fauna in the uplands evolved in physical environments which have always been harsh, and have evolved to cope, and in some cases to thrive in these conditions. The evolutionary adaptations for the physical conditions may in some cases confer useful attributes to tolerate pollutant inputs. However, in many cases they may be the cause of current problems. Taking the case of some lower plants, which thrive at relatively low temperatures in the wet and cool uplands of the UK, there are mosses which rely almost exclusively on the atmosphere for their nitrogen, and other nutrient supply. In the current chemical climate of the UK uplands, the nitrogen supply has been increased from ~3kg N ha^{-1} in the late 19th century to current values in the range 10 to 40 kg N ha^{-1} (Goulding *et al.*, 1998). The decline in abundance of some lower plants has been attributed to enhanced nitrogen inputs from the atmosphere (Woodin *et al.*, 1985; Baddeley *et al.*, 1994) and the foliar N concentrations of a range of species have been shown to be associated with atmospheric inputs (Pitcairn *et al.*, 1995; Leith *et al.*, 1999). Thus, the adaptations to the physical climate have proved a disadvantage in the presence of excess nitrogen deposition.

In the case of freshwaters, some of the low ionic strength freshwaters draining slowly weathering mineralogy, such as granites, are particularly sensitive to the effects of acidification. The uplands, as described earlier, greatly enhance the annual input of the acidifying chemical species, and thus the uplands become the most sensitive targets in the landscape, as the diatom records show.

The number of chemical species showing hot spots for deposition in the uplands extends across the spectrum of current environmental problems, including:

- acidifying species (SO_2, SO_4, NO_3, NH_4);
- eutrophying species (NO_y, NH_x);
- heavy metals (Cd, Pb, Zn, Cu);
- Persistent Organic Pollutants (DDE, PCBs, HCH, PAHs); and
- photochemical oxidants (O_3, H_2O_2).

It is a consistent property of the mountains to generate larger deposition of the full range of pollutants contained in aerosols and precipitation. For concentrations in cloud water and precipitation, the mountains are generally exposed to larger values, again through the orographic processes described, but in the case of convective rain, concentrations may decrease at the higher elevations with larger precipitation rates. So, there are important exceptions, and the altitude-wet deposition and concentration relationships developed in the cool maritime climates of the UK do not apply in continental climates, e.g. of the Alps.

The degree to which larger exposure to pollutants is reflected in effects on vegetation, soils or freshwater ecology clearly depends on the dose, the sensitivity of the organism and whether the pollutant accumulates or depletes an important regulator of the chemical environment, such as the acid neutralizing capacity. However, it is clear that mountain ecosystems are excellent locations to detect early signs of developing environmental problems, and a network of stations to define the long-term trends in the chemical climate of the uplands would provide an early warning for a much wider distribution of ecosystems in the landscape. Mountains are one of the best places to look for developing problems.

Acknowledgements

The authors acknowledge the support of the Department of the Environment, Food and Rural Affairs for support for the research on which this chapter is based at our respective Institutions. RWB is in addition grateful to the EU for additional funding through the AL:PE, MOLAR and EMERGE projects. Two referees commented on the manuscript.

References

Agustí-Panareda, A. & Thompson, R. (2002). Reconstructing air temperature at eleven remote alpine and arctic lakes in Europe from 1781 to 1997 AD. *Journal of Paleolimnology*, **28**, 7-23.

Baddeley, J.A., Woodin, S.J. & Alexander, I.J. (1994). Effects of increased nitrogen and phosphorus availability on the photosynthesis and nutrient relations of three arctic dwarf shrub from Svalbard. *Functional Ecology*, **84**, 189-196.

Battarbee, R.W. (1984). Diatom analysis and the acidification of lakes. *Philosophical Transactions of the Royal Society of London*, **B305**, 451-477.

Battarbee, R.W. (1990). The causes of lake acidification with special reference to the role of acid deposition. *Philosophical Transactions of the Royal Society of London*, **B327**, 339-347.

Battarbee, R.W., Thompson, R., Catalan, J., Grytnes, J.A. & Birks, H.J.B. (2002). Climate variability and ecosystem dynamics of remote alpine and arctic lakes: the MOLAR project. *Journal of Paleolimnology*, **28**, 1-6.

Carrera, G., Fernandez, P., Grimalt, J.O., Ventura, M., Camarero, L., Catalan, J., Nickus, U., Thies, U. & Psenner, R. (2002). Atmospheric deposition of organochlorine compounds to remote high mountain lakes of Europe. *Environmental Science and Technology*, **36**, 2581-2588.

Catalan, J., Pla, S., Rieradevall, M., Felip, M., Ventura, M., Buchaca, T., Camarero, L., Brancelj, A., Appleby, P.G., Lami, A., Grytnes, J.A. Agustí-Panareda, A. & Thompson, R. (2002a). Lake Redó ecosystem response to an increasing warming the Pyrenees during the twentieth century. *Journal of Paleolimnology*, **28**, 129-145.

Catalan, J., Ventura, M., Brancelj, A. Granados, I., Thies, H., Nickus, U., Korhola A., Lotter, A.F., Barbieri, A., Stuchlik, E., Lien, L., Bitusik, P., Buchaca, T., Camarero, L., Goudsmit, G.H., Kopacek, J., Lemcke, G., Livingstone, D.M., Muller, B., Rautio, M., Sisko, M., Sorvari, S., Sporka, F., Strunecky, O. & Toro, M. (2002b). Seasonal ecosystem variability in remote mountain lakes: implications for detecting climatic signals in sediment records. *Journal of Paleolimnology*, **28**, 25-46.

Chandler, T.J. & Gregory, S. (1976). *The Climate of the British Isles*. Longmans, London.

Fernandez, P., Vilanova, R.M. & Grimalt, J.O. (1999). Sediment fluxes of polycyclic aromatic hydrocarbons in European high altitude mountain fluxes. *Environmental Science and Technology*, **33**, 3716-3722.

Fernandez P., Grimalt, J.O. & Vilanova, R.M. (2002). Atmospheric gas-particle partitioning of polycyclic aromatic hydrocarbons in high mountain regions of Europe. *Environmental Science and Technology*, **36**, 1162-1168.

Fowler, D. (1984). Transfer to terrestrial surfaces. *Philosophical Transactions of the Royal Society of London, Series B*, **305**, 281-297. [Also published in *The Ecological Effects of Deposited Sulphur and Nitrogen Compounds*, ed. by J.W.L. Beament & A.D. Bradshaw. Royal Society, London. pp. 23-39.]

Fowler, D., Flechard, C., Skiba, U. Coyle, M. & Cape, J.N. (1998). The atmospheric budget of oxidized nitrogen and its role in ozone formation and deposition. *New Phytologist,* **139**, 11-23.

Goulding, K.W.T., Bailey, N.J., Bradbury, N.J., Hargreaves, P., Howe, M., Murphy, D.V., Poulton, P.R. & Willison, T.W. (1998). Nitrogen deposition and its contribution to nitrogen cycling and associated soil processes. *New Phytologist,* **139**, 49-58.

Grimalt, J.O., Fernandez, P., Berdie, L., Vilanova, R.M., Catalan, J., Psenner, R., Hofer, R., Appleby, P.G., Rosseland, B.O., Lien, L., Massabuau, J.C. & Battarbee, R.W. (2001). Selective trapping of organochlorine compounds in mountain lakes of temperate areas. *Environmental Science and Technology,* **35**, 2690-2697.

Guenther, A.B., Hewitt, C.N., Erickson, D., Fall, R., Geron, C., Graedel, T., Harley, P.C., Klinger, L., Lerdau, M., Mckay, W.A., Pierce, T., Scholes, B., Steinbrecher, R., Tallamraju, R., Taylor, J. & Zimmerman, P. (1995). A global model of natural volatile organic compound emissions. *Journal of Geophysical Research,* **100**, 8873-8892.

Hjellbrekke, A.G. & Tarrason, L. (2001). Mapping concentrations in Europe combining measurements and acid deposition models. *Water Air and Soil Pollution,* **130**, 1529-1534.

Hori, T. (1953). *Studies on Fogs in Relation to Fog-Preventing Forest.* Tanne Trading Co., Sapporo, Japan.

Houghton, J.T., Ding, Y., Griggs, D.J., Noguer, M., van der Linden, P.J., Dai, X., Maskell, K. & Johnson, C.A. (2002). *Climate Change 2001: The Scientific Basis. Contribution of Working Group I to the Third Assessment Report of the Inter-governmental Panel on Climate Change.* Cambridge University Press, Cambridge.

Jones, V.J., Flower, R.J., Appleby, P.G., Natkanski, J., Richardson, N., Rippey, B., Stevenson, A.C. & Battarbee, R.W. (1993). Palaeolimnological evidence for the acidification and atmospheric contamination of lochs in the Cairngorm and Lochnagar areas of Scotland. *Journal of Ecology,* **81**, 3-24.

Koinig, K.A., Kamenik, C., Schmidt, R., Agusti-Panareda, A., Appleby, P., Lami, A., Prazakova, M., Rose, N., Schnell, O.A., Tessadri, R., Thompson, R. & Psenner, R. (2002). Environmental changes in an alpine lake (Gossenkollensee, Austria) over the last two centuries: the influence of air temperature on biological parameters. *Journal of Paleolimnology,* **28**, 147-160.

Leith, I.D., Hicks, W.K., Fowler, D. & Woodin, S.J. (1999). Differential responses of UK upland plants to nitrogen deposition. *New Phytologist,* **141**, 277-289.

Monteith, D.T. & Evans, C. (in press). Recovery from acidification in the UK: Evidence from 15 years of acid waters monitoring. *Environmental Pollution.*

NEGTAP (2001). *Transboundary Air Pollution: Acidification, Eutrophication and Ground-level Ozone in the UK.* Report of the National Expert Group on Transboundary Air Pollution (NEGTAP) for the UK Department for Environment, Food and Rural Affairs, Scottish Executive, The National Assembly for Wales/Cynulliad Cenedlaethol Cymru and the Department of the Environment for Northern Ireland. Centre for Ecology & Hydrology, Edinburgh.

Nemitz, E., Gallagher, M.W., Duyzer, J.H. & Fowler, D. (2002). Micrometeorological measurements of particle deposition velocities to moorland vegetation. *Quarterly Journal of the Royal Meteorological Society,* **128**, 2281-2300.

Pitcairn, C.E.R., Fowler, D. & Grace, J. (1995). Deposition of fixed atmospheric nitrogen and foliar nitrogen content of bryophytes and *Calluna vulgaris* (L.) Hull. *Environmental Pollution,* **88**, 193-205.

PORG (1997). *Ozone in the United Kingdom. Fourth Report of the Photochemical Oxidants Review Group.* Department of the Environment, Transport and the Regions, London.

Prather, M., Gauss, M., Berntsen, T., Isaksen, I., Sundet, J., Bey, I., Brasseur, G., Dentener, F., Derwent, R., Stevenson, D., Grenfell, L., Hauglustaine, D., Horowitz, L., Jacob, D., Mickley, L., Lawrence, M., von Kuhlmann, R., Muller, J.F., Pitari, G., Rogers, H., Johnson, M., Pyle, J., Law, K., van Weele, M. & Wild, O. (2003). Fresh air in the 21st century? *Geophysical Research Letters*, **30**, 100. doi:10.1029/2002GLO16285

Psenner, R. & Schmidt, R. (1992). Climate-driven pH control of remote alpine lakes and effects of acid deposition. *Nature*, **356**, 781-783.

Raddum, G.G. & Fjellheim, A. (2002). Species composition of freshwater invertebrates in relation to chemical and physical factors in high mountains in southwestern Norway. *Water Air and Soil Pollution: Focus*, **2**, 311-328.

Rognerud, S., Grimalt, J.O., Rosseland, B.O., Fernandez, P., Hofer, R., Lackner, B., Lauritzen, L., Lien, L., Massabuau, J.C. & Ribes, A. (2002). Mercury and organochlorine contamination in brown trout (*Salmo trutta*) and arctic charr (*Salvelinus alpinus*) from high mountain lakes in Europe and the Svalbard archipelago. *Water Air and Soil Pollution: Focus*, **2**, 209-232.

Rose, N.L., Backus, S., Karlsson, H. & Muir, D.C.G. (2001). An historical record of toxaphene and its congeners in a remote lake in Western Europe. *Environmental Science & Technology*, **35**, 1312-1319.

Rose, N.L., Shilland, E., Yang, H., Berg, T., Camarero, L., Harriman, R., Koinig, K., Lien, L., Nickus, U., Stuchlik, E., Thies, H. & Ventura, M. (2002). Deposition and storage of spheroidal carbonaceous fly-ash particles in European mountain lake sediments and catchment soils. *Water Air and Soil Pollution: Focus*, **2**, 251-260.

Rosseland, B.O., Grimalt, J., Lien, L., Hofer, R., Massabuau, J.-C., Morisson, B., Rodriguez, A., Moiseenko, T., Galas, J. & Birks, J.B. (1997). Fish. Population structure and concentrations of heavy metals and organic micropollutants. In *AL:PE, Acidification of Mountain Lakes: Palaeolimnology and Ecology. Part 2 Remote Mountain Lakes as Indicators of Air Pollution and Climate Change*, ed. by B. Wathne, S. Patrick & N. Cameron. EU-Research Report, NIVA Report 3638-97. Norsk institutt for vannforskning, Oslo. Chapter 4, pp. 4-1 to 4-73.

Sheppard, L.J., Cape, J.N. & Leith, I.D. (1993). Influence of acidic mist on frost hardiness and nutrient concentrations in red spruce seedlings. 1. Exposure of the foliage and the rooting environment. *New Phytologist*, **124**, 595-605.

Sovari, S. & Korhola, A. (1998). Recent diatom assemblage changes in subarctic Lake Saanajärvi, NW Finnish Lapland, and their palaeoenvironmental implications. *Journal of Paleolimnology*, **20**, 205-215.

Volz, A. & Kley, D. (1988). Evaluation of the Montsouris series of ozone measurements made in the nineteenth century. *Nature*, **332**, 240-242.

Wathne, B., Patrick, S. & Cameron, N. (Eds) (1997). *AL:PE, Acidification of Mountain Lakes: Palaeolimnology and Ecology. Part 2 Remote Mountain Lakes as Indicators of Air Pollution and Climate Change*. EU-Research Report, NIVA Report 3638-97. Norsk Institute for vannforskning, Oslo.

Woodin, S.J., Press, M.C. & Lee, J.A. (1985). Nitrate reductase activity in *Sphagnum fuscum* in relation to wet deposition of nitrate from the atmosphere. *New Phytologist*, **99**, 381-388.

Yang, H.D., Rose, N.L., Battarbee, R.W. & Boyle, J.F. (2002). Mercury and lead budgets for Lochnagar, a Scottish mountain lake and its catchment. *Environmental Science and Technology*, **36,** 1383-1388.

7 Contemporary and historical pollutant status of Scottish mountain lochs

M. Kernan, N.L. Rose, L. Camarero, B. Piña & J. Grimalt

Summary

1. The distribution of spheroidal carbonaceous particles, lead and persistent organic pollutant concentration in the surface sediments of upland lochs in Scotland shows no spatial coherence. This is likely to be a result of the combination of distance from major emission sources and variation in sediment accumulation rate.
2. The observed distribution of contaminants appears to be a result of a combination of regional, national and international sources.
3. Sediment extracts from the study sites and fish extracts from Lochnagar show clear oestrogenic potential, raising concerns for fish and higher predators.
4. The lack of decline of trace metals in the full-basin sediment record of Lochnagar, in contrast to that of emissions and atmospheric deposition, is thought to be due to the enhanced release of previously deposited metals to the catchment, possibly exacerbated by climate changes.

7.1 Introduction

Remote mountain lakes are sensitive environments characterised by unique biological communities and are considered excellent early warning indicators of environmental change (see Fowler & Battarbee, this volume). Although mountain lakes are often perceived to be pristine environments, they are vulnerable to atmospheric deposition of acidifying compounds (Jones *et al.,* 1993), persistent organic pollutants (POPs) (Fernandez *et al.,* 1996), trace metals (Camarero *et al.,* 1995), and the impacts of climate change (Psenner & Schmidt, 1992), whilst their sediments provide a reliable record of pollutant deposition. A recent survey of mountain lakes (EU-EMERGE project – European Mountain lake Ecosystems: Regionalisation, diagnostics & socio-economic evaluation), covering most of the geographical and environmental gradients across Europe, examined the distribution patterns of a range of POPs and trace metals in surface sediments. This chapter examines the variability of selected pollutants in the surface sediments of upland lochs in Scotland sampled as part of the pan-European study. The historical perspective is also considered in this context using data from Lochnagar, a high altitude site within the Cairngorms National Park.

Kernan, M., Rose, M.L., Camarero, L., Piña, B. & Grimalt, J. (2005). Contemporary and historical pollutant status of Scottish mountain lochs. In *Mountains of Northern Europe: Conservation, Management, People and Nature,* ed. by D.B.A. Thompson, M.F. Price & C.A. Galbraith. TSO Scotland, Edinburgh. pp. 89-98.

7.2 Sites and methods

A Geographical Information System (GIS) was used to identify all high altitude lochs north of the Central Lowlands, with a surface area greater than 0.5 ha and above the extent of the estimated maximum Holocene treeline (ranging from 700-800 m in Cairngorm and the Grampians to 500 m in the north-west Highlands) (see Birks, 1988). This identified 399 sites from which a sub-set of 27 was selected for detailed sampling, covering gradients of bed-rock geology, altitude, surface area and geographical distribution (Table 7.1). Of these, six are within Special Areas of Conservation (SACs), designated specifically for freshwater habitats under EC Habitats and Species Directive 92/43/EEC. Figure 7.1 shows examples of three lochs. Surface sediment cores were taken from the deepest part of each loch using a Glew corer (Glew, 1991) and the surficial sample (0-0.5 cm) was removed using teflon utensils previously rinsed with Milli-Q water and acetone. Samples were frozen and subsequently analysed using standard protocols, for trace metals (Camarero *et al.*, 2003), POPs (Grimalt *et al.*, 2004) and spheroidal carbonaceous particles (SCPs) (Rose, 1994). SCPs are produced only from the high temperature combustion of fossil fuels and provide an unambiguous record of atmospheric deposition from this source. A suite of metals and POPs was analysed but here, for brevity, we focus on lead (Pb) and polychlorinated biphenyls (PCBs).

Figure 7.1 Three images relating to the work presented in this chapter.

(a) Lochnagar is a corrie loch located towards the centre of a granitic massif which comprises much of the Balmoral Forest in Aberdeenshire and within the newly established Cairngorms National Park. The loch lies at an altitude of 785 m with a steep north-east facing corrie wall which rises to the summit at 1,155 m. The catchment is comprised mostly of bare rock and extensive fields of large boulders and coarse screes but closer to the shore, and particularly to the north-east and north-west, there are blanket peat deposits which have become heavily eroded. The sparse vegetation is dominated by *Calluna* and *Vaccinium* species. The loch itself is 26 m deep with an impoverished aquatic flora and fauna. (Photo: ECRC).

(b) Loch Coir' a' Ghrunnda is a small corrie lochan located at the western edge of the Cuillin Hills, Isle of Skye, at an altitude of 750 m above sea level. The catchment is characterised by a high proportion of bare rock with very little soil and vegetation cover. (Photo: ECRC).

(c) Loch Coire Mhic Fearchair is a high altitude loch (600 m above sea level) in Glen Torridon, north-west Scotland. (Photo: ECRC).

7.3 Results and discussion

Figure 7.2 shows the spatial distributions of SCP, Pb and sum of PCB concentrations in the surface sediments of the selected mountain lakes. These pollutants, whilst all atmospherically transported and deposited, represent different industrial sources but exhibit no coherent spatial distribution. The absence of a spatial concentration gradient may be due to the fact that similar levels of atmospheric contamination have occurred at these sites due to their

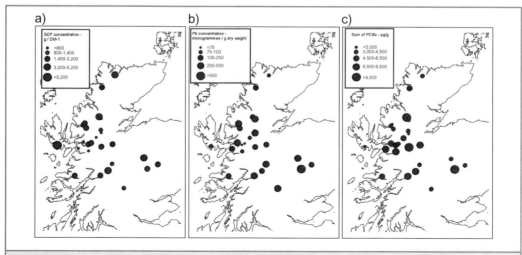

Figure 7.2 a) Distribution of SCPs; b) Pb; and c) sum of all PCBs across high altitude survey sites.

Table 7.1 Details of 27 high altitude lochs surveyed for POPs (Persistent Organic Pollutants) and trace metal concentrations in sediments.

Name	Easting	Northing	Altitude (m)	Area (ha)	County	SAC	SAC name	Underlying geology
Loch Coir' a' Ghrunnda	145161	820239	698	2.87	Highland	N		gabbro & allied types
Lochan Bac an Lochain	176361	765029	577	1.76	Highland	N		mica-schist, semi-pelitic schist & mixed schist
Un-named	182918	814895	717	2.52	Highland	N		quartz-feldspar-granulite
Loch Coire na Caime	192210	858193	527	3.21	Highland	Y	Loch Maree Complex	sandstone and grit
Loch Bhuic Moir	192250	826128	536	3.55	Highland	N		quartz-feldspar-granulite
Loch Coire Mhic Fhearchair	194199	860897	592	10.05	Highland	Y	Loch Maree Complex	sandstone and grit
Un-named	196940	852471	665	9.65	Highland	N		sandstone and grit
Loch a' Chleirich	200700	822606	747	1.44	Highland	N		quartz-feldspar-granulite
Un-named	201216	810559	665	1.90	Highland	N		mica-schist, semi-pelitic schist & mixed schist
Loch an Fhraoich-choire	205609	825042	757	4.42	Highland	N		mica-schist, semi-pelitic schist & mixed schist
Loch Beag	214627	835483	658	2.45	Highland	Y	Strathglass Complex	undifferentiated gneiss
Loch a' Mhadaidh	219927	873206	567	31.10	Highland	N		mica-schist, semi-pelitic schist & mixed schist
Un-named	222514	765437	736	9.32	Highland	N		quartzite, grit, interstratified quartzose-mica-schist
Loch Toll Lochan	223023	848876	518	7.54	Highland	N		mica-schist, semi-pelitic schist & mixed schist
Loch an Fhuar-thuill Mhoir	223911	844025	768	5.36	Highland	N		quartz-feldspar-granulite
Loch Gorm	223952	869554	538	21.61	Highland	N		quartz-feldspar-granulite
Loch Carn a' Chaochain	224067	818147	657	2.29	Highland	N		quartz-feldspar-granulite
Loch a' Choire Dhairg	225182	927244	526	4.22	Highland	N		undifferentiated gneiss
Loch Bealach na h-Uidhe	226445	925587	534	2.90	Highland	N		pipe-rock & basal quartzite
Lochan Coire an Lochain	236320	774324	738	5.06	Highland	N		slate, phyllite & mica-schist
Lochan Coire Choille-rais	243339	786715	808	7.40	Highland	Y	Creag Meagaidh	quartz-feldspar-granulite
Un-named	244352	822794	538	1.92	Highland	N		quartz-feldspar-granulite
Un-named	248499	950528	536	3.82	Highland	N		quartz-feldspar-granulite
Lochan nan Cat	264493	742643	717	12.37	Perth & Kinross	Y	Ben Lawers	slate, phyllite & mica-schist
Lochan Uaine	300118	798093	956	3.78	Aberdeenshire	Y	Cairngorms	granite, syenite, granophyre & allied types
Loch nan Eun	306374	778117	787	14.46	Perth & Kinross	N		quartzite, grit, interstratified quartzose-mica-schist
Lochnagar	325221	785991	788	9.90	Aberdeenshire	N		granite, syenite, granophyre & allied types

distance from major local sources. This is supported by data from southern Scotland and northern England, where SCP surface sediment concentrations have been observed an order of magnitude higher than the maximum observed here (Rose *et al.*, 1995), and sum of PCB concentrations are also 2-5 times higher (e.g. Sanders *et al.*, 1992). The combination of similar levels of atmospheric deposition and variations in sediment accumulation rate (albeit that the lakes are comparable in many aspects) leads to the lack of spatial pattern.

The source of deposited contaminants would thus appear to be the result of a combination of long-range transport from diverse and diffuse sources. This assertion is supported by data from atmospheric transport models. Using one of the survey sites, Lochnagar (the most easterly site in Figure 7.2), as an example, the EMEP (European Monitoring and Evaluation Programme) Lagrangian Deposition Model (A. Benedictow, pers. comm.) predicts that c. 30% of the sulphur deposition (and by implication other industrially-derived and atmospherically transported pollutants) received by the site are from non-UK sources including Germany (6%), Poland (5%), Republic of Ireland (5%), and the Czech Republic (4%) (A. Benedictow, NILU; pers. comm.). Furthermore, the Hull Acid Rain Model (HARM) predicts that of the UK-derived sulphur deposited at Lochnagar, over 60% is from power station and refinery sources in England and Wales (J. Nicholson, University of Edinburgh, pers. comm.). 'Local' Scottish sources thus only account for c. 27%, less than that derived from overseas, and even these are mainly located along the Forth and Clyde, more than 100 km from Lochnagar.

Although concentrations of pollutants in the sediments of these remote lochs are comparatively low when compared with other UK sites, they appear higher when compared to other sites in mountain regions across Europe (the median Pb concentration for Scotland is 162 ppm compared with 98 ppm for mainland sites). This is despite the fact that deposition levels tend to be lower than in other mountain areas across Europe (Camarero *et al.*, submitted). The reason for this may be that the moorland and peat catchments in upland Scotland have accumulated previously deposited metals which are then transported to the sediment via catchment erosion or the leaching of organic materials to the lochs (Yang *et al.*, 2002).

These pollutant levels are not without ecological significance. Analysis of sediments from Lochnagar shows the presence of significant oestrogenic potential. This is the potential to cause endocrine disruption in organisms. Endocrine disruptors (such as those that mimic the hormone estradiol) can bind to hormone receptors and interfere with the natural functioning of these hormones. This can result in deleterious effects on, for example, the development of the reproductive tracts of fish and other organisms. It was calculated that the sediment sample contained endocrine disruptors at a concentration equivalent to 5 to 20 ppb of the female steroid hormone estradiol. Although this is a very high value for a mountain lake, similar levels have been found in other sites in Scotland and in southern Norway. Analysis of brown trout (*Salmo trutta*) from Lochnagar showed a very uneven distribution of oestrogenic activity with one individual (LN1) showing the equivalent of 28 ppb of estradiol. Although chemical analyses gave no direct indication as to the source of the chemical compound responsible for the oestrogenic activity, LN1 also exhibited a high concentration of highly chlorinated PCB congeners, as did all highly oestrogenic fish analysed across Europe. The clear oestrogenic potential found at these sites is thus thought to be due to atmospherically deposited volatile hydrocarbons and organochlorine compounds which are easily bioaccumulated in fish fat. The deleterious effects of endocrine disruption occur at concentration levels commonly found within this dataset (García-Reyero, 2005) raising concerns for fish and higher predators in these remote lochs, and in particular, at less remote sites where fish may be taken for human consumption.

The spatial distributions shown in Figure 7.2 provide a good indication of the contemporary status of Scottish mountain lochs, given the geographical coverage of the survey sites, but do not allow any judgement to be made about whether the contamination of these remote areas is improving or worsening. The analysis of dated sediment cores provides this information at a site-specific scale. As an example, Figure 7.3 shows the ^{210}Pb-dated sediment profile of SCP concentration for Lochnagar. This profile is fairly typical for Scottish sites and shows that a maximum concentration is observed in the late 1970s since when there has been a significant reduction such that contemporary concentrations are at their lowest since the 1930s.

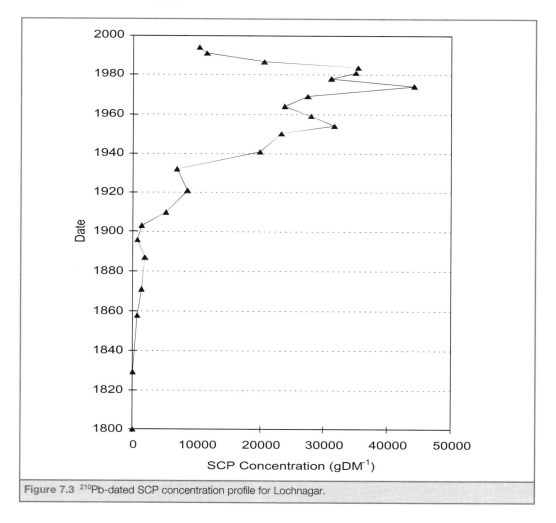

Figure 7.3 ^{210}Pb-dated SCP concentration profile for Lochnagar.

Such temporal information is important as it allows the rate and direction of change to be determined, and, coupled with information on the ecological impacts of pollutants in these systems, allows biologically significant targets to be set. These will show whether emissions policies have had an impact and will allow estimations to be made as to what else needs to be done to reduce contamination to a desired level in the future. However, recent studies have shown that at some remote lochs, whilst concentration profiles of some metals (e.g. Hg, Pb) are declining, the total flux of metal in the region of accumulating sediment within the lake has plateaued and is not showing the declines expected as a result of emissions reductions. There are a number of hypotheses for this observed phenomenon, but most suggest that this is due to enhanced inputs of previously deposited metals stored in the catchment. This could simply be the result of a delay in metals passing through the catchment, increased soil erosion, climate-enhanced decomposition of organic matter leading to elevated leaching of metals adsorbed to dissolved organic carbon (DOC) during wetter winters, or the enhanced scavenging by algae within the water column as a result of increased ice-free periods (Rose *et al.*, 2004). Whatever the cause, historically deposited pollutants stored in catchment soils have the potential to be a major source of pollutants to upland fresh waters across the UK into the future and may prevent 'recovery' of contamination levels for many decades, negating the impacts of increased emissions reductions. Such effects would be exacerbated by the impacts of predicted climate change scenarios. The sediment data described therefore provide a useful baseline against which to monitor future changes and assess the restoration of pollutant inputs to levels below which ecological affects may be seen to occur. Further work remains to understand more fully the nature of catchments as a potential source of variability and how the role of climate may impact the future input of pollutants to these important sites and the restoration of these pollutants to pre-industrial levels.

Acknowledgments

The EMERGE project was funded by the EU (EU contract EVK1-CT-1999-00032).

References

Birks, H.J.B. (1988). Long-term ecological change in the British uplands. In *Ecological Change in the Uplands*, Special Publication Number 7 of The British Ecological Society, ed. by M.B. Usher & D.B.A. Thompson. Blackwell Scientific Publications, Oxford. pp. 37-56.

Camarero, L. (2003). Spreading of trace metals and metalloids pollution in lake sediments over the Pyrénées. *Journal de Physique IV France*, **107**, 249-253.

Camarero, L., Botev, I., Muri, G., Psenner, R., Rose, N.L., Stuchlik, E. (submitted). Pan-European survey of heavy metals in alpine/arctic lake sediments: a record of diffuse atmospheric pollution at a continental scale. *Freshwater Biology*.

Camarero, L., Catalan, J., Pla, S,. Rieradevall, M., Jiménez, M., Prat, N., Rodríguez, M., Encina, L., Cruz-Pizarro, L., Sánchez Castillo, P., Carrillo, P., Toro, M., Grimalt, J., Berdié, L., Fernández, P. & Vilanova, R. (1995). Remote mountain lakes as indicators of diffuse acidic and organic pollution in the Iberian Peninsula (AL:PE Studies). *Water, Air and Soil Pollution*, **85**, 487-492.

Férnandez, P., Vilanova, R. & Grimalt, J.O. (1996). PAH distributions in sediments from high mountain lakes. *Polycyclic Aromatic Compounds*, **9**, 121-128.

García-Reyero, N., Piña, B., Grimalt, J.O., Fernández, P., Fonts, R., Polvillo, O. & Martrat, B. (2005). Estrogenic activity in sediments from European Mountain Lakes. *Environmental. Science & Technology,* **39**, 1427-1435.

Glew, J.R. (1991). Miniature gravity corer for recovering short sediment cores. *Journal of Palaeolimnology,* **5**, 285-287.

Grimalt, J.O., van Drooge, B.L., Ribes, A., Vilanova, R.M., Férnandez, P. & Appleby, P. (2004). Persistent organochlorine compounds in soils and sediments of European high mountain lakes. *Chemosphere,* **54**, 1549-1561.

Jones, V.J., Flower, R.J., Appleby, P.G., Natkanski, J., Richardson. N., Rippey, B., Stevenson, A.C. & Battarbee, R.W. (1993). Palaeolimnological evidence for the acidification and atmospheric contamination of lochs in the Cairngorm and Lochnagar areas of Scotland. *Journal of Ecology,* **81**, 3-24.

Psenner, R. & Schmidt, R. (1992). Climate-driven pH control of remote alpine lakes and effects of acid deposition. *Nature,* **356**, 781-783.

Rose, N.L. (1994). A note on further refinements to a procedure for the extraction of carbonaceous fly-ash particles from sediments. *Journal of Palaeolimnology,* **11**, 201-204.

Rose, N.L., Monteith, D.T., Kettle, H., Thompson, R., Yang, H. & Muir, D.C.G. (2004). A consideration of potential confounding factors limiting chemical and biological recovery at Lochnagar, a remote mountain loch in Scotland. *Journal of Limnology,* **63**, 63-76.

Rose, N.L., Harlock, S., Appleby, P.G. & Battarbee, R.W. (1995). Dating of recent lake sediments in the United Kingdom and Ireland using spheroidal carbonaceous particle (SCP) concentration profiles. *The Holocene,* **5**, 328–335.

Sanders, G., Jones, K.C., Hamilton-Taylor, J. & Dörr, H. (1992). Historical inputs of polychlorinated biphenyls and other organochlorines to a dated lacustrine sediment core in rural England. *Environmental Science & Technology,* **26**, 1815–1821.

Yang, H., Rose, N.L., Battarbee, R.W. & Boyle, J. (2002). Mercury and lead budgets for Lochnagar, a Scottish mountain lake and its catchment. *Environmental Science & Technology,* **36**, 1383-1388.

8 Climate change and effects on Scottish and Welsh montane ecosystems: a conservation perspective

N.E. Ellis & J.E.G. Good

Summary

1. Climate change is currently occurring so rapidly that many plant and animal species in the montane environment are being adversely affected. Examples of change are given.
2. Both biotic and abiotic factors within the montane environment are affected. The impact on montane species therefore needs to be considered at the ecosystem level.
3. Contingency plans for the protection of the most vulnerable species and ecosystems need to be made. However, this requires the urgent acquisition of autecological baseline data in order to be able to make assessments of potential impacts under various climate change scenarios.

8.1 A changing climate

Between 1861 and 2000, the global mean annual temperature rose by 0.6°C, with the Northern Hemisphere warming at a rate about twice as great as in the Southern Hemisphere, and minimum temperatures rising at about twice the rate of maximum temperatures (Folland *et al.*, 2001). Snow cover has decreased by about 10% globally since the late 1960s (Folland *et al.*, 2001). Over the next 50 years, models indicate that the Earth may warm by 0.6-2.6°C above the 1960-1991 baseline (Cubasch *et al.*, 2001), with Northern Europe expected to warm by a further 40% above the global average in the winter months (Giorgi *et al.*, 2001). At least one climate change scenario shows a further increase of about 0.1°C per 100 m elevation in winter across Scotland (Hulme *et al.*, 2001). Montane (alpine) ecosystems across Britain are therefore likely to be particularly adversely affected.

Over the last century, precipitation increased by 5-10% in the Northern Hemisphere, associated with an estimated 2-4% increase in the frequency of heavy rain events, and summers became drier (Folland *et al.*, 2001). Projections for Northern Europe indicate that annual precipitation may increase by 5-20% by 2100, but almost entirely falling in the winter months (Giorgi *et al.*, 2001). The seasonal balance of precipitation has therefore polarised.

Ellis, N.E. & Good, J.E.G. (2005) Climate change and effects on Scottish and Welsh montane ecosystems: a conservation perspective *In Mountains of Northern Europe: Conservation, Management, People and Nature,* ed. by D.B.A. Thompson, M.F. Price & C.A. Galbraith. TSO Scotland, Edinburgh. pp. 99-102.

8.2 Implications for montane ecosystems

A mean annual increase of 1°C is associated with an isotherm shift of 200-275 m uphill. Organisms inhabiting arctic-alpine habitats above 600 m in 1960-1991 will therefore need to move to 800-875 m to experience a similar climate. This would result in about 90% and 96% reductions in arctic-alpine extent in Scotland and Wales, respectively. In Scotland, declines in arctic-alpine species' populations are likely to occur, e.g. for ptarmigan (*Lagopus mutus*), dotterel (*Charadrius morinellus*), snow bunting (*Plectrophenax nivalis*) and mountain hare (*Lepus timidus*). Seven globally scarce arctic-alpine lichens, including *Catopyrenium psoromoides,* may be threatened; four have recently become extinct, believed to be the result of climate change, including *Caloplaca nivalis,* a lichen which had its own Biodiversity Action Plan (Gilbert, 2004). Three species typical of upland areas, Norwegian mugwort (*Artemisia norvegica* var *scotica*), cloudberry (*Rubus chamaemorus*) and mountain ringlet *(Erebia epiphron),* are predicted to lose climate space across the whole of Britain and Ireland under one of the 2050s' climate change scenarios (Harrison *et al.,* 2001). The same model shows that wild azalea (*Loiseleuria procumbens*) may lose all climate space from Wales under the same climate change scenario. It is therefore likely that this would also apply to other mountain species in Wales, like Snowdon lily (*Lloydia serotina*), purple saxifrage (*Saxifraga oppositifolia*), moss campion (*Silene acaulis*), mountain avens (*Dryas octopetala*) and tufted saxifrage (*Saxifraga cespitosa*), with Scotland remaining a temporary refuge in Britain.

Figure 8.1 Warming of 1°C above the 1960-1991 average is expected to reduce the area suitable for Arctic-alpine species by 60 and 90% in Wales and Scotland respectively. Examples of Arctic-alpine species are shown: (a) purple saxifrage (Photo: SNH); (b) trailing (wild) azalea (Photo: SNH); (c) tufted saxifrage (Photo: SNH); (d) dotterel on nest (Photo: Des Thompson); and (e) snow bunting with food. (Photo: Sandy Tewnion).

Warmer temperatures enhance the productivity of soil ecosystems through the greater mobilisation of nutrients. With the continuing addition of nitrogenous airborne pollutants, the typically nutrient-poor soils of the uplands are likely to gain fertility, therefore favouring lower-altitude species (see Britton *et al.*, this volume). Many scarce or rare plant species of the montane environment are slow-growing and long-lived, often not reproducing sexually until they are several years old, and then not every year, e.g. Snowdon lily and moss campion. These and other plant species, such as tufted saxifrage, also seem to have poor dispersal capabilities. This makes them not only vulnerable to being 'trapped' in sub-optimal climatic environments from which they are unable to 'escape', but also to being overshadowed by faster-growing plant species encroaching from lower altitudes.

Whilst heavier winter precipitation might fall as snow and lead to the development of larger, more persistent snowpatches at higher altitude sites, warmer spring temperatures will accelerate melting. In an analysis of Scottish data, Harrison, Winterbottom & Johnson (2001) concluded that snowbeds at altitudes below 800 m were unlikely to survive over most summer months, although larger, deeper snowbeds at higher altitudes may persist over the occasional summer over this century. Species-rich bryophyte and invertebrate communities associated with snowbeds might therefore be expected to continue to flourish at the highest altitudes, whereas those that occur at the margins of snow-lie are likely to disappear.

8.3 Knowledge required to manage montane ecosystems under climate change

Appropriate management of montane ecosystems may need to consider assisted reproduction and dispersal (*in situ* and *ex situ*), including several methods of translocation in the case of globally-threatened species (e.g. seed transfer), possibly to new, more northerly, higher altitude and/or north-facing locations, including those in other countries. It is therefore important to determine the degree and pattern of shifts in suitable climate space for montane species. Whilst impacts on individual species' potential for re-distribution may be described under different scenarios of climate change, this can only be done for species for which there are adequate distribution data (see Harrison, Berry & Dawson, 2001).

Impacts on individual species are also likely to include alterations in species communities following differing rates of re-distribution under climate change, as well as changes in the abiotic environment. Defining these impacts will depend upon each species' dispersal capability and general ecology, including their reproductive strategy. Unfortunately, such information is often lacking, which hinders understanding of species' responses to projected climate scenarios and therefore the preparation of appropriate management plans.

8.4 Conclusion

Before contingency plans for the protection of the most vulnerable montane species and ecosystems are written, there is a need to collate fundamental autecological information for many montane species. Whilst the abundance of some species may be enhanced by restoring populations to current climatically-suitable locations, some existing locations are likely to become unsuitable in time, therefore restricting these species' range to ever more northerly and higher-altitude locations. The listing of interest features within montane designated sites for nature conservation therefore needs to be reconsidered in order to identify species most likely to disappear or at least change in distribution, especially those that might need careful management. Species likely to be globally threatened may require *ex-situ* action and/or international co-operation, possibly as early as the next 15-20 years (Hossell *et al.*, 2003). There are philosophical and methodological issues in this area which merit considerable debate and research.

References

Cubasch, U., Meehl, G.A., Boer, G.J., Stouffer, R.J., Dix, N., Noda, A., Senior, C.A., Raper, S. & Yap, K.S. (2001). Projections of future climate change. In *Climate Change 2001: The Scientific Basis. Contribution of Working Group I to the Third Assessment Report of the Inter-governmental Panel on Climate Change*, ed. by J.T. Houghton, Y. Ding, D.J. Griggs, M. Noguer, P.J. van der Linden, X. Dai, K. Maskell & C.A. Johnson. Cambridge University Press, Cambridge. pp. 525-582.

Folland, C.K., Karl, T.R., Christy, J.R., Clarke, R.A., Gruza, G.V., Jouzel, J., Mann, M.E., Oerlemans, J., Salinger, M.J. & Wang, S.-W. (2001). Observed climate variability and change. In *Climate Change 2001: The Scientific Basis. Contribution of Working Group I to the Third Assessment Report of the Inter-governmental Panel on Climate Change*, ed. by J.T. Houghton, Y. Ding, D.J. Griggs, M. Noguer, P.J. van der Linden, X. Dai, K. Maskell & C.A. Johnson. Cambridge University Press, Cambridge. pp. 99-192.

Giorgi, F., Hewitson, B., Christensen, J., Hulme, M., Von Storch, H., Whetton, P., Jones, R., Mearns, L. & Fu, C. (2001). Regional climate information - evaluation and projection. In *Climate Change 2001: The Scientific Basis. Contribution of Working Group I to the Third Assessment Report of the Inter-governmental Panel on Climate Change*, ed. by J.T. Houghton, Y. Ding, D.J. Griggs, M. Noguer, P.J. van der Linden, X. Dai, K. Maskell & C.A. Johnson. Cambridge University Press, Cambridge. pp. 583-638.

Gilbert, O. (2004). *Naturally Scottish: Lichens.* Scottish Natural Heritage, Perth.

Harrison, P.A., Berry, P.M. & Dawson, T. (Ed.) (2001). *Climate Change and Nature Conservation in Britain and Ireland: Modelling Natural Resource Responses to Climate Change (the MONARCH project).* UK Climate Impacts Programme, Oxford.

Harrison, S.J., Winterbottom, A. & Johnson, R. (2001). *Climate Changes and Changing Snowfall Patterns in Scotland.* Scottish Executive Central Research Unit, Edinburgh.

Hossell, J.E., Ellis, N.E., Harley, M. & Hepburn, I.R. (2003). Climate change and nature conservation: implications for policy and practice. *Journal for Nature Conservation*, **11**, 67-73.

Hulme, M., Crossley, J. & Xianfu, L. (2001). *An Exploration of Regional Climate Change Scenarios for Scotland.* Scottish Executive Central Research Unit, Edinburgh.

9 Modelling future climates in the Scottish Highlands - an approach integrating local climatic variables and regional climate model outputs

John Coll, Stuart W. Gibb & S. John Harrison

Summary

1. Despite ongoing advances in climate modelling, considerable uncertainties remain in relation to climate model outputs and performance, particularly at regional scales. This is especially true for a topographically diverse region such as the Scottish Highlands.

2. In this work we evaluate a new approach to obtaining more locally relevant representations of future climates for key climatic variables. We utilise a network of observed values for temperature and precipitation to evaluate the performance of a Regional Climate Model (RCM) for selected annual and seasonal values of these variables. Analyses were conducted that variably combined RCM outputs with quality-controlled station data and some temperature lapse rate modelling experiments.

3. In the case of temperature, results indicated poor RCM performance for the region. However, upon the application of lapse rate models, which accounted for differences between observed station data values and corresponding RCM grid values, far better agreement was obtained. It is concluded that this approach is a viable tool in improving the altitudinal modelling of future changes of temperature at the local scale.

4. In the case of precipitation, considerable disparities were observed between observed precipitation values and RCM-simulated representations. Further model development and evaluation are required to address this issue.

5. Given the current limitations of present climate system modelling in topographically diverse regions, we suggest that the approach combining observed data and RCM outputs can provide a better estimate of the range of possible future climates at the local scale. However, as with many aspects of climate change research, considerable uncertainty remains. Some aspects of this are discussed, both in a wider context and in relation to how best to conduct future upland climate change impact assessments at the local scale.

Coll, J., Gibb, S.W. & Harrison, S.J. (2005). Modelling future climates in the Scottish Highlands - an approach integrating local climatic variables and regional climate model outputs. In *Mountains of Northern Europe: Conservation, Management, People and Nature*, ed. by D.B.A. Thompson, M.F. Price & C.A. Galbraith. TSO Scotland, Edinburgh. pp. 103-120.

9.1 Introduction

There is scientific consensus that the global climate is warming at a rate unprecedented in recent times, with warming trends being especially evident over northern hemisphere land areas and at high latitudes (Albritton *et al.*, 2001). However, there are significant uncertainties in current predictions of future change, especially at regionally and locally relevant scales. Consequently, planners are faced with a wide range of predicted changes from different models of unknown relative quality due to large, but unquantified, uncertainties in the modelling process (Cubasch *et al.*, 2001). This is especially true for regions in north-western Europe and the North Atlantic, which are particularly challenging in terms of climate system understanding.

In N.W. Europe and the N. Atlantic regions, climate shifts over the past ~100,000 years have been frequent and relatively sudden, and have been associated with changes to the large-scale meridional overturning motion of the thermohaline circulation (THC) of the Atlantic Ocean (Stocker & Schmittner, 1997; Hall & Stouffer, 2001). With increasing freshwater inputs to Arctic waters associated with recent warming, and projected increases for the future, there remains the possibility that the THC may weaken during this century and actually induce a cooling over parts of northern Europe (Hansen *et al.*, 2004). However, while the concept of a weakened THC is supported by some simple numerical models (Rahmstorf, 2000), there are disagreements among more complex Global Climate Models (GCMs) in describing the future development of the THC (Hansen *et al.*, 2004). These differences in the stability of the THC constitute one of the sources of uncertainty in climate projections for the North Atlantic region (Vellinga, 2004).

As a result, considerable research effort is being directed at improving understanding of the processes which influence North Atlantic climate variability and the implications of these for future change in this climatically dynamic region (Blunier & Brook, 2001; Marshall *et al.*, 2001; Srokosz & Gommenginger, 2002). In particular, specific research efforts are being directed at both monitoring and modelling prospective changes to important oceanic gyres associated with warming and increased inputs of freshwater at high latitudes, including the THC (Schiermeier, 2004a; Alley, 2004).

Despite disagreements in the magnitude of warming predicted by various GCMs, there is agreement that the observed warming trends are set to continue throughout the 21st century. For example, on hemispheric scales the GCMs show broad similarities in their patterns of change, such as the land warming faster than the ocean, and warming in high latitudes faster than in equatorial regions (Albritton *et al.*, 2001; Jenkins *et al.*, 2003; Fowler & Battarbee, this volume).

9.1.1 Maritime upland sensitivity to climate change

Situated on the seaward western edge of north-western Europe and subject to maritime and to a lesser extent continental influences, the climate of the Scottish Highlands is typified by spatial and temporal variability. Superimposed on these synoptic controls, orographic effects produce a locally variable climate across the region. With more than 4,500 km² of the land surface being higher than 600 m above sea level, such altitudinal gradients of change are an important control in the spatial pattern of climate across Scotland (Harrison & Kirkpatrick, 2001).

This variable climate contributes greatly to the biodiversity of the uplands, with a diverse mix of Atlantic, Arctic, Arctic-alpine and boreal elements occurring within a limited geographical area, and including many species on the edge of their global distributional range (Birks, 1997). Within this continuum of microclimates, most high altitude plant species are adapted to slow growth and are less competitive, with survival in the increasingly severe abiotic environment largely determined by the altitudinal range over which a species occurs. It is this combination of highly variable local climates, together with many designated habitats of high conservation value being located in the uplands, which necessitates the development of more locally relevant methods of climate change impact assessment.

9.1.2 Sensitivity to temperature changes

Amid general concerns that mountain regions are particularly susceptible to warming (Gottfried *et al.*, 1999; Becker & Bugmann, 2001; Klanderud & Birks, 2003; Fowler & Battarbee, this volume), there is particular concern that upland maritime regions such as the Highlands of Scotland may be especially vulnerable. Here even a relatively modest warming has the potential to induce far-reaching changes in the latitudinal and altitudinal distribution of habitats (Gordon *et al.*, 1998; Hill *et al.*, 1999). For example, it has been estimated that a 1°C increase in annual temperature would be associated with an uphill isotherm shift of 200-275 m, resulting in a 90% reduction in area of Scotland's arctic-alpine habitat and with substantial implications for vulnerable species (Kerr & Ellis, 2002; Nagy, 2003; Ellis & Good, this volume).

With many mountain plants intolerant to competition, fast-growing lowland species with broad altitudinal and ecological ranges are predicted to expand at the cost of slow-growing competition-intolerant species with narrow habitat demands (Saetersdal & Birks, 1997; Pauli *et al.*, 2003; Ellis & Good, this volume). In addition, with soil productivity enhanced by warming temperatures, the altitudinal expansion of lower altitude species will be increasingly favoured (Ellis & Good, this volume). An advancing tree line for example, or a denser forest below the tree line, would have important implications for the carbon cycle (increasing the terrestrial carbon sink) and for biodiversity of the alpine ecotone, possibly ousting rare species and disrupting alpine and arctic plant communities (Grace *et al.*, 2002; Kullman, 2004).

While it might be expected that oceanic mountains would be buffered against climatic change by their more limited annual temperature range, by comparison with some of the higher European mountain regions, the nival zone here is insufficient in extent to accommodate any potential upward migration of species (Crawford, 2000a,b). Consequently under global warming scenarios, species with low migration potential are unlikely to achieve the relatively small distance migration which ensured their survival during preceding warmer periods of the Holocene. Winter climatic warming is a particular threat to montane floras in the more oceanic west of the region, since the risks of premature bud burst and warm winter-induced carbohydrate consumption are greater (Crawford, 2000b).

The Hadley Centre Regional Climate Model (HadRM3) suggests future increases in oceanicity (particularly in the autumn and winter half-year) for the more continental hills of the east. Changes in the pattern of snowfall and duration of snow-lie (Harrison *et al.*, 2001) occurring as a consequence of these changes will also affect community structure in these upland regions. The viability of upland species which are dependent on snow cover for survival will be particularly impacted, e.g. lemon scented fern (*Dryopteris oreopteris*), great wood rush (*Luzula sylvatica*), and hard fern (*Blechnum spicant*) which tend to be found in the more continental east of Scotland and not in the west (Poore & McVean, 1957). However, spring and summer changes in climate are also likely to have an impact, with HadRM3 simulating substantial warming and drying over eastern areas as the century progresses.

Aside from changes in mean seasonal values, climate models predict increasingly frequent extreme weather events which would impact upland plant communities as the century progresses. For example, recent modelling results suggest that future heat waves will increase in the second half of the century (Meehl & Tebaldi, 2004) as anthropogenic interference in the climate system becomes increasingly noticeable (Stott *et al.*, 2004): an influence which will increasingly impact the more continentally influenced uplands of the east.

With the magnitude and rate of changes to seasonal temperature and precipitation regimes indicated by GCM scenario outputs for the coming century likely to have an adverse impact on many higher plant and animal species in the montane environment, there is a need for contingency conservation and management measures to protect the most vulnerable species and ecosystems (Ellis & Good, this volume). This requirement is augmented by concerns that the required dispersal rates are greater than many species can manage. However, the response of mountain plant species to climate change remains relatively uncertain due to a lack of accurate observational data on long-term changes and knowledge about adaptation (Klanderud & Birks, 2003; EEA, 2004). This recognition drives the requirement for data acquisition on the abundance and distribution of the most climate-vulnerable montane species, their reproductive abilities, dispersal capabilities and their general ecology, particularly in relation to their minimum ecosystem requirements (Ellis & Good, this volume). In addition, there is a need for developing methods which will deliver more locally relevant representations of future change to key climatic variables in order that land managers can make better informed decisions on stewardship of vulnerable communities and habitats.

9.1.3 Sensitivity to precipitation changes

Maritime upland vulnerability and adaptive responses to a changing climate extend beyond the direct effects of temperature: seasonal changes to precipitation regimes will also contribute to shifts in altitudinal zonations of vegetation. For example, a warmer and wetter climate over the coming decades would alter the deposition of nitrogen, organic pollutants and heavy metals further affecting ecosystems already sensitive to changes in their physical and chemical climates (Fowler & Battarbee, this volume). With Regional Climate Model (RCM) outputs indicating substantial changes to seasonal rainfall regimes and an increase in extreme rainfall events (Christensen & Christensen, 2002; Huntingford *et al.*, 2003; Frei *et al.*, 2003), there is a suggestion that future changes to the intensity and return period of such events could trigger the accelerated release of historically deposited pollutants in catchment soils in affected uplands. Poor historical land use practices have left substantial tracts of the region vulnerable to accelerated erosion (Grieve *et al.*, 1995; Grieve & Hipkin, 1996) and mobilisation of pollutants if the pulsed inputs from low frequency, high magnitude extreme rainfall events are set to increase in the future.

Some of these changes may already be underway. Increases in seasonal precipitation intensity associated with changes to the North Atlantic Oscillation (NAO) have been greatest in the parts of the UK which experience significant orographic rainfall (Osborn *et al.*, 2000). These have been particularly evident in the east of Scotland where the 50 year autumn and winter extreme event during the period 1961-1990 has become an 8 year event during the 1990s (Fowler & Kilsby, 2003). These changes have been implicated in a marked increase in gullying and erosion in upland areas despite reductions in grazing pressure (Kerr & Ellis, 2002), and are concurrent with a sharp increase in the concentration of dissolved organic carbon (DOC) in upland freshwaters between 1989 and 2001 (Freeman *et al.*, 2001). It has also been reported that excessive stream discharges are dislodging spawning grounds of young fish and freshwater pearl mussel (*Margaritifera margaritifera*) beds, particularly in north-west Scotland (Hastie *et al.*, 2001). It is this suite of anticipated changes which drives the need to obtain more locally representative estimates of what future climate may be. However, these are difficult to obtain from the present generation of climate model outputs alone.

9.1.4 Climate model limitations in topographically diverse regions and recent model developments

Even with the evolution of ever more complex and sophisticated GCMs, issues remain concerning their robustness (Chase *et al.*, 2004), and their reproduction of the detail of regional climates remains rather poor (Zorita & von Storch, 1999; Gonzalez-Rouco *et al.*, 2000; Jones & Reid, 2001). For the North Atlantic region, GCMs generate widely varying results and have problems in simulating the present climate of the region (Saelthun & Barkved, 2003). This is, in part, due to a limited parameterisation of many specific climate processes. Consequently, these are represented in different (but plausible) ways in the various GCMs, resulting in different outputs (Jenkins & Lowe, 2003). Thus, when nine current GCMs are inter-compared at the UK regional scale, there are substantial differences in simulated changes to seasonal precipitation for example (Jenkins & Lowe, 2003). As a result, more recent approaches have aimed at reducing some of the uncertainties (Murphy *et al.*, 2004) and increasing confidence in the projections of future climate (Kerr, 2004; Stocker, 2004).

With RCMs, the horizontal resolution is increased up to the mesoscale over a limited area of interest allowing a much more accurate description of topography, coastlines and lakes (Christensen, 2001; Jenkins *et al.*, 2003). However, RCMs take their boundary values from the coarse grid GCMs, and the effects of these coarse resolution boundary values can propagate within the RCMs, thereby perpetuating the uncertainties originating in the large-scale models (Carter *et al.*, 1999; Jenkins & Lowe, 2003; Moberg & Jones, 2003; Saelthun & Barkved, 2003; Rowell, 2004; Schiermeier, 2004b). For example, the United Kingdom Climate Impact Programme (UKCIP02) climate change scenarios were generated using the HadRM3 RCM, driven by predictions from the HadCM3 GCM. However, HadCM3's representation of some aspects of observed climate is not realistic, e.g. storm tracks over north-west Europe are displaced too far south (Hulme *et al.*, 2002). Therefore, in order to improve the coupling of the models, an intermediate, 150 km resolution, atmosphere-only GCM was used (HadAM3H) to provide the lateral boundary conditions (Hulme *et al.*, 2002). This double-nesting approach improves the quality of the simulated European climate – the position of the main storm tracks, for example, are better located – and also allows the UKCIP02 scenarios to present information with greater spatial detail: 50 km rather than 250 to 300 km (Hulme *et al.*, 2002). However, in an area of complex topography such as the Highlands where the close spatial juxtaposition of sea and land areas gives rise to highly localised climatic effects, it is unlikely that even such a sophisticated model will capture the detail of local climatic processes here.

9.2 Objectives and approach

In this chapter we aim to establish if variably combined climate model outputs and station data can be used to improve local representations of future changes to temperature and precipitation. We then examine the implications of experiments for modelling future changes in our upland environments.

A number of distinct analyses have been undertaken.

- An observed station network with data quality controlled for the 1961-1990 baseline period is used to assess HadRM3's performance in simulating the 1961-1990 baseline for selected seasonal mean values.
- Temperature lapse rate values are used to correct the mismatch in elevations between HadRM3 grid cells and the stations to improve the validity of the above assessment. This is also undertaken to establish if the approach can be modified and refined to project future temperature change outputs from HadRM3 upslope.
- HadRM3 future time slice outputs and station observed 1961-1990 baseline data are variably combined for selected stations and results compared.

9.3 Methods

Precipitation (Precip) and temperature data (maximum, T_{max} and minimum, T_{min}) data from stations across the Highlands with a fundamentally intact record for 1961-1990 were extracted from the British Atmospheric Data Centre (BADC) database. Due to missing data for some years and gaps due to logger/instrument failure, intact baseline years available ranged from 22-30 years (this reflecting some of the wider problems of data availability and spatial cover of instrumental records for the region). To counter the problem of gaps in the observed 1961-1990 baseline record, annual values of T_{max}, T_{min} and Precip for all of the stations used were cross-correlated (Temperature, n = 19; Precipitation, n = 55). Station numbers here differ from those given below because the existing Meteorological Office station records were not used and island station data were included. Those stations with data sets exhibiting the highest Correlation Coefficients (Pearson) and lowest p-values were identified and used to infill gap years in the 1961-1990 record (T_{max}, r = 0.725-0.983, p <0.001; T_{min}, r = 0.533-0.930, p = 0.000-0.006; Precip, r = 0.575-0.960, p <0.001-0.005). Following these measures, 30 year baseline means were constructed for 1 and 2 below (it should be noted that for 1961 complete monthly rain gauge records were unobtainable and the precipitation baseline was constructed for 29 years).

1. Annual and seasonal maximum and minimum temperatures for 16 stations ranging in elevation from 4 m to 283 m. Four further station records were added from existing Meteorological Office 1961-1990 station data to improve spatial cover.
2. Annual and seasonal total precipitation for 50 monthly rain-gauge sites ranging in elevation from 27 m to 536 m.

3. Island station records were excluded from this analysis, since even at a 50 km x 50 km resolution the Western and Northern Isles are not treated as 'land' in HadRM3.

The Hadley Centre's RCM (HadRM3) 50 km x 50 km land grid cells for the 'Highlands' were identified (Table 9.1) and outputs from the UKCIP02 database extracted for:

1. annual and seasonal maximum and minimum temperature and precipitation values for the simulated 1961-1990 baseline and future time slices (2020s to 2080s); and
2. HadRM3 grid cell latitude and longitude co-ordinates.

Table 9.1 HadRM3 'Highland' land grid cell identifiers. Cell latitude and longitude co-ordinates refer to the centre of each 50 km x 50 km grid, elevation is constant across the cell (Source: Hulme *et al*., 2002).

HadRM3 Grid Cell ID	Cell Latitude (°N)	Cell Longitude (°E)	Cell Elevation (m)
92	58.55	-4.48	100.35
93	58.63	-3.66	48.35
108	58.03	-5.13	252.02
109	58.11	-4.32	202.93
110	58.20	-3.50	68.58
124	57.50	-5.75	277.78
125	57.60	-4.96	384.65
126	57.68	-4.16	151.23
142	57.07	-5.58	445.07
143	57.17	-4.79	438.53
144	57.25	-4.00	412.19
145	57.33	-3.20	320.73
146	57.41	-2.40	113.38
160	56.64	-5.41	460.54
161	56.73	-4.63	618.85
162	56.82	-3.84	479.38
163	56.90	-3.06	339.39
178	56.21	-5.24	333.26
179	56.30	-4.47	365.22
180	56.39	-3.69	201.68
181	56.47	-2.91	77.14

A matrix of resolution of 0.1° latitude and longitude was constructed for the region and a nearest neighbour method used to interpolate stations to the corresponding HadRM3 grid cell. This 0.1° resolution equates to ~11.3 km and ~11.1 km in distance for latitude and longitude, respectively. Observed seasonal means of temperature and precipitation for the 1961-1990 baseline were subtracted from the corresponding HadRM3 grid 1961-1990 simulation and the differential plotted (Figures 9.1a,b and 9.2). Given the considerable

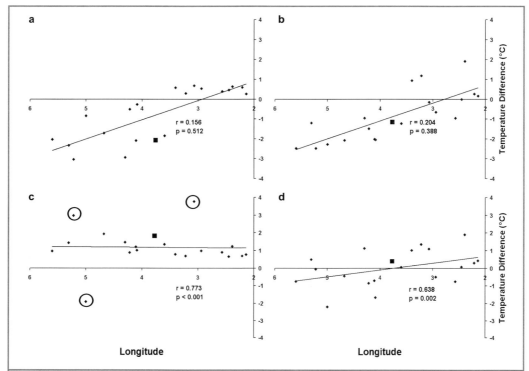

Figure 9.1 Temperature differences (°C) HadRM3 – Observed Station Values (n = 20), Pearson correlation coefficient (r) and probability value (p). (a) Winter maximum temperature (T_{max}). (Circled points denote outliers, see text for explanation). (b) Autumn minimum temperature (T_{min}). (c) Winter T_{max} lapse adjusted to corresponding HadRM3 grid elevation. (d) Autumn T_{min} lapse adjusted to corresponding HadRM3 grid elevation. (■ Denotes Faskally).

mismatch in actual altitude and that of the corresponding HadRM3 grid cell for many of the stations, seasonal lapse rate values 'typical' of the Highlands (Table 9.2) were used to relate station altitudes (nearest 10 m) to the corresponding HadRM3 grid (Figure 9.1c,d). Only seasonal mean values are presented here.

Two stations near Pitlochry (Latitude ~56.7°, Longitude ~3.8°) with a full observational record of daily temperature (Faskally, 94 m) and monthly precipitation (Bruar Intake, 381 m) for the 1961-1990 baseline were selected to illustrate the range of possible future climate for the area and the observed data variably combined with HadRM3 data for the 2080s Low & 2080s High Scenarios (Figure 9.3).

9.4   Results and Discussion

9.4.1   Temperature

For all stations, there is a clear west to east gradient of departures in both annual and seasonal maxima and minima from the corresponding HadRM3 grid cells. However, HadRM3's simulation of autumn minima for example has more scatter (Figure 9.1b, r = 0.156, p = 0.512) west to east than for the simulation of winter maxima (Figure 9.1a, r = 0.204, p = 0.388). For the region as a whole, HadRM3's closest simulation of the observed baseline is for the eastern lowlands, whereas the largest departures are associated with the western

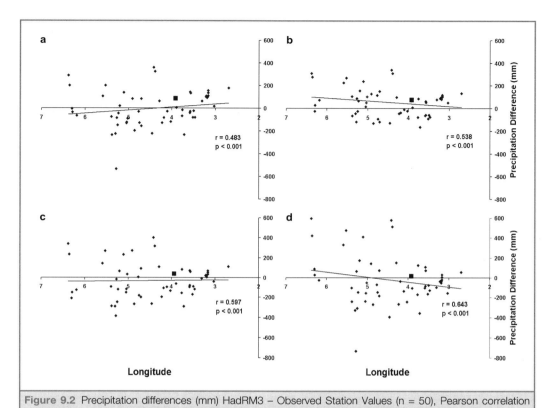

Figure 9.2 Precipitation differences (mm) HadRM3 – Observed Station Values (n = 50), Pearson correlation coefficient (r) and probability value (p). (a) Spring precipitation totals. (b) Summer precipitation totals. (c) Autumn precipitation totals. (d) Winter precipitation totals. (■ Denotes Bruar Intake).

Table 9.2 Seasonal range of temperature lapse rate values.

Season	Lapse Rate °C/1,000 m	
	Maximum Temperature (T_{max})	Minimum Temperature (T_{min})
Spring	9.8-10.2	6.0–8.0
Summer	9.5-9.8	5.0–7.0
Autumn	9.2-10.0	3.0–5.0
Winter	9.5-10.2	3.0–8.0

lowlands and central mountainous areas. At least part of this pattern is attributable to the difference in altitude between the stations and their corresponding HadRM3 grid cell which are generally greater in western and central inland areas of the region. When observed data are modified to the corresponding HadRM3 grid using the appropriate lapse rate, greatly improved correlations were observed for both winter T_{max} (Figure 9.1c; r = 0.773, p <0.001) and autumn T_{min} (Figure 9.1d; r = 0.638, p = 0.002). Residual scatter of data (particularly evident for winter T_{max} values) suggests a parameterisation problem with some HadRM3 grid cells.

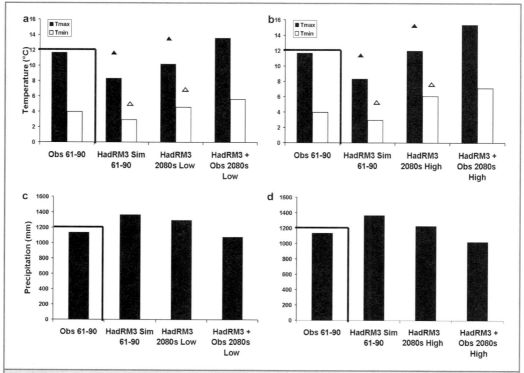

Figure 9.3 (a) Mean annual temperature (°C) for Faskally; observed 1961-90, HadRM3 simulated 1961-90, HadRM3 2080s Low scenario, 2080s Low added to observed 1961-1990. (b) Same procedure for 2080s High scenario. (c) Mean annual precipitation total (Precip, mm) for Bruar Intake; observed 1961-90, HadRM3 simulated 1961-90, HadRM3 simulated 1961-90, HadRM3 2080s High scenario, 2080s High added to observed 1961-1990. (d) Same procedure for 2080s High scenario. (▲, Denotes HadRM3 grid values lapse adjusted to station elevation, see text for explanation).

It may be noted that when the outlying points of Onich, Cape Wrath and Mylnefield (west to east respectively and circled in Figure 9.1c) are omitted, greatly improved correlation coefficients are observed (r = 0.981, p <0.001). Given these are coastal sites, this may suggest that HadRM3 and the driving HadCM3 may have a problem with the land-sea interface. Alternatively, there may be a problem with the lapse rate models at these locations as other near coastal sites are not outlying to the same extent; in either case, further work is required.

9.4.2 Precipitation

There is a less discernible pattern in the difference between HadRM3's simulation of seasonal precipitation and observed data in moving from west to east (Figure 9.2). Figures 9.2c and 9.2d illustrate that the largest differentials are observed for winter and autumn, despite the better correlations in data sets observed for these seasons (r = 0.643 and 0.597 respectively; Figure 9.2a,b).

Across the region there is a pattern of positive departures in annual and seasonal total precipitation associated with the elevational mismatch between HadRM3 grids and the stations, as would be expected. However, as with temperature data, the magnitude of some of the differences suggest parameterisation problems with some of the HadRM3 grid cells resulting in a variable pattern of over- and under-representation of model outputs with respect to observed values across the region.

This problem is not unique to HadRM3 and the driving HadCM3. Despite progressive increases in model resolution, it is not yet possible to use GCMs to resolve atmospheric processes such as storms, fronts and their interaction with topography in sufficient detail to predict rainfall behaviour (Huntingford *et al.*, 2003). To illustrate this, in an inter-comparison of precipitation outputs from five RCMs and daily rainfall records across the European Alps, HadRM3 was only one of three RCMs found to be under-representing mean summer precipitation by up to 25% (Frei *et al.*, 2003). While no attempt has been made here to adjust station precipitation data by elevation to the corresponding HadRM3 grid elevation, the residual scatter of data for eastern sites (where station and HadRM3 grid elevations more closely correspond) would suggest that HadRM3 parameterisations are having difficulty resolving local orographic effects for all seasons, including summer (Figure 9.2b).

Ongoing work indicates that locally observed precipitation-elevation relationships can be used to construct simple models which may be used to tune HadRM3 to generate outputs more representative of observed values. Advances in this approach are challenged by the highly localised orographic controls characteristic of the region.

Generally, climate models produce outputs that areally average climate at the grid box scale. Consequently, the validation of climate model simulations creates substantial demands for observed climate datasets which are both geographically and historically extensive (Osborn & Hulme, 1997). For a data-sparse region such as the Highlands, the inter-comparison between the HadRM3 grid cells and stations on the geographically sparse datasets used here is limited in extent. Since the HadRM3 grids are realised at elevations substantially different from those of available station data, this only compounds the general problem of regional evaluation of climate model outputs here. For the purposes of future climate change research, both generally and for the uplands in particular, this work flags the need for a more extensive network of recording stations at varying elevations.

9.4.3 Variably combining observed data and HadRM3 future time slices

The annual temperature plots for Faskally in Figures 9.3a and 9.3b illustrate that the differences between the observed 1961-90 values (Obs 61-90) and HadRM3's simulation (HadRM3 Sim) of the same period have implications for the application of HadRM3 future time slices. HadRM3's apparent under-representation of both mean T_{max} and T_{min} for the baseline period is again an artefact of the mismatch between station and grid elevation (HadRM3 Grid 162 = 479m, compared to Faskally at 94 m). One approach in creating a range of possible future climates is simply to add the future time slice outputs to the observed baseline (HadRM3 + Obs, both plots). If the HadRM3 cell outputs are lapse-adjusted to the station elevation (values denoted by triangles, Figure 9.3a,b), there is a closer accordance of values. However, this adjustment indicates that HadRM3 is still simulating higher T_{max} and T_{min} values with respect to the observed baseline (HadRM3 Sim 61-90). When future time slice outputs are applied to the grid lapse-adjusted values, these under- and over-representations are carried through (HadRM3 2080s, both plots) by comparison with the same outputs added to the observed baseline (HadRM3 + Obs, both plots).

Treating the annual precipitation data for Bruar Intake in the same way creates a similar range of possible futures (Figure 9.3c,d, notations as above). In this case, the higher annual values simulated by HadRM3 are attributable to grid cell 162 being ~100 m higher than Bruar Intake. Even if time slice outputs unadjusted for elevation for the two 2080s scenarios selected are added to the observed baseline, future drying is indicated (HadRM3 + Obs, both plots). Clearly, the sign of future precipitation change is different from that of temperature for the selected location. This reflects HadRM3's prediction of future drying for the eastern Highlands. An obvious caveat is that this sign change is an artefact of the location of the station selected and future sign changes for precipitation at other stations are different.

At a 50 km x 50 km resolution, HadRM3's representation of 'land' for the Highlands is truncated to the west and east by comparison with reality and the Western and Northern Isles are not treated as land. Consequently, island station records were dropped from this analysis. New scenarios and data outputs for island sites using a version of the Hadley Regional Model at a resolution of 25 km x 25 km (Jenkins *et al.*, 2003) should offer new possibilities for purposes of impact assessment at island sites. However, even at this resolution the smaller features which produce local climates, such as hills and valleys on the islands, cannot be accounted for (Jenkins *et al.*, 2003). Some of the data presented here and ongoing work indicate that observed temperature lapse rates for the region can be used to improve agreement between model-simulated and observed values and that simple elevational models can be constructed to project HadRM3 future time slice outputs up and down slope.

As with many aspects of regional climate change assessment, there is inherent uncertainty in the work. This includes uncertainty in

- future emissions pathways (reflecting the inability to foresee future socio-economic trajectories);
- the response of the climate system to these emissions (due to imperfect models); and
- natural climate variability and downscaling to a higher resolution (Huntingford *et al.*, 2003; Jenkins & Lowe, 2003; Willows & Connell, 2003).

Added to these, the formulation of RCMs adds further uncertainty to climate change at the national scale; and for example, while this effect is small for mean surface air temperature in HadRM3, it is more substantial for precipitation (Rowell, 2004). Also, given the existence of substantial biases in current RCMs, future scenario integrations should be treated with care until further research aimed at reducing systematic errors is undertaken (Moberg & Jones, 2003). For the Highlands, where there is significant year-to-year and decade-to-decade variability of climate, the imposition of this natural variability on an underlying anthropogenic trend in the future is a further complication. In the context of this work, there is a considerable assumption involved in considering that future relationships of T and Precip with elevation will remain the same in a warmer and seasonally wetter world; and that the frequency and type of synoptic frontal systems, which in turn influence lapse rates and orographic precipitation, will remain the same.

9.5 Extending the temperature lapse rate modelling

As indicated above, ongoing studies extend the work reported here on temperature lapse rates. As well as being used to improve the agreement between the HadRM3 Sim and Obs

61-90 values, lapse rate model performance has been tested against some observed upland station records at differing elevations across the region and found to be performing credibly across the range of annual and seasonal values applied (Coll, unpublished work). Following the model validation, the lapse rate models have been used to project both Obs 61-90 and HadRM3 Sim values to an elevation of 1,300 m for 'representative' western and eastern upland areas (Lochaber and Cairngorm/Caenlochan).

Additively perturbing the baseline models with selected UKCIP02 scenario outputs for seasonal temperature changes from HadRM3 (2050s Medium-Low and 2080s High) causes very substantial upward shifts in seasonal isotherms both east and west (Coll, unpublished work). If temperature is taken as the only driver, the climate space currently associated with present vegetation zones in the uplands is subject to substantial altitudinal migration (particularly for seasonal minima), even by the 2050s for the selected scenario. For example, in the east, the 1961-1990 mean spring minimum of −1.0°C associated with the present treeline limit of ~600 m migrates to 820 m under the 2050s Medium-Low scenario when the corresponding HadRM3 grid output is used to perturb the observed baseline data for Balmoral. In the west, the mean 1961-1990 autumn minimum of 4.6°C associated with the current treeline limit of ~300 m shifts to 830 m under the same 2050s scenario when the corresponding HadRM3 grid output is used to perturb the observed baseline data for Onich. In this latter case, the lapse rate models are carrying through the dramatic increases in autumn minima simulated by HadRM3 for the western Highlands.

To date, a particular focus has been on altitudinal changes to winter maxima and spring minima (specifically the elevation associated with the 0°C isotherm) in order to infer changes to snow-lie and the implications for vulnerable communities at a localised scale (Coll, unpublished work). However, the lapse rate models can integrate any combination of observed and GCM or RCM data, allowing the possibility of modelling an ensemble of possible future changes at any given elevation for any selected value of T_{max} and T_{min} at the local scale. Such an approach may constitute a useful site assessment tool for policy-makers, since a wide range of possible temperature futures could be represented for purposes of contingency planning. Given the problems with the current generation of GCMs and RCMs, it is also suggested that such an approach would be better able to envelope the range of possible futures for planning purposes at the local scale.

9.6 Conclusions and future work

Our work indicates that despite the sophistication of HadRM3 and its improved spatial performance over its predecessor, HadCM2, its 50 km x 50 km grid resolution does not adequately capture the spatial variability of climate in the Highlands and Islands of Scotland. Mountains give rise to a varied climate as local orographic effects are superimposed on complex terrain and, as indicated here, the representation of mountainous areas in GCMs and even the finer-scaled RCMs do not adequately capture local climate detail. While advances in climate system modelling have been substantial, further advances are required before RCM outputs meet the local-scale projections required by policy-makers (Schiermeier, 2004b). At present, especially for a region such as the Highlands, it is difficult to conduct valid assessments using outputs from the present generation of climate models alone. Despite the difficulties and uncertainties surrounding climate model outputs, it is possible to adopt a pragmatic approach for topographically diverse regions. By variably combining RCM outputs and station data, the results presented here indicate that (for temperature at least) possible future changes to climate can be better resolved at the local scale. With ongoing refinement, the approach can be used to better inform stewardship decisions for plant communities of high conservation value by providing an improved range of local projections.

Improvements to the next Hadley Centre Global Environmental Model version 1 (HadGEM1) will also enhance our capability for carrying out new climate experiments (Johns *et al.*, 2004). Plans to double-nest a regional model (HadRAM1) in this improved GCM at a spatial resolution of 10 km (Hulme *et al.*, 2002) create future scope for using climate model outputs at an improved spatial resolution to refine the methods developed here.

These advances will facilitate improved assessment of the impacts of future climate changes to key natural and managed systems spanning the range of habitat elevations. However, arriving at improved assesments of local future climate is only one of the many challenges involved in managing our uplands amid the suite of likely future environmental and social drivers of change. Many of the challenges involved in managing future climatically driven changes are superimposed on already complex historical and contemporary drivers of ecological and land use change, often in poorly understood environments (see Thompson *et al.*, this volume).

There is a clear need for improved understanding of the nature and magnitude of future climate change at regional scales. The environmental impacts and socio-economic consequences of this change must also be robustly assessed and managed. It is only through programmes of extensive monitoring and intensive and highly inter-disciplinary effort that the research community can address these needs.

Acknowledgements

We would like to thank the BADC and UKCIP02 for providing access to the Met. Office surface station data and HadRM3 output data, and Dr Uwe Peterman for developing the programme used to process daily temperature records. We would also like to thank the two anonymous referees for providing helpful comments on the original manuscript.

References

Albritton, D.L., Meira Filho, L.G., Cubasch, U., Dai, X., Ding, Y., Griggs, D.J., Hewitson, B., Houghton, J.T., Isaksen, I., Karl, T., McFarland, M., Meleshko, V.P., Mitchell, J.F.B., Noguer, M., Nyenzi, B.S., Oppenheimer, M., Penner, J.E., Pollonais, S., Stocker, T. & Trenberth, K.E. (2001). Technical Summary of the Working Group I Report. In *Climate Change 2001: The Scientific Basis. Contribution of Working Group I to the Third Assessment Report of the Inter-governmental Panel on Climate Change,* ed. by J.T. Houghton, Y. Ding, D.J. Griggs, M. Noguer, P.J. van der Linden, X. Dai, K. Maskell & C.A. Johnson. Cambridge University Press, Cambridge. pp. 21-83.

Alley, R.B. (2004). A slowing cog in the North Atlantic Ocean's climate machine. *Science*, **304**, 371-372.

Becker, A. & Bugmann, H. (Eds) (2001). Global Change and Mountain Regions (The Mountain Research Initiative), IGBP Report 49. International Geosphere-Biosphere Programme, Royal Swedish Academy of Sciences, Stockholm.

Birks, H.J.B. (1997). Scottish biodiversity in a historical context. In *Biodiversity in Scotland: Status, Trends and Initiatives,* ed. by L.V. Fleming, A.C. Newton, J.A. Vickery & M.B. Usher. The Stationery Office, Edinburgh. pp. 21-35.

Blunier, T. & Brook, E.J. (2001). Timing of millennial-scale climate change in Antarctica and Greenland during the last glacial period. *Science*, **291**, 109-112.

Carter, T.R., Hulme, M. & Viner, D. (Eds) (1999). Representing Uncertainty in Climate Change Scenarios and Impact Studies. ECLAT-2, Report No. 1, Helsinki Workshop, 14-16 April 1999. CRU, Norwich.

Chase, T.N., Pielke, R.A., Herman, B. & Zeng, X. (2004). Likelihood of rapidly increasing surface temperatures unaccompanied by strong warming in the free troposphere. *Climate Research*, **25**, 185-190.

Christensen, J.H. (2001). Major characteristics of the global climate system and Global Climate Models. In *Integrated Regional Impact Studies in the European North*, ed. by M.A. Lange. Institute for Geophysics, University of Munster, Munster. pp. 99-120.

Christensen, J.H. & Christensen, O.B. (2002). Severe summertime flooding in Europe. *Nature*, **421**, 805.

Crawford, R.M.M. (2000a). Ecological hazards of oceanic environments (Tansley Review No. 114). *New Phytologist*, **147**, 257-281.

Crawford, R.M.M. (2000b). Plant community responses to Scotland's changing environment. Botanical *Journal of Scotland*, **53**, 77-105.

Cubasch, U. Meehl, G.A., Boer, G.J., Stouffer, R.J., Dix, N., Noda, A., Senior, C.A., Raper, S. & Yap, K.S. (2001). Projections of future climate change. In *Climate Change 2001: The Scientific Basis. Contribution of Working Group I to the Third Assessment Report of the Inter-governmental Panel on Climate Change*, ed. by J.T. Houghton, Y. Ding, D.J. Griggs, M. Noguer, P.J. van der Linden, X. Dai, K. Maskell & C.A. Johnson. Cambridge University Press, Cambridge. pp. 525-582.

EEA (2004). *Impacts of Europe's Changing Climate: an Indicator-Based Assessment*. European Environment Agency, Copenhagen.

Fowler, H.J. & Kilsby, C.G. (2003). A regional frequency analysis of United Kingdom extreme rainfall from 1961 to 2000. *International Journal of Climatology*, **23**, 1313-1334.

Freeman, C., Evans, C.D., Monteith, T.D., Reynolds, B. & Fenner, N. (2001). Export of organic carbon from peat soils. *Nature*, **412**, 785.

Frei, C., Christensen, J.H., Deque, M., Jacob, D., Jones, R.G. & Vidale, P.L. (2003). Daily precipitation statistics in regional climate models: evaluation and intercomparison for the European Alps. *Journal of Geophysical Research*, **108**, 1-22.

Gonzalez-Rouco, J.F., Heyen, H. & Valero, E. (2000). Agreement between observed rainfall trends and climate change simulations in the south-west of Europe. *Journal of Climate*, **13**, 3057-3065.

Gordon, J.E., Thompson, D.B.A., Haynes, V.M., Brazier, V. & Macdonald, R. (1998). Environmental sensitivity and conservation management in the Cairngorm Mountains, Scotland. *Ambio*, **27**, 335-344.

Gottfried, M., Pauli, H., Reiter, K. & Grabherr, G. (1999). A fine-scaled predictive model for changes in species distribution patterns of high mountain plants induced by climate warming. *Diversity and Distributions*, **5**, 241-251.

Grace, J., Berninger, F. & Nagy, L. (2002). Impacts of climate change on the tree line (review paper). *Annals of Botany*, **90**, 537-544.

Grieve, I.C., Davidson, D.A. & Gordon, J.E. (1995). Nature, extent and severity of soil erosion in upland Scotland. *Land Degradation and Rehabilitation*, **6**, 41-55.

Grieve, I.C. & Hipkin, J.A. (1996). Soil erosion and sustainability. In S*oils, Sustainability and the Natural Heritage*, ed. by A.G. Taylor, J.E. Gordon & M.B. Usher. HMSO, Edinburgh. pp. 236-248.

Hall, A. & Stouffer, R.J. (2001). An abrupt climate event in a coupled ocean-atmosphere simulation without external forcing. *Nature*, **409**, 171-174.

Hansen, B., Osterhus, S., Quadfasel, D. & Turrell, W. (2004). Already the day after tomorrow? *Science*, **305**, 953-954.

Harrison, J., Winterbottom, S. & Johnson, R. (2001). *Climate Change and Changing Snowfall Patterns in Scotland.* Scottish Executive, Edinburgh.

Harrison, S.J. & Kirkpatrick, A.H. (2001). Climatic change and its potential implications for environments in Scotland. In *Earth Science and The Natural Heritage: Interactions and Integrated Management,* ed. by J.E. Gordon & K.F. Leys. The Stationery Office, Edinburgh. pp. 296-305.

Hastie, L.C., Boon, P.J., Young, M.R. & Way, S. (2001). The effects of a major flood on an endangered freshwater mussel population. *Biological Conservation,* **98**, 107-115

Hill, M.O., Downing, T.E., Berry, P.M., Coppins, B.J., Hammond, P.S., Marquiss, M., Roy, D.B., Telfer, M.G. & Welch, D. (1999). Climate changes and Scotland's natural heritage: an environmental audit. Scottish Natural Heritage Research, Survey and Monitoring Report 132.

Hulme, M., Jenkins, G.J., Lu, X., Turnpenny, J.R., Mitchell, T.D., Jones, R.G., Lowe, J., Murphy, J.M., Hassell, D., Boorman, P., McDonald, R. & Hill, S. (2002). *Climate Change Scenarios for the United Kingdom: The UKCIP02* Scientific Report. Tyndall Centre for Climate Change Research, School of Environmental Sciences, University of East Anglia, Norwich.

Huntingford, C., Jones, R.G., Prudhomme, C., Lamb, R., Gash, J.H.C. & Jones, D.A. (2003). Regional climate-model predictions of extreme rainfall for a changing climate. *Quarterly Journal of the Royal Meteorological Society,* **129**, 1607-1621.

Jenkins, G., Cooper, C., Hassell, D. & Jones, R. (2003). *Scenarios of Climate Change for Islands within the BIC Region.* British-Irish Council (Environment) and Department for Environment, Food and Rural Affairs, London.

Jenkins, G. & Lowe, K. (2003). Handling uncertainties in the UKCIP02 scenarios of climate change. Hadley Centre Technical Note 44. Hadley Centre for Climate Change Research, Exeter.

Johns, T., Durman, C., Banks, H., Roberts, M., McLaren, A., Ridley, J., Senior, C., Williams, K., Jones, A., Keen, A., Rickard, G., Cusack, S., Joshi, M., Ringer, M., Dong, B., Spencer, H., Hill, R., Gregory, J., Pardaens, A., Lowe, J., Bodas-Salcedo, A., Stark, S. & Searl, Y. (2004). HadGEM1 - Model description and analysis of preliminary experiments for the IPCC Fourth Assessment Report. Hadley Centre Technical Note 55. Hadley Centre for Climate Change Research, Exeter.

Jones, P.D. & Reid, P.A. (2001). Assessing future changes in extreme precipitation over Britain using Regional Climate Model integrations. *International Journal of Climatology,* **21**, 1337-1556.

Kerr, A. & Ellis, N. (2002). Managing the impacts of climate change on the natural environment. In *The State of Scotland's Environment and Natural Heritage,* ed. by M.B. Usher, E.C. Mackey & J.C. Curran. The Stationery Office Edinburgh. pp. 239-250.

Kerr, R.A. (2004). Three degrees of consensus. *Science,* **305**, 932-934.

Klanderud, K. & Birks, H.J.B. (2003). Recent increases in species richness and shifts in altitudinal distributions of Norwegian mountain plants. *The Holocene,* **13**, 1-6.

Kullman, L. (2004). The changing face of the Alpine world. *The Global Change Newsletter of the International Geosphere-Biosphere Programme,* **57**, 12-14.

Marshall, J., Kushnir, Y., Battisti, D., Chang, P., Cjaza, A., Dickson, R., Hurrell, J., McCartney, M., Saravanan, R. & Visbeck, M. (2001). North Atlantic climate variability: phenomena, impacts and mechanisms. *International Journal of Climatology,* **21**, 1863-1898.

Meehl, G.A. & Tebaldi, C. (2004). More intense, more frequent, and longer lasting heatwaves in the 21st century. *Science,* **305**, 994-997.

Moberg, A. & Jones, P.D. (2003). Regional climate model simulation of daily maximum and minimum near-surface temperatures across Europe 1961-1990: comparisons with observed station data. Climate Research Unit Research, University of East Anglia, Norwich. www.cru.uea.ac.uk/cru/posters/.

Murphy, J.M., Sexton, D.M.H., Barnett, D.N., Jones, G.S., Webb, W.J., Collins, M. & Stainforth, D.A. (2004). Quantification of modelling uncertainties in a large ensemble of climate change simulations. *Nature*, **430**, 768-772.

Nagy, L. (2003). The high mountain vegetation of Scotland. In *Alpine Biodiversity in Europe*, ed. by L. Nagy, G. Grabherr, C. Körner & D.B.A. Thompson. Springer-Verlag, Berlin. pp. 39-46.

Osborn, T.J. & Hulme, M. (1997). Development of a relationship between station and grid-box rainday frequencies for climate model evaluation. *Journal of Climate,* **10**, 1885-1908.

Osborn, T.J., Hulme, M., Jones, P.D. & Basnett, T.A. (2000). Observed trends in the daily intensity of United Kingdom precipitation. *International Journal of Climatology*, **20**, 347-364.

Pauli, H., Gottfried, M., Dirnbock, T., Dullinger, S. & Grabherr, G. (2003). Assessing the long-term dynamics of endemic plants at summit habitats. In *Alpine Biodiversity in Europe*, ed. by L. Nagy, G. Grabherr, C. Körner & D.B.A. Thompson. Springer-Verlag, Berlin. pp. 195-207.

Poore, M.E.D. & McVean, D.N. (1957). a new approach to Scottish mountain vegetation. *Journal of Ecology*, **45**, 401-439.

Rahmstorf, S. (2000). The thermohaline ocean circulation: a system with dangerous thresholds? An editorial comment. *Climatic Change,* **46**, 247-256.

Rowell, D.P. (2004). An Initial Estimate of the Uncertainty in UK Predicted Climate Change Resulting from RCM Formulation. Hadley Centre Technical Note 49. Hadley Centre for Climate Change Research, Exeter.

Saelthun, N.R. & Barkved, L.J. (2003). *Climate Change Scenarios for the SCANNET Region.* Norwegian Institute for Water Research, Oslo.

Saetersdal, M. & Birks, H.J.B. (1997). A comparative ecological study of Norwegian mountain plants in relation to possible future climatic change. *Journal of Biogeography*, **24,** 127-152.

Schiermeier, Q. (2004a). Gulf Stream probed for early warnings of system failure. *Nature*, **427**, 769.

Schiermeier, Q. (2004b). Modellers deplore 'short-termism' on climate. *Nature*, **428**, 769.

Srokosz, J. & Gommenginger, C. (2002). *Science Plan: to Improve our Ability to Quantify the Probability and Magnitude of Future Rapid Change in Climate.* Natural Environment Research Council and Southampton Oceanography Centre, Southampton.

Stocker, T.F. (2004). Models change their tune. *Nature*, **430**, 737-738.

Stocker, T.F. & Schmittner, A. (1997). Influence of CO_2 emission rates on the stability of the thermohaline circulation. *Nature*, **388**, 862-865.

Stott, P.A., Stone, D.A. & Allen, M.R. (2004). Human contribution to the European heatwave of 2003. *Nature*, **432**, 610-614.

Vellinga, M. (2004). Robustness of climate response in HadCM3 to various perturbations of the Atlantic meridional overturning circulation. Hadley Centre Technical Note 48. Hadley Centre for Climate Change Research, Exeter.

Willows, R. & Connell, R. (Eds) (2003). *Climate Adaptation: Risk, Uncertainty And Decision Making.* United Kingdom Climate Impacts Programme, Oxford.

Zorita, E. & von Storch, H. (1999). The analog method as a simple statistical downscaling technique: comparison with more complicated methods. *Journal of Climate*, **12**, 2474-2489.

10 The effects of nitrogen deposition on mountain vegetation: the importance of geology

A.J. Britton, J. Fisher & G. Baillie

Summary

1. Patterns of nutrient limitation were investigated for *Calluna vulgaris* and *Racomitrium lanuginosum* at a range of montane heathland sites across Scotland.
2. Nitrogen availability alone was rarely found to be limiting for growth of these species.
3. Patterns of nutrient limitation were different between species, reflecting the source of their nutrients (rainfall vs. soils).
4. Nutrient limitation in *Calluna* was closely linked to phosphorus availability in the soil, which varied according to underlying geology.
5. We suggest that a single plant community may have a variable response to N deposition across its range, which could be predicted from underlying geology and the ecology of the constituent species.

10.1 Introduction

Due to the high annual rainfall which they experience, mountain areas are especially vulnerable to wet deposition of pollutants such as nitrogen (N) and sulphur (NEGTAP, 2001; Fowler & Batterbee, this volume). Long periods of cloud cover can further increase pollution loads when additional nitrogen and sulphur are deposited directly on the vegetation in cloud water (occult deposition) or are washed out of polluted cloud by rain falling from above (a process known as seeder-feeder enhancement). Increases in N deposition, in particular, are a cause for concern as they lead to increased nutrient availability and can affect plant growth and interactions between species. Fast growing species are generally most able to benefit from increased nutrient supply, giving them a competitive advantage over slower growing species. Ultimately, this can lead to changes in community composition such as those seen in polluted regions of Europe. Examples of such changes include the expansion of *Deschampsia flexuosa* and *Molinia caerulea* on Dutch heaths, over 30% of which are thought to have changed from heathland to grassland (Bobbink *et al.*, 1998), and the expansion of graminoid species in Scottish montane (alpine) heaths (van der Wal *et al.*, 2003). Such large-scale changes have important consequences for biodiversity and conservation, as characteristic or rare species often disappear from the community.

Britton, A.J, Fisher, J. & Baillie G, (2005). The effects of nitrogen deposition on mountain vegetation: the importance of geology. In *Mountains of Northern Europe: Conservation, Management, People and Nature*, ed. by D.B.A. Thompson, M.F. Price & C.A. Galbraith. TSO Scotland, Edinburgh. pp. 121-126.

In mountain ecosystems, nutrient availability is not the only limitation on plant growth. Effects of N deposition may be altered by interactions with harsh climatic conditions, which limit the plants' ability to utilise additional nutrients. N deposition effects may also vary between soil types, due to differences in availability of other limiting nutrients such as phosphorus (P). The varied topography and geology of mountain areas results in a complex mosaic of soil types, and this increases the difficulty of predicting how mountain vegetation might respond to increased N deposition.

10.2 Methods

In this study we investigated potential variation in sensitivity to N deposition between two species with differing ecology and between populations of the same species at sites with varying soil properties. Two mountain heathland species, heather (*Calluna vulgaris*) and woolly hair moss (*Racomitrium lanuginosum*) were selected for study as they are widespread species with contrasting life-form and ecology. We used N:P of plant tissue as an indicator of nutrient limitation status following the method of Koerselman & Meuleman (1996). Their study of a range of species and communities suggested that N:P values below 14 generally imply N limitation, while N:P values above 16 imply P limitation. Where N:P values are between 14 and 16, nutrient limitation can be variable (N, P or co-limitation).

The two selected species were sampled from 15 prostrate *Calluna* heathland stands (National Vegetation Classification communities H13 and H14; see Rodwell (1991)) distributed across the Scottish Highlands, overlying a range of different rock types. At each site five samples each of *Calluna* and *Racomitrium* were collected. Five samples of the soil surface horizon (to 10 cm depth) were also collected from each site. Total N and total P content were measured by the Kjeldahl method in the terminal 2 cm portion of the *Calluna* and *Racomitrium* shoots (Wall *et al.*, 1975) and used to calculate N:P. Soil samples were analysed to determine total P content (Smith & Bain, 1982). Plant tissue N:P was then compared between species and sites and related to soil chemistry.

10.3 Species differences

Patterns of nutrient limitation were different for the two species studied (see Figure 10.1). *Calluna* N:P varied widely, with N, P and co-limitation all occurring; however, P or N and

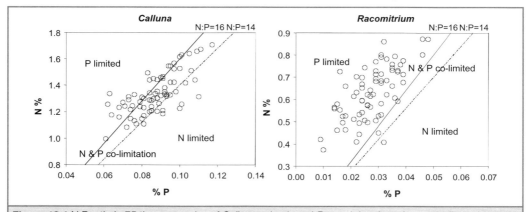

Figure 10.1 N:P ratio in 75 tissue samples of *Calluna vulgaris* and *Racomitrium lanuginosum* collected from 15 prostrate *Calluna* (H13/14) heathland sites in the Scottish Highlands.

P co-limitation were most common. *Racomitrium* samples showed a different pattern, most samples having high N:P values indicating P limitation. These differences may reflect the different nutrient acquisition strategies of the two species. *Racomitrium* derives most of its nutrients from rainfall (Pitcairn *et al.*, 1995) and reflects the high N:P seen in precipitation. *Calluna* derives its nutrients from the soil and the greater variability seen in its N:P may reflect variations in soil nutrient availability between sites.

10.4 Site differences

Variation in nutrient limitation between sites was investigated further for *Calluna* (see Figure 10.2). *Calluna* tissue N:P was found to be related to the P content of underlying soil, P limitation occurring on soils with the lowest P contents. The majority of sites in the mid-range for soil P content had *Calluna* N:P suggesting co-limitation, while only one site with high soil P appeared N-limited. When the influence of geology was examined, P limitation was seen to be associated with soils derived from very nutrient-poor rocks such as sandstone and quartzite, while the N-limited site occurred on diorite, an intermediate-basic rock giving rise to richer soils.

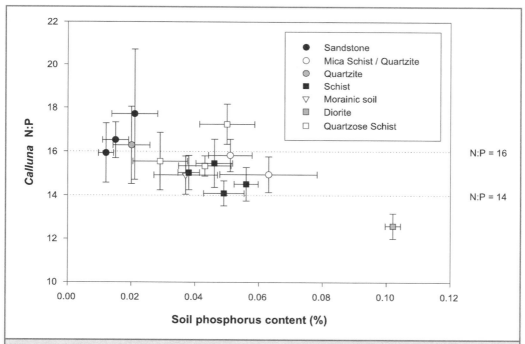

Figure 10.2 N:P ratio in tissue of *Calluna vulgaris* from 15 prostrate *Calluna* (H13/14) heathland sites in the Scottish Highlands vs. soil phosphorus content (%P) and underlying geology. Geological information is derived from British Geological Survey 1:50 000, mapping.

10.5 Conclusions

For the two species investigated here it, seems that N availability is rarely limiting for growth. This is somewhat contradictory to the general view that N is usually limiting in arctic and alpine ecosystems and that the resulting tight N cycling and retention make these systems sensitive to N deposition (Bobbink *et al.*, 1998). However, there have been few previous studies of this type of dwarf shrub-dominated vegetation on acid soils typical of the Scottish montane zone.

Patterns of nutrient limitation differed between the two species and reflected the source of their nutrient supply. While *Racomitrium* had a high N:P similar to that found in rainfall, *Calluna* was much more variable, reflecting the site to site variation seen in soil properties. This suggests that species occurring within the same community may be subject to different limitations on growth, depending on the nature of their nutrient supply, and hence respond differently to nitrogen deposition. Of the two species investigated here, the N:P values suggest that *Calluna* is more likely to be able to use additional nitrogen for growth, although this will vary between sites. The high N:P values found in *Racomitrium* suggest that it is unlikely to show a positive growth response to N deposition and may show little response unless toxic levels of N accumulate in tissues, resulting in reduced growth and ultimately death. This type of response has indeed been observed in N addition experiments on *Racomitrium* heath (Pearce & van der Wal, 2002; Pearce *et al.*, 2003). Further studies are needed to determine whether this type of response is common amongst montane bryophyte species.

Variation of N:P between sites was only found for *Calluna,* and the variation appeared to be linked to the measured soil P content and underlying geology. Only one site overlying diorite had N:P values indicating N limitation. This type of rock is of limited spatial extent in Scotland, the majority of Scottish mountains being made up of sandstones, quartzites and schists on which *Calluna* appeared generally P limted or N and P co-limited. The consequences of this for conservation of montane heaths are difficult to predict. While the scarcity of N limitation suggests that growth stimulation in response to nitrogen deposition is likely to be uncommon, the result may be N accumulation in plant tissues when N is taken up but not used. Such increases in tissue N content have been linked to a variety of effects, including reduced ability to tolerate frost and/or drought and increased palatability to herbivores (Bobbink *et al.*, 1998). However, most existing information on such impacts relates to lowland or upland heaths which experience very different climatic constraints. Further studies are needed to discover exactly how montane species will respond to nitrogen deposition and at what level of nitrogen deposition effects may be seen. What is clear from this study is that nitrogen deposition impacts are likely to be complex, with both species and spatial variation in response. Currently, communities are assumed to behave in a similar way throughout their range, but inclusion of such spatial and inter-specific variation may help to refine our predictions of the impact of N deposition on mountain vegetation.

Acknowledgements

This work was supported by the Scottish Executive Environment and Rural Affairs Department.

References

Bobbink, R., Hornung, M. & Roelofs, J.G.M. (1998). The effects of air-borne nitrogen pollutants on species diversity in natural and semi-natural European vegetation. *Journal of Ecology*, **86**, 717-738.

Koerselman, W. & Meuleman, A.F.M. (1996). The vegetation N:P ratio: a new tool to detect the nature of nutrient limitation. *Journal of Applied Ecology*, **33**, 1441-1450.

NEGTAP (2001). *Transboundary Air Pollution: Acidification, Eutrophication and Ground-Level Ozone in the UK*. Department of Environment, Food & Rural Affairs, London.

Pearce, I.S.K. & van der Wal, R. (2002). Effects of nitrogen deposition on growth and survival of montane *Racomitrium lanuginosum* heath. *Biological Conservation*, **104**, 83-89.

Pearce, I.S.K., Woodin, S.J. & van der Wal, R. (2003). Physiological and growth responses of the montane bryophyte *Racomitrium lanuginosum* to atmospheric nitrogen deposition. *New Phytologist*, **160**, 145-155.

Pitcairn, C.E.R., Fowler, D. & Grace, J. (1995). Deposition of fixed atmospheric nitrogen and foliar nitrogen content of bryophytes and *Calluna vulgaris* (L.) Hull. *Environmental Pollution*, **88**, 193-205.

Rodwell, J.S. (1991). *British Plant Communities, Vol 2. Mires and Heaths*. Cambridge University Press, Cambridge.

Smith, B.F.L. & Bain, D.C. (1982). A sodium hydroxide fusion method for the determination of total phosphate in soils. *Communications in Soil Science and Plant Analysis*, **13**, 185-190.

van der Wal, R., Pearce, I., Brooker, R., Scott, D. Welch, D. & Woodin, S. (2003). Interplay between nitrogen deposition and grazing causes habitat degradation. *Ecology Letters*, **6**, 141-146.

Wall, L.L., Gehrke, C.W., Neuner, T.E., Cathey, R.D. & Rexroad, P.R. (1975). Total protein nitrogen: evaluation and comparison of four different methods. *Journal of the Association of Official Analytical Chemists*, **58**, 811-817.

11 People, recreation and the mountains with reference to the Scottish Highlands

John W. Mackay

Summary

1. This chapter reviews the history of recreational enjoyment of mountains, focusing on the Scottish Highlands.

2. A broad range of recreational activities occurs in the Scottish Highlands, including walking, shunting, climbing and skiing.

3. There has been only limited survey of recreational use of the hills, and summary tables for the most recent are given.

4. Relatively new access legislation is described, which provides for a general right of access in Scotland.

11.1 Introduction

Mountains attract people for enjoyment. They are major tourist destinations around the globe, and offer people a range of attractions: a cooler and healthier escape from hotter or humid lowlands, and a setting of physical challenge for outdoor pursuits. The life and work of mountain people can provide visitors with a sense of cultural stability, albeit that these are often fragile communities, and the open character of mountain lands, with few constraints on access, offers great freedom of action. Above all, these can be truly inspiring settings – places of great aesthetic value, founded on their wildness and naturalness, and the drama and wonder of their elevation and verticality.

Mountain tourism is not quite at the mass-market end of the tourist spectrum, except where special facilities for uplift or accommodation – winter or summer – allow larger numbers of people to reach high ground. In practice, most mountain-based tourism is made up of a series of niche markets. Many people are content to look at mountains, touring or staying there and mainly walking in and around them at lower elevations; traditional family holidays still prevail; outdoor activities expand, driven by fashion and changing tastes, and there are opportunities for nature tourism. Downhill skiing has been an influential type of recreation because of its commercial drive and demand for access and development, often used in summer as well as winter. Guided holidays in the remote mountain ranges – sometimes termed 'soft adventure' – have also grown. Finally, at the more adventurous end of the scale, there are challenges in mountains which test human skills and endeavour to the utmost.

Mackay, J.W. (2005). People, recreation and the mountains with reference to the Scottish Highlands. In *Mountains of Northern Europe: Conservation, Management, People and Nature,* ed. by D.B.A. Thompson, M.F. Price & C.A. Galbraith. TSO Scotland, Edinburgh. pp. 127-136.

This diversity makes it difficult to generalise about the present state and direction of recreation in mountain areas. The pattern of visitor activities will vary greatly from country to country, indeed within countries, according to local accessibility to the mountains, and the scale of commercial or public-sector drive in developing mountain-based tourism as a force in the economy of these areas. There is also a strong commercial drive behind the active pursuits, which has international influence and, progressively, this leads general trends in mountain-based tourism in the world's accessible mountain ranges. The northern European mountains are not immune to these homogenising effects.

The north-west European mountains also have certain common attributes which influence their recreational use. They fall into that class of mountains which lie adjacent to wealthy and urbanised societies, and where open-air recreation is a valued outlet for populations which now have greater leisure aspirations. These are spacious, often empty landscapes of great aesthetic appeal. The traditional land uses survive, but they depend mainly on subsidy to fill the gap between the low productive capacity of the land and the costs of a modern lifestyle which people living there wish to, and should, share. Peripheral location contributes to weakness in the economy; tensions arise as the traditional land uses are overtaken by more service-based employment; and there is a pernicious trend to outward migration of younger people. Finally, this is a harsh environment, even at low altitudes, from the truly Arctic to the wet and exposed Atlantic fringe – terrain which can be unforgiving as a living environment, and of low carrying capacity for almost all uses of the land, including recreation.

11.2 Mountain recreation – where have we come from?

The adoption by the general public of the enjoyment of mountains as an end in itself is a relatively young phenomenon. Early local ascents of mountains will have happened but remain unrecorded, and the best early accounts of climbs and visits to mountainous areas come from travellers and natural scientists, both being groups of visitors who were given to leaving written accounts of their journeys. That these early travellers had the means to visit mountainous areas was a consequence of expanding wealth in society, allied to growing curiosity in travel, exploration and natural history. From the 1950s onwards, an expanding middle-class with spare resources, both financial and free time, added to the momentum of going to the hills as a holiday destination or as a weekend activity. The other critical door to open here in fostering more participation was that of improved accessibility – by steamer, rail, bus, even the humble cycle, and eventually leading to the contemporary domination of access to the hills by the motor car.

Active exploration of the northern European mountains was overshadowed in the late 18th and 19th centuries by interest in the Alps, but the record of early ascents in both Scandinavia and Scotland indicates that peaks were being climbed regularly by the turn of the 19th century. Later, the Alpine influence spread northwards in both Scotland and in Scandinavia by climbers such as Slingsby, but the Scandinavian tradition in open-air recreation evolved through its own distinctive cultural strand of a close engagement between people and their right to use the land.

While there are differences then from country to country in relation to accessibility, land tenure and access law, as well as in social structure and culture, there are also some common themes which have fostered the evolution of mountain-based recreation in the main mountain ranges. Some of these factors still apply today and, in outline, are as follows.

- The emergence of a rich literature describing mountain travel and ascents acted as an early stimulus to visiting mountains by a wider public and to participation in adventurous pursuits. This influence continues to lead and inspire – indeed no other outdoor recreation has quite the same volume of descriptive writing.
- Travel is inherent in mountain-based recreation – often long-distance travel – and this has helped to disseminate both a new outdoor culture and technique in the use of new equipment, which has also fostered new skills through example.
- Social change, with the evolution of an expanded middle class having spare leisure time, has been a critical stimulus to the growth of participation in open-air recreation. Surveys of mountain-based pursuits indicate that there is a dominance of the middle and professional classes in the ranks of participants.
- Improved accessibility has fostered participation, and the dominance now in Britain of day or short-break visits to the mountains has been led by continued improvements to the public road network, which enables committed hill-goers to travel further in a day or to have longer at their destination.
- The early development of climbing clubs and similar associations also provided a stimulus – often acting as associations of like-minded participants to foster skills and knowledge, or as a means of making collective arrangements for travel or accommodation, and also for social purposes. While clubs are in some ways less significant today, their role in fostering participation was important from the late 19th century and through most of the 20th century.
- Fashion and taste have affected mountain-based sports in the same way as they influence other facets of human behaviour. This can be fashion for new trends in participation, or for the opening up of new destinations, or for new technical development, and commerce has played a main role in such change.
- While mountaineering is not a competition sport, climbing can be. But the reality is that a competitive urge has always suffused mountaineering, whether it be the drive to early first-ascents in the Alps (and elsewhere), in aspiring to or exceeding elite performance, or in the completion of a round of ascents of predetermined worthiness – say, in Scotland, the 284 Munros (mountain summits higher than 3,000 ft/914m). The exploratory drive to be there first, to have done more, to have done it faster, is innate to the activity and is itself a stimulus, drawing onwards standards of practice by the participants.

The social psychology of the drive to engage in the more adventurous activities is not too different from the drives and motivations to behaviour in other facets of daily life. Leisure (of all kinds) provides an outlet for social interaction in which some of the conventional restraints of home and work are released, thus allowing people a greater freedom of action. Key elements in this complex area of human behaviour are:

- self-projection, through peer group approval and self-esteem, gained through achievement and promotion of what has been achieved by the individual;
- self-appreciation by the individual, through better understanding of their competence and personal capacity, arising from recreational activities which can test physical and psychological limits; and

- self-expression, in the undertaking of activities which have risk and offer opportunity for free and novel action, and which can be challenging in a physical sense.

It would be wrong to emphasise that adventure is all that counts in the mountains or that social achievement is the only driver. More modest but more universal influences on participation lie in the physical well-being gained from hard exercise and, crucially, the stimulus of simply being in places of great aesthetic attraction. Most people will go to the mountains mainly to look at them or to engage in general recreational activities, and participants at the highest levels of challenge and attainment have always been relatively few in number. But this small sector has always been important in being in the forefront of giving a lead to performance, with some individuals acting as role-models.

Alongside mountaineering and related pursuits, field sports have been an important land use in the mountains. Hunting has, since Victorian times, been a fashionable part of land use in the Scottish Highlands, where deer stalking and the shooting of moorland birds, notably red grouse (*Lagopus lagopus scoticus*) became the leisure pursuits of choice for the very wealthy. The role of hunting – often in other countries being pursued more in forests around the mountains than on the open hill – varies from nation to nation, according to culture and tradition, and whether or not the legal basis of hunting is a public right, or a facet of private ownership, as in Britain. Special interest in natural history has long drawn people to the mountains, and downhill skiing – which links back to cross-country travel on ski, and also to the early days of mountaineering – has become a significant, commercially-led part of mountain-based recreation. Other challenging extreme sports which use steep ground, swift rivers and rugged relief continue to emerge, but overall participation in these kinds of activities remains tiny and their practice, albeit now as a global trend, is mainly located in the more touristic settings.

11.3 Who participates?

It was asserted earlier that the middle and professional classes participate most in the active pursuits in the mountains. Good surveys of recreational use of the hills are sparse. The gathering of data on public recreation dispersed over extensive areas can be difficult and costly, and the conventional mode of survey in Scotland is the cordon survey, capturing respondents at key access points as they return from their trip out.

Survey at the end of the day is the best means of recording the details of the outing, but the timings of return from the hill can be extended and late respondents can be disinclined to participate in lengthy face-to-face outdoor interviews, especially if the weather is inclement. Hence, there is a preference for gathering data by postal-return questionnaire, either left on parked cars or delivered directly to respondents on site, or distributed later using addresses gathered at self-registration survey points. Response levels vary but, on the best surveys, comforting levels of up to 70% return have been attained and this, along with the quality and consistency of information on returned questionnaires, gives some confidence in survey quality. Some sampling imperfections will remain. Of Scottish surveys of this kind, the most useful are surveys mounted in three areas of the Highlands: East Grampians (1995; Mather, 1998), Glen Shiel (1996; Herries, 1998) and the Cairngorms (1997-98; Taylor & MacGregor, 1999). Earlier mountain-recreation surveys are probably now too dated, have sampling flaws or, if done as student exercises, are too limited in scope.

Some comparative data from these three Scottish surveys are given in Tables 11.1-11.3. While broadly similar in survey mode, differences in these data reflect that they come from three different settings and are for time-periods which do not exactly match, survey to survey. Thus there are no participants in short local walks recorded in Glen Shiel because terrain and lack of obvious opportunity inhibit visitors looking for modest excursions away from the road. This was a survey limited to late summer and autumn. The mode of survey here was entirely based on self-return questionnaires left on cars, and was strongly focused on those visiting the Glen for hill walking. The other two surveys cover areas where there is a wider range of recreation opportunity among, as well as on, the hills, and here there is a bigger day-trip component on account of closer proximity to urban areas, especially for the East Grampians. The main comparisons between these surveys, as set down below, are based on selecting responses for the more active pursuits on the hill, the survey data being taken from summer and/or autumn visits. While there are the aforementioned differences of setting, the patterns of activity do seem to be realistic for each area. Some data draw from the response of the participants while others refer to a respondent's report on the characteristics of the group he or she was in.

Table 11.1 Main purpose of visit in Scotland. Source: Herries (1998), Mather (1998), Taylor & MacGregor (1999).

Main Activity	% respondents		
	Glen Shiel	East Grampians	Cairngorms
Hillwalking	97	70	52
Low level walking 1h +	1	18	34
Mountain biking	-	2	3
Wildlife watching	1	1	3
Rock climbing – scrambling	1	3	3
Others (photography, hill running, study, outdoor training)	2	-	4

People participating in the more active pursuits comprise a well-educated group, spread across the adult age classes, of average age around 40 years, but distinctly lacking in younger cohorts, and dominated by the male gender. In winter survey data for the Cairngorms and East Grampian areas, the male dominance is greater and the age balance inclines more to the younger groups, there being more climbers at this time of the year. Residents of Scotland dominate the samples, but there are noticeable numbers of visitors from abroad – the Glen Shiel survey had respondents from Germany, France, the Netherlands, the USA and Canada. Day visits predominate closer to the main urban catchments, and here the proportion of repeat visits was also higher. Overnight stays in the hills (mainly camping) are modest in number.

Of the activities pursued, walking on high or low ground dominates. In those areas where mountain biking is a possibility the numbers are not high, but there may be some under-representation here, with cyclists being more difficult to catch for interview, and some walkers claim cycling as a secondary activity (4% in the East Grampian Survey), presumably using a bike to get closer to the hills. Other main purposes of the day out are only marginally represented – such as wildlife watching at around 1% – but many more

Table 11.2 Mountain recreation visitor characteristics in Scotland, based on questionnaire survey. Source: Herries (1998), Mather (1998), Taylor & MacGregor (1999).

Characteristic[1]		% respondents		
		Glen Shiel	East Grampians	Cairngorms
Gender	Male	69	76	63
	Female	30	23	37
Age Class	<18	2	2	2
	19–24	8	7	5
	25–34	29	28	21
	35–44	26	27	19
	45–54	25	21	31
	53–64	9	12	16
	65+	2	3	6
Home residence	Scotland	60	72[2]	54
	England/Wales	37	24	38
	Elsewhere	3	4	8
Club Member	Yes	29	ca.25	23
	No	70	75	78
Educational background	Degree/diploma			61
	Other higher qualification			9
	Higher/A level			4
	Others			26
Socio-economic class	1		25	
	2		40	
	3		16	
	4		1	
	5		<1	
	Unclassified		19[3]	
How long recreating in the hills – years	<5	19	23	9
	6–10	25	18	19
	11–20	28	26	19
	21–30	16	16	21
	31–40	8	17[4]	14
	40+	3	-	13

[1] Gaps indicate no comparable data. [2] All visitors. [3] Mainly students, retired, armed forces and unemployed. [4] This figure covers the 31+ age group.

respondents (39% in East Grampian) claimed this as a secondary activity. Going to the hills is a sociable activity in the main, but not so much a club activity as in the past, with only a minority of respondents (around 30%) claiming club membership. More data than can be set out here were gathered on participants and their visit, including attitudinal data.

Our understanding of the longer-term trends in the more active mountain-based pursuits is limited, apart from a clear identification of substantial growth from the early 1960s as mobility and wealth became the main drivers of more participation. Data on overall levels of participation in mountain-based activities are difficult to interpret. From

Table 11.3 Some features of the visit in Scotland, based on questionnaire surveys. Source: Herries (1998), Mather (1998), Taylor & MacGregor (1999).

| Feature | | % respondents | | |
		Glen Shiel	East Grampians	Cairngorms
Type of visit	Day visit from home	29	52	25
	Away from home	71	48	74
Overnight in the hills		2	5	13
Number in group	1	20	20	24
	2	49	47	50
	3	13	11	10
	4	8	9	7
	5–7	8	9	9*
Length of day out (mean)		7 hrs 3 mins	5 hrs 50 mins (Hill walkers)	5 hrs 42 mins
First visit	Yes	27	22	16
	No	72	78	84
If no, how many visits in previous year?	0	32		21
	1	28		19
	2–5	35		32
	5+	4		28
Groups having a dog	Yes	11	12	14
	No	88	87	86

* This figure covers 5+ people in a group.

general recreational surveys, we know that participation in mountaineering or hill climbing ranges from 2-5% in reply to a question about whether the respondent had participated in the previous month, with the level of response varying according to the wording of the question or the time of year. General surveys of walks taken in the Scottish countryside find that up to 10% of walks have been taken in a mountain or moorland setting – that is, more people do walk in and close to the mountains rather than up them. There appears to have been a progressive rise in participation over the past decade, and it is difficult to predict a future for active mountain-based recreation which does not continue present patterns and perhaps even some augmented levels of participation.

The level of longer-term commitment by participants to their recreational activities is high – the surveys cited above indicate a long time-span to respondents' involvement in mountain-based recreation, once started, and many participants espouse a strong attachment to their recreation. While the longevity of participation may signal that this is a robust market sector, there are some trends within hillwalking and mountaineering which may cause some attrition to the basic values and tenets of these pursuits. There is more in the way of commerce-led influence in the outdoors and a trend towards a more self-centred approach to how participants approach their recreation (now more activity- than environment-led). We know little of these trends or of any implications arising.

11.4 Mountain-based recreation and the land

The recreational communities, indeed wider society, place high value on the mountains for their aesthetic or conservation values and, in Scotland, wealthy people may pay large sums to acquire such land as a kind of private wilderness for hunting, evidently being drawn there by the same qualities that attract public recreation. There is economic value to be extracted from mountains in the form of minerals, hunting, agricultural grazing and timber on the lower slopes. Mountains also provide society with wider social benefits as water catchments and high-quality natural heritage features and, increasingly, renewable energy.

Mountain-based recreation by the mainly town and city-based public is an overlay on existing land uses, often a dominant overlay as the main use of the land, albeit not the owner's or manager's use. Some of the pre-existing land uses, such as hunting, fishing or the gathering of natural fruits, have evolved from their traditional role as part of local people's economy of survival to become predominantly recreational in their own right. Thus hunting (and fishing) has moved to become primarily a sport, either mainly a matter of private trade (as in Scotland) or as regulated in other countries. Tensions can arise from these conflicting uses over whose recreational needs and rights have primacy.

But generally, these are lands of inherently low productivity, harsh and uncompromising and offering poor returns for the traditional, pastoral land uses, to the extent that agriculture can only survive through heavy subsidy. Grazing on high ground tends to be extensive and often communal in management and, in some cases, in collective ownership. Tourism and general service provision to local communities have thus become a mainstay of many mountain economies, sometimes with a discordance between those who own or manage the land and those who benefit commercially from tourism, and who have no management responsibilities for the land they use. In the most mountainous areas, the number of people in Highland Scotland directly claiming a living from the land is now a small proportion of those employed locally.

Public recreation on the mountains can thus raise new management needs. Given that most public open-air recreation is without payment or is a public right, no charges can arise although there are costs, and these have mainly been allocated as a public responsibility assigned to both local authorities and national bodies in all northern nations (for example in Britain through the Countryside (Scotland) Act 1967 and the Countryside Act 1968). The voluntary land-owning sector has also played an important role in such management, which has not always received the right level of attention and priority by the public authorities. Regulation of activities and their practice have generally been avoided as difficult to enforce and overbearing, and the emphasis is mainly on self-regulation through responsible actions by participants. But this still leaves a large task of physical management, yet to be fully addressed.

Ownership of mountainous lands has varied forms – private as a form of exclusive use and often for non-economic reasons, and community and state ownership are widespread, including public sector owners with sectoral interests in resource management and development, especially for water and timber. More recently in the Scottish Highlands ,ownership by Trusts and, non-govermental organisations (NGOs) has grown, with these bodies acting as beneficial holders of valued land in the wider public interest, and there is a small trend towards community acquisition of land.

The reality of extensive-area ownership, few physical barriers except of the natural kind, low resident populations, and low-value usage has been that access for open-air recreation has evolved to be relatively unimpeded. What was once locally-based access – either customary or by right – to use the land for the needs of daily life has been transmuted into general access for all. The law has supported this in some countries. While exclusivity of private ownership has been asserted in the Scottish uplands, from time to time, this was overtaken by the *force majeure* of the public taking access, and now by recent legislation to create a general right of access (Anon., 2003). This new Scottish legislation is notable in that it confirms a general right of access to land and inland water, subject to responsibilities falling on those exercising rights and on land managers, which parallels the Scandinavian *allemannsretten* and which stands close to the arrangements in Norway. Conservation has also become a force on the uplands, as these are the most extensive areas of natural habitat in the industrial economies but, in many areas of the Scandinavian and British uplands, enjoyment – public and private – is now the main use of the land (see Thompson *et al.*, this volume).

11.5 The evolving debate on access

It is characteristic of these northern mountains that an important part of their attraction for recreation comes from the austere quality of landscapes which are of great aesthetic merit and which, in turn, command great affection from those who enjoy them. This affection can create a strong sense of ownership for many people, apart from those who legally own and manage the land. Local people may espouse a close interest in the land through community residence and use, and others from afar can have a sense of possession of the land, arising from their commitment to the recreational values enjoyed there. In Scotland, the Access Forum has developed the Scottish Outdoor Access Code (Scottish Natural Heritage, 2005.) to promote formal guidance on the right to responsible access.

Tensions can arise between these national, local and individual owners' perspectives. Locally some of these tensions have strong historical and socio-political roots, as in the case of the Highland clearances. Tensions are also fostered by the continued economic vulnerability of communities in the mountains and their inability to effect control over the external forces of change which affect them. In mountain areas there can be a strong expression of localism, which can provide the basis for significant conflict with external perspectives on the use and care of these lands. These external perspectives often assert that resources of national significance – landscape and nature conservation, in particular – are not being well enough safeguarded locally, and this has been at the heart of a wide range of *cause célèbre* conflicts over development, or land use proposals, which have arisen in mountainous regions around the globe over the past century. Perhaps the 1906-13 debate over the Hetch-Hetchy Dam in the Californian Sierra, to augment the water supply for San Francisco, was the first notable such case (e.g. Nash, 1982).

There has been a century of debate since that seminal case in the high Sierra, and this has led to the evolution of protected-area mechanisms of different degrees of sophistication and effectiveness, which often bear heavily on mountainous lands. Indeed, the growth of these conservation systems has itself been a matter of conflict and a contributor to divisive tensions between local and national interests. It is a characteristic of protection of this kind that it focuses on the objects of protection – nature, landscape, buildings – and that it fails to give sufficient recognition of the values of enjoyment, which underpin the primary motive to protect. Indeed it is the case that recreational values of this kind have no sound basis of safeguard, and have no institutional framework or, indeed, any strong underpinning philosophy to deploy in the process of debate about whether proposed change to valued recreational settings is appropriate.

Yet recreational values are the driver to people's visits to the mountains and at the heart of them lie the aesthetic qualities to be found there. There must be a commonality of interest here between local and national perspectives, which has yet to be explored and agreed. Places of home and work are the landscapes of daily life, and for those who visit the mountains, often from urban settings, these are landscapes of escape and refreshment. At one level this may seem a trivial and one-sided comparison but, in a rapidly changing society, there is a more profound interdependence between these different sectors of society at both economic and social levels. This should be a basis for agreeing a common interest in mountainous lands which, through their value to wider society, have much more than local significance.

References

Anonymous (2003). *Land Reform (Scotland) Act (Part 1)*. HMSO.

Herries, J. (1998). Glen Shiel Hillwalking Survey. Scottish Natural Heritage Research, Survey & Monitoring Report No. 106.

Mather, A.S. (1998). East Grampians & Lochnagar Visitor Survey, 1995-overview. Scottish Natural Heritage Research, Survey & Monitoring Report No. 104.

Nash, R. (1982). *Wilderness and the American Mind. Chapter 10. 3rd edition*. Yale University Press, New Haven, CT.

Scottish Natural Heritage (2005). *Scottish Outdoor Access Code*. Scottish Natural Heritage, Perth.

Taylor, J. & MacGregor, C. (1999). Cairngorms Mountain Recreation Survey, 1997-98. Scottish Natural Heritage Research, Survey & Monitoring Report No. 162.

PART 3:
Land Use Change

The quiltwork pattern of muirburn, near Abington, Scotland, UK
(Photo: Patricia and Angus Macdonald).

PART 3:

Land Use Change

For researchers, one of the greatest challenges in working on mountain environments is to tease apart the human-related influences on mountain landscapes, habitats, species, and the great range of earth and biological processes. Some take a historical approach, essentially documenting the changes in, for instance, vegetation composition; some make long-term observations, detailing changes in response to variation in numbers of herbivores or concentrations of pollutants; and a few undertake experiments. In this part of the book, we have nine chapters which get to the heart of understanding human influences on mountain areas.

Simmons (Chapter 12) presents a narrative environmental history of the mountains of Northern Europe over the last 10,000 years. He defines four periods, characterised by: hunters and gatherers; pre-industrial agriculture; industrialisation; and the post-industrial period (from 1950 to the present time). A trend emerges of a movement towards the integration of mountain areas with other parts of individual countries; indeed, there has been a steady growth in the recognition of mountain areas within local, national, regional and international policies. Simmons raises important questions regarding the lessons learned from understanding the history of these areas.

Three chapters deal with the more recent history of mountain areas. Olsson (Chapter 13) examines the history of livestock grazing in Norwegian mountains. She gives a detailed analysis of the decline in the extent of semi-natural habitats as a result of heavy grazing pressure, and the trend towards agricultural intensification. This leads to a broad consideration of the role of livestock grazing in sustainable food production in mountain areas. Mary Edwards (Chapter 14) compares landscape history and biodiversity conservation in mountain areas of Norway and Britain, viewing both as essentially 'cultural' landscapes. She presents some very revealing comparisons between her two study areas, in Sjodalen, mid-Norway and Coed Gallywd, North Wales. In Wales for example, the uplands have undergone major physical and biological changes, to the extent that they are more impoverished biologically than the uplands of Norway, evidently due to the great difference in scale of cultural influences, and earlier phases of forest clearance and burning. In Iceland, Thomson (Chapter 15) has studied the history of vegetation change before 1900. Over the past millennium, Iceland has had approximately 40% of its vegetation and soil cover removed, and large areas of the remaining vegetation are eroding; much of this degradation has been attributed to livestock grazing. Using a combination of photographs and a computer simulation model, Thomson has elegantly unravelled the interactions between climate and land management in shaping the appearance of the landscape. An important message here is that one needs to be flexible in managing livestock grazing systems under conditions which change climatically; grazing has a greater impact on mountain vegetation under colder conditions, when the growing season and vegetation production are reduced, and livestock fodder requirements increase.

Morrocco's study of terrain sensitivity on high mountain plateaux in the Scottish Highlands (Chapter 16) follows on well from the previous study of Icelandic mountain landscapes. Morrocco has undertaken a PhD study of the sensitivity to disturbance of soils and vegetation in high mountain landscapes. He has developed some important new techniques that enable researchers and land managers to compare, for the first time, the 'shear strength' of different plant communities. This work is important in developing land management practices for existing vegetation susceptible to disturbance, but also for planning habitat restoration measures. Lilly *et al.* (Chapter 17) have developed a map showing the inherent risk of soil erosion across the Scottish Uplands due to overland flow. This work offers an important tool to guide management of priority peatland habitats, which are highly vulnerable to erosion. High-altitude mountain slopes are also prone to erosion where associated with downhill skiing developments. Cadell & Matthews (Chapter 18) have developed the use of remotely sensed imagery to identify areas at risk of being damaged. Again, this new technique has important land management applications.

The remaining chapters in this part of the book address, in different ways, the consequences of land use practices. Fredman & Heberlein (Chapter 19) review current patterns and recent trends in mountain tourism in Sweden, and compare their results with other mountain regions in Northern Europe. They show how some historical events have, by chance, influenced the popularity of some tourism activities. For instance, when Ingemar Stenmark won his first skiing slalom world cup for Sweden in 1976, and then dominated the sport for almost a decade, he started a national downhill skiing boom!

Finally, Bayfield *et al.* report on three international environmental monitoring networks established to detect and report on the sorts of natural and human-related influences on mountain environments described above. They report that a long-term ecosystem research network for Europe was established in 2005, with monitoring seeking to understand the social and economic drivers and pressures which influence land use change.

The growth of international monitoring networks is encouraging, and it is heartening to see participants eager to embrace the multi-disciplinary approach needed to understand the complexities of change in mountain environments.

12 The mountains of Northern Europe: towards an environmental history for the last ten thousand years

I.G. Simmons

Summary

1. The post-Holocene environmental history of the mountains of Northern Europe can be divided into four economic-ecological periods: hunters and gatherers; pre-industrial agriculture; industrialisation; and post-industrial.

2. Over this period, mountain areas have been increasingly integrated into national economies and global processes.

3. This chapter provides an initial framework for reviewing the environmental history of Northen Europe's mountain areas by considering social and environmental factors.

12.1 Introduction

The emerging discipline of environmental history has ambitious aims. Though developed initially by historians, especially in the USA, it has been espoused by workers from several backgrounds (especially where historical geography had been poorly developed). It affords the opportunity and challenge of synthesising information from a range of discourses. The natural sciences are fundamental, but data from economic and political history and other social sciences are also needed. The role played by the humanities in evaluating the way in which the cognition of the environment evolves is often significant. In Europe, there is now a European Society for Environmental History, and bibliographies of work in the spirit of the discipline are being published. However, no extended account of the environmental history of Europe has yet appeared and, to my knowledge, no synthesis of the environmental history of the northern mountains has been crafted from the many pieces of work in individual fields and different places. Consequently, this chapter can at best lay out some of the groundwork and suggest an agenda: it is for other workers to take up the task of bringing together the numerous items of relevant scholarship that already exist, to weave the kind of story that can be nested inside broader-scale treatments like that of Price (1981).

12.2 The strands in the skein

There can be no ignoring the geophysical base. The mountains of Northern Europe are not immensely high by world standards, but many rise from sea-level, which contributes to their impressiveness and their lapse rates alike. The highest summits at 2,500 m (Norway), are

Simmons, I.G. (2005). The mountains of Northern Europe: towards an environmental history for the last ten thousand years. In *Mountains of Northern Europe: Conservation, Management, People and Nature,* ed. by D.B.A. Thompson, M.F. Price & C.A. Galbraith. TSO Scotland, Edinburgh. pp. 141-150.

lower than the Pyrenees (3,900 m), the Alps (4,500 m) and the Caucasus at 5,600 m, for example, and Wales's highest mountain Snowdon (Yr Wyddfa) is 1085 m above sea level. The rise from sea-level is indicative of the fact that many of the mountain areas receive a maritime influence. Active glaciers are confined to Iceland, Svalbard, Sweden and Norway. There are differences in the rates of recovery after the Pleistocene ice-loading and so relative sea levels are not changing uniformly. Active volcanism is confined to Iceland, and the fall of tephra during the Holocene is unevenly distributed, though as more is looked for, more is found. These changes have to be understood in the context of Holocene climatic change where, in the mountains, even small changes are almost always significant. Thus the course of post-glacial climates (including the currently modelled scenarios of future change and impact) is rarely without significance (Frenzel, 1993, 1994, 1996; Malmer & Wallen, 1996; Grove, 2001; Seppa & Birks, 2001).

Into this dynamic set of processes, we inject the human communities: as the ice melts and land surfaces appear, then so do the humans, who moreover do not confine themselves to the valleys but are aware of the resource potential of the upper areas and develop technologies to enhance this potential and the usage of these environments. As examples we might cite the following.

a) The control of fire at the landscape scale in manipulating plant communities and hunting animals. These techniques are mostly associated with gatherer-hunters but are by no means confined to them since, for example, low shrub management is often carried out with fire. Fire history, determined from early Holocene palaeoecological data, proposes that many zones in the mountains were thus manipulated from very early times of human occupancy (Rasanen, 2001). The Holocene succession has therefore been vulnerable to deflection by human action, as in the attempts to prevent deciduous forest taking over the open moorland in some regions of England (Simmons, 1996).

b) The development of an agriculture which was adapted to an efficient use of the mountain resources, with a preference for oats (Gramineae) on the wetter western margins. There has also been the use of domesticated animals to transform grasses, low shrub ecosystems and forest floors into milk and meat, hides and wool, and to integrate this culture with valley land use for meadow and hay production. The use of wild plants has often been intensive (e.g. Gudmundsson, 1996).

c) Industrialization based on steam power brought immense changes to mountain areas, transforming them both directly (e.g. the rate of timber processing in steam-powered sawmills) and indirectly (e.g. the import of coal and recreationists by rail). Not least, rural electrification must have slowed down the apparently inexorable drift of people from the land in the 20th century.

12.3 The shapes of change

Mountain regions are not totally isolated from the surrounding zones of their nation and continent. This can be seen most obviously in their energy relations: until the 19th century most were solar-powered entirely, either by the current products of photosynthesis or recently stored organic materials like wood and animals. After industrialism was introduced, the outside world became more influential: it could provide coal and also demanded the flooding of valleys to make reservoirs for hydropower or flood control. Fossil fuels generally made it easier for the mobile to leave; the demographics of emigration and city growth have both been affected by the movements of the young.

Thus no mountain is an island, so to speak. In that sense we can apply to Northern Europe's mountains a simple classification of economic-ecological periods which is more or less universal. The Holocene can be divided into four periods on the understanding that the transitions between them are asynchronous and while sudden in some regions, are gradual in others.

12.3.1 Hunters and gatherers

In Europe, as elsewhere, gatherer-hunters were the initial colonists of land freed from ice and of coasts undergoing isostatic recovery. On the other hand, they lost territories to rising seas in the early Holocene, and the occupants of the British Isles found themselves separated from their European ancestors and relatives. Early occupants of mountain areas tended to use caves and rock shelters as their base but, once into the rapidly-warming part of the Holocene, other settlement sites were favoured. Resource preferences learned in open country were still favoured even as the forests colonised the slopes of the mountains. Pollen analysis has shown that fire was used to burn along the upper edge of the higher woodlands (Tallis & Switsur, 1990). Preventing more mature forests (with an unfamiliar fauna) from colonising these slopes may have been temporarily successful in a few places but, on maritime uplands, it allowed the growth of blanket mire. Eventually, Mesolithic populations became adapted to the new biome and were successful in it (see Olsson, this volume). Hunters and gatherers have accumulated a romantic image in which they had little impact on nature – but there are many places in which scientific evidence tells a different story.

12.3.2 Pre-industrial agriculture

This term encompasses all of the verities of rural economies in the time between the adoption of life-ways based on domesticated plants and animals, and the availability of significant quantities of fossil fuels and their associated technologies (Birks *et al.*, 1988). Thus the Neolithic to the 19th century can be included, though not without recognising some changes in between. All of these populations faced the challenges of getting enough nutrition from their surroundings, or else producing sufficient surplus to trade with the lowlands. Where possible, sedentary communities produced some cereals as a staple (oats *Avena sativa* and rye *Secate cereale* being the best adapted to upland climates), to go with dairy products. A far northern adaptation saw reindeer (*Rangifer tarandus*) herding as a distinct economy (see Hallanaro & Usher, this volume).

Where cereal cultivation was possible, the clearance of fields and their maintenance by drainage and dunging was central to the feeding of families. Laborious removal of stones and boulders, hand-dug ridge-and-furrow, and seasonal confinement of animals on the fields were all central to the maintenance of fertility levels. The uphill movement of animals onto open land or enclosed community pastures was also important since it brought manure to lower levels on a daily basis. If beasts were stalled through the winter, forest materials were brought down to feed them as a supplement to the stored hay from the valley meadows, another precious resource. Seasonal transhumance (see Olsson, this volume) has the effect of relieving pressure on the valley ecology (not least, perhaps, taking the animals to a distance where they are unlikely to get into the ripening cereal crop) and of producing a surplus of storable dairy items, such as cheese, which are available for trade. The role of the woods and the use of fire as management tools were clearly central (Esseen *et al.*, 1997: Nilsson, 1997; Lindbladh & Nilsson, 1999; Ericsson *et al.*, 2000).

All of this adds up to a considerable environmental impact. The scenery which came to delight travellers in the 19th century was created out of environments which were not replete in easily tapped biological productivity. There was, however, a combination of lightly managed vegetation types (both forest and hyper-forest zones) and closely manipulated cropland and meadows which could support a modest density of people who used wood and peat as fuels. The basis was mostly that of local self-sufficiency, but with the possibility of trading in dairy and wood products. Such communities had few buffers against bad times. The onset of climatic deterioration could not easily be managed except by emigration; avalanches and floods required permanent avoidance of their paths. Famine was relatively common. Supplementing the local diet by hunting was no doubt common, but large areas of upland were often sequestered by the aristocracy as reserves of game.

In a long-term perspective, we can see the foundations of today's landscapes being laid out, with upper vegetation zones little modified, a forest zone with wild species whose composition and physiognomy has been altered by millennia of human use, and a valley zone of considerable human imprint. The biological productivity and the woodland distributions had been notably altered (Heal & Perkins, 1978; Harkness, 1982; Hallanaro & Usher, this volume). Moreover, distinct mountain cultures developed.

12.3.3 Industrialisation

The essential features of fossil fuel-based economies (notably steam power and the factory system) were slower to arrive than in the lowlands. In many areas, too, the availability of wood meant that coal was not such a cheap and convenient fuel. Likewise, mountain areas were unlikely to have dense enough populations to make factory work an economic replacement for domestic production of textiles or wood products. Mining had for centuries drawn in men, often requiring them to walk many kilometres on a weekly or monthly basis from their homes to the mine. Even so, in the period between 1800 and 1950, there was a considerable change in mountain economies: some caused by the drawing-in of new knowledge and technologies, and some by the discovery of the mountains by those who lived outside them.

There came pressures for more momentous changes from the outside world. In part, these were severely practical: mountains had water that the urban-industrial zones needed and moreover might use for generating electricity and diminishing the impact of floods. There was flooding of valleys behind large dams, with associated consequences for the downstream ecology of the rivers and, often, the catchment areas of the impoundments. Demands for quarry and mine products increased dramatically. One parallel change was the use of the steam- and coal-powered engines of fishing vessels and ferries in the mountain areas that abut the sea, as in Norway and Scotland. The 19th century saw many small-scale fishing enterprises linked to faster transport and refrigeration grow in remote areas. But perhaps the most easily overlooked influence is that of sheep (*Ovis aries*), which provided meat to lowland areas either for direct consumption or for fattening, as well as wool and some cheese. The ecology of sheep grazing is well understood: not for nothing did John

Muir call them (in the context of the Sierra Nevada of California, admittedly) 'hooved locusts'.

At the same time as these results of the 19th century industrial anabasis of Europe were occurring (intermittently spurred on by wars), a new group of mountain users began to grow: those for whom the mountains were not frightening places to be shunned, but unstead localities of the greatest delight in terms of the passive enjoyment of scenery, a non-industrial way of life, and the challenge of scaling the peaks or traversing the wildernesses (see Mackay, this volume). Initially, the summer season brought most people but, gradually, the attractions of winter penetrated the well-off from the cities of the plains. With this group of recreationists arrived demands for protection of scenery and structures as well as the call for more urban facilities (see Mackay, this volume). Sporting use of some uplands increased the extent of selected habitats, such as *Calluna vulgaris*-dominated heather moor (Thompson *et al.*, 1995).

The key connection between production and recreation is transport. In particular the railways, even in much of Northern Europe, played a primary role in getting materials to sites of industrial development, in the extraction of materials, and in taking people to the mountains for recreation. The narrow-gauge mountain railway was never quite so popular outside Switzerland, though there is one on Snowdon, and a more recently constructed one on Cairn Gorm, and demands for funicular construction seem still quite active here and there. Yet in 1950, the older ways of mountain transport were still to be seen in active 'real' use rather than elements of a self-conscious tourist scene.

12.3.4 'Post-industrial time': 1950-present

Although it is more difficult to demarcate and label the transition than for other periods, European economies seem to have undergone a phase-shift after 1950. Oil began to replace coal, nuclear power grew to some extent, road-building bloomed and private transport made significant gains at the expense of railways, though the latter have begun to claw back some trade in recent years. Above all, leisure time and disposable incomes grew markedly.

In the mountains, the impact has been an intensification of many pre-existing trends. There was a continuing outflow of young people and several products could better be supplied by more intensive methods on lowland farms. Where the farming methods in the mountains have underlain a particularly valued scene as in the Alps and Carpathians, for example, the inhabitants are paid to maintain the old ways. This coincides with the main explosion of use of the uplands in the form of tourism. New roads, and expenditure on keeping them open in winter, have contributed greatly to the exponential growth of winter sports. Cheap energy has also allowed inexpensive ski-lift development and even artificial snow-making as well as road-building, including tunnel construction as in Norway. A major outcome has been the local removal of forests to facilitate downhill skiing, with consequences for soils and flora and most likely for avalanche frequency.

As in many regions in the West, the development of this industry (for no other word can now be applied to mountain recreation) has brought a reaction from environmentalist individuals and groups. They want to see the mountains kept as wild as possible and the native species of plant and animal disturbed as little as possible. Many conflicts between developers, both private and municipal, and conservationists have occurred and will no doubt continue, especially in the face of the pressures upon northern biota that are likely with global warming. Many designations have been invented to enshrine the special appeal of the mountains and their ecosystems ('National Park' is just one of several: see Part Four of this volume) but it is rare that any of them have been able to resist all the pressures for development that the post-1950 period has brought. The predictions of global warming have evoked a number of syntheses of likely impacts on the mountains at a time when other pressures are also strong (for example, Pitt & Nilsson, 1991; Stone, 1992; Beniston, 1994; Price, 1994, 1999; Denniston, 1995; Guisan, 1995; Huntley, 1997; Price & Kim, 1999; Becker & Bugmann, 2001; Ellis & Good, this volume; Fowler & Battarbee, this volume). Recent advances in modelling and GIS have been rapid as well (Council of Europe, 1985; Cernusca *et al.*, 1998; Price & Heywood, 1994).

12.3.5 Overview
To extract a single trend from all of that time and all of those activities is probably over-ambitious. Nevertheless, it is possible to see a long-term movement towards the integration of mountain areas with the rest of their national economies and indeed into global processes. When the mountains were lightly populated and relatively isolated (and their people often acquired a reputation for independence and rebellion), their interaction with other zones was limited. Now they are much more fully part of the regional, national and international policies and, like other ecosystems, their integrity depends upon the integrity of human societies that exert so much influence. One use of all of this history is to put today's policies in a historical context, including those of conservation (Wallis de Vries, 1995; Berglund *et al.*, 1996; Bjørse & Bradshaw, 1998; Pykala, 2000).

12.4 Beyond the narrative
The compilation of a narrative of the kind outlined above, with proper attention to accuracies of time and place, is perhaps a first step in an environmental history. Getting the narrative right involves imposing order on myriad details, with the danger that false generalisations are made in the cause of telling a coherent story. This is especially so when the chronicle is part of an approach to policy, which usually demands a straightforward version. This is liable to be misleading since it usually contains science and a little history – but not much on the human cognition of environmental variables.

The intellectual advance then comes, it can be argued, with interrogating any narrative to see if it throws light on other questions. I pose three questions by way of example.

a) How great are the differences and similarities between different regions, on regional, national and European scales? How do we account for these in terms of different mixes of nature and national cultures in space and time? Were there periods of especially rapid, long-lasting or irreversible change? Have climatic changes been absolutely dominant in bringing about changes in 'natural' and human-directed land cover (Thompson *et al.*, 1995; Rose *et al.*, 1997; Rundgren, 1998; Segerström *et al.*, 1996; Grove, 2001; Nesje *et al.*, 2001; Simpson *et al.*, 2001)?

b) Does any of this history inform the immediate future? Was the 19th century such a huge break that nothing before it has any relevance to today? If this is argued for lowland areas, is it less true of the mountains?

c) Does history provide lenses through which the longer-term future makes any sense?

Does the concept of 'sustainability' have any meaning or is chance (often in the form of technological innovation) always going to destabilise any situation? Is it the case, for example, that before the 19th century there was an integrated upland/valley, grazing/forest exploitation/slope/valley protection system, which only came apart under 'outside' influences (Oestlund *et al.*, 1997), or is that a Romantic construction read back by a later age? And if it was environmentally stable, was it a fulfilling life for the humans who lived there?

12.5 A broad conceptual context

Our task is to deal with 10,000 years of interaction between social and natural processes that interact iteratively and where dominance is never constant. Some of the processes in society are towards individualism and these produce social fragmentation that leads to similar environmental processes: common ownership breaking up into individual ownership, for example, and producing new management regimes for natural and semi-natural ecosystems. By contrast, some of the socio-economic processes tend to coalescence and so the opposite occurs: the introduction of species from another continent would be an example, though perhaps mountains stand aloof from some of the more spectacular coalescences, albeit not from imported influenza viruses. But they do exhibit uniform ski runs and plantation-style forestry.

Mountains have been marginal for most of their human history but that seems now to be less the case: there is now a greater degree of contest over their future (Pollard, 1997; Smout, 2000; Simmons, 2003). They are both different and the same. Their environmental histories have singular qualities but are linked ever more strongly to the rest of the world. In terms of our cognition of them, it is the differences which fascinate us and which this volume is no doubt dedicated to perpetuating, but we must not be blind to their integration into mainstream economies and cultures.

References

Becker, A. & Bugmann, H. (Ed.) (2001). *Global Change and Mountain Regions: the Mountain Research Initiative*. International Human Dimensions Programme on Global Environmental Change) No. 13. IGBP Secretariat, Royal Swedish Academy of Sciences, Stockholm.

Beniston, M. (1994). *Mountain Environments in Changing Climates*. Routledge, London.

Berglund, B.E., Barnekow, L., Hammarlund, D., Sandgren, P. & Snowball, I.F. (1996). Holocene forest dynamics and climate changes in the Abisko area, northern Sweden - the Sonesson model of vegetation history reconsidered and confirmed. *Ecological Bulletins*, **45**, 15-30.

Birks, H.H., Birks, H.J.B., Kaland, P.E. & Moe, D. (Ed.) (1988). *The Cultural Landscape - Past, Present and Future*. Cambridge University Press, Cambridge.

Bjørse, G. & Bradshaw, R. (1998). 2000 years of forest dynamics in southern Sweden: suggestions for forest management. *Forest Ecology and Management*, **104**, 15-26.

Cernusca, A., Bahn, M., Chemini, C., Graber, W., Seigwolf, R., Tappeiner, U. & Tenhunen, J. (1998). ECOMONT: a combined approach of field measurements and process-based modelling for assessing effects of land-use changes in mountain Landscapes. *Ecological Modelling*, **113**, 167-178.

Council of Europe (1985). *Ecological Charter for Mountain Regions in Europe*. European Committee for the Conservation of Nature and Natural Resources Council of Europe, Strasbourg.

Denniston, D. (1995). *High Priorities: Conserving Mountain Ecosystems and Cultures.* Worldwatch Paper No. 123. Worldwatch Institute, Washington, DC.

Ericsson, S., Østlund, L, & Axelsson, A.L. (2000). A forest of grazing and logging: deforestation and reforestation history of a boreal landscape in central Sweden. *New Forests,* **19**, 227-240.

Esseen, P.-A., Ehnstrom, B., Ericson, L. & Sjoberg, K. (1997). Boreal forests. *Ecological Bulletin,* **46**, 16-47.

Frenzel, B., Eronen, M., Vorren, K.-D. & Gläser, B. (Eds) (1993). Oscillations of the Alpine and Polar tree Limits in the Holocene. 5th EPC/ESF Workshop, Innsbruck, Österreich, 20-22 October 1990. Paläoklimaforschung/Palaeoclimate Research; 9. Gustav Fischer, Stuttgart.

Frenzel, B., Pfister, C. & Gläser, B. (Ed.) (1994). Climatic trends and anomalies in Europe 1675-1715. 11th ESF/EPC Workshop, Bern, Schweiz, 3-5 September 1992. Paläoklimaforschung/Palaeoclimate Research; 13. Gustav Fischer, Stuttgart.

Frenzel, B., Vorren, K.-D., Birks, H.H. & Alm, T. (Ed.) (1996). Holocene treeline oscillations, dendrochronology and palaeoclimate. Nordic Symposium of EPC/ESF, Skibotn, Norwegen, 6-8 September 1993. Paläoklimaforschung/Palaeoclimate Research; 20. Fischer, Stuttgart.

Grove, J.M. (2001). The onset of the Little Ice Age. In *History and Climate: Memories of the Future?,* ed. by P.D. Jones, A.E.J. Ogilvie, T.D. Davies & K.R. Briffa. Kluwer, Dordrecht. pp 135-185.

Gudmundsson, G. (1996). Gathering and processing of lyme-grass (*Elymus arenarius* L.) in Iceland: an ethnohistorical account. *Vegetation History and Archaeobotany,* **5**, 13-24.

Guisan, A. (1995). *Potential Ecological Impacts of Climate Change in the Alps and the Fennoscandian Mountains.* Intergovernmental Panel on Climate Change Working Group II. Publication hors-série des Conservatoire et jardin botaniques de la ville de Genève; no. 8. Conservatoire et Jardin botanique de la Ville, Geneva.

Harkness, C.E. (1982). Mapping changes in the extent of woodland in upland areas, 1885-1975. University of Birmingham Department of Geography Moorland Change Project Surveys of Moorland and Roughland Change No. 12. Department of Geography, University of Birmingham, Birmingham.

Heal, O.W. & Perkins, D.F. (Ed.) (1978). *Production Ecology of British Moors and Montane Grasslands.* Ecological Studies v. 27. Springer-Verlag, New York.

Huntley, B. (1997). The responses of vegetation to past and future climatic changes. *Ecological Studies,* **124**, 290-311.

Lindbladh, M. & Nilsson, S.G. (1999). Forest and trees in the cultural landscape at Stenbrohult, S. Sweden - vegetation history based on biological and historical archives. *Svensk Botanisk Tidskrift,* **93**, 19-30.

Malmer, N. & Wallen, B. (1996). Peat formation and mass balance in subarctic ombrotrophic peatlands around Abisko, northern Scandinavia. *Ecological Bulletins,* **45**, 79-92.

Nesje, A., Matthews, J.A., Dahl, O., Berrisford, M.S. & Andersson, C. (2001). Holocene glacier fluctuations of Flatebreen and winter-precipitation changes in the Jostedalsbreen region, western Norway, based on glaciolacustrine sediment records. *The Holocene,* **11**, 267-280.

Nilsson, S.G. (1997). Forests in the temperate-boreal transition: natural and man-made features. *Ecological Bulletins,* **46**, 61-71.

Øestlund, L., Zackrisson, O. & Axelsson, A.-L. (1997). The history and transformation of a Scandinavian boreal forest landscape since the 19th century. *Canadian Journal of Forest Research,* **27**, 1198-1206.

Pitt, D.C. & Nilsson, S. (1991). *Mountain World in Danger: Climate Change in the Mountains and Forests of Europe.* Earthscan in association with Alp Action and International Institute for Applied Systems Analysis (IIASA), London.

Pollard, S. (1997). *Marginal Europe: the Contribution of Marginal Lands Since the Middle Ages.* Clarendon Press, Oxford.

Price, L.W. (1981). *Mountains and Man: A Study of Process and Environment.* University of California Press,

Berkeley and London.

Price M.F. (1994). *Mountain Research in Europe: an Overview of MAB Research from the Pyrenees to Siberia.* UNESCO Man and the Biosphere Series v. 14. Parthenon Publishing, Carnforth, UK/New York.

Price, M. (Ed.) (1999). *Global Change in the Mountains.* Parthenon Publishing, New York and London,

Price, M.F. & Heywood, D.I. (1994). *Mountain Environments and Geographic Information Systems.* Taylor & Francis, London.

Price, M.F. & Kim, E.-G. (1999). Priorities for sustainable mountain development in Europe. *International Journal of Sustainable Development and World Ecology,* **6**, 203-219.

Pykala, J. (2000). Mitigating human effects on European biodiversity through traditional animal husbandry. *Conservation Biology,* **14**, 705-712.

Rasanen, S. (2001). Tracing and interpreting fine-scale human impact in northern Fennoscandia with the aid of modern pollen analogues. *Vegetation History and Archaeobotany,* **10**, 211-218.

Rose, J., Whiteman, C.A., Lee, J., Branch, N.P., Harkness, D.D. & Walden, J. (1997). Mid- and late-Holocene vegetation, surface weathering and glaciation, Fjallsjoekull, southeast Iceland. *The Holocene,* **7**, 457-472.

Rundgren, M. (1998). Early-Holocene vegetation of northern Iceland: pollen and plant macrofossil evidence from the Skagi peninsula. *The Holocene,* **8**, 553-564.

Segerström, U., Hoernberg, G. & Bradshaw, R. (1996). The 9000-year history of vegetation development and disturbance patterns of a swamp-forest in Dalarna, northern Sweden. *The Holocene,* **6**, 37-48.

Seppa, H. & Birks, H.J.B. (2001). July mean temperature and annual precipitation trends during the Holocene in the Fennoscandian tree-line area: pollen based climate reconstructions. *The Holocene,* **11**, 527-539.

Simmons, I.G. (1996). *The Environmental Impact of Later Mesolithic Cultures.* Edinburgh University Press, Edinburgh.

Simmons, I.G. (2003). *The Moorlands of England and Wales. An Environmental History 1000 BC–AD 2000.* Edinburgh University Press, Edinburgh.

Simpson, I.A., Dugmore, A.J., Thomson, A. & Vésteinsson, O. (2001). Crossing the thresholds: human ecology and historical patterns of landscape degradation. *Catena,* **42**, 175-192.

Smout, T.C. (2000). *Nature Contested. Environmental History in Scotland and Northern England since 1600.* Edinburgh University Press, Edinburgh.

Stone, P.B. (1992). *The State of the World's Mountains. A Global Report Edited on Behalf of Mountain Agenda.* Zed Books, London.

Tallis, J.H. & Switsur, V.R. (1990). Forest and moorland in the south Pennine uplands in the mid-Flandrian period. II. The hillslope forests. *Journal of Ecology,* **78**, 857-853.

Thompson, D.B.A., Hester, A.J. & Usher, M.B. (Ed.) (1995). *Heaths and Moorland: Cultural Landscapes.* HMSO, Edinburgh.

Wallis de Vries, M.F. (1995). Large herbivores and the design of large-scale nature reserves in Western Europe. *Conservation Biology,* **9**, 25-33.

13 The use and management of Norwegian mountains reflected in biodiversity values - what are the options for future food production?

E. Gunilla A. Olsson

Summary

1. The human use of mountains in Scandinavia by livestock grazing is documented back to the Bronze Age, some 4,500 years ago.
2. Long-term livestock grazing has shaped and maintained the mountains' semi-natural habitats. The mountains' semi-natural grasslands on base-rich bedrock have a high species diversity. Such habitats function today as refuges for species that are threatened due to habitat extinction in lowlands.
3. The changes in agrarian production over the last 50 years have led to major changes in landscapes and ecosystems and negatively affected many ecological components. The disappearance of semi-natural habitats is a significant example.
4. The rationalisation and industrialisation of agriculture detached the links between local and regional ecosystems and food production when the inputs for high-yielding agriculture, transported over long distances, became available through the availability of cheap fossil fuel.
5. Sustainable development in all parts of the world demands a halt to the transfer of high-quality protein from the south for maintaining high-yielding livestock production in the north. Food production has to fall back on regional and local ecosystems to a larger extent than now.
6. A framework for sustainable development of valuable resources in the mountain areas in Norway and many other countries is needed for extensive grazing of livestock. This would maintain the biodiversity qualities of landscapes and ecosystems and contribute to ethical livestock rearing and the production of human food of the highest quality, and thus sustain food production for the future.

13.1 Introduction

For how long have humans used Norway's mountain resources? As the continental ice sheet retreated some 10,000 years ago, humans immigrated into Scandinavia and lived at a subsistence level on the available biological resources. The hunting of wild herbivores was a crucial subsistence activity (Barth, 1979). Already in prehistoric times – at the transition

---

Olsson, E.G.A. (2005). The use and management of Norwegian mountains reflected in biodiversity values - what are the options for future food production? In *Mountains of Northern Europe: Conservation, Management, People and Nature,* ed. by D.B.A. Thompson, M.F. Price & C.A. Galbraith. TSO Scotland, Edinburgh. pp. 151-162.

between the Neolithic and Bronze Ages – we have indications of livestock grazing in the Norwegian mountains (Moe *et al.*, 1988). From palaeoecological studies in other mountains in central Norway, the existence of semi-natural grasslands has been documented from the late Bronze Age, 3000 BP in Hordaland (Kaland, 1992), Sogn og Fjordane (Kvamme *et al.*, 1992) and Hedmark, Innerdalen (Paus & Jevne, 1987). As the existence of semi-natural grasslands is an indication of livestock grazing, the time perspective of the human use of mountains by their livestock is 3,000 to 5,000 years (see also Simmons, this volume). This time period can be compared to the time of existence of wild spruce (*Picea abies*) in eastern Norway, which is some 1,500 years (Hafsten, 1992). The human use of the mountains (through their livestock) is thus some 2,000 years older than the spontaneous existence of spruce in Norway (Figure 13.1). Using the same logic of old age as the spruce, semi-natural vegetation shaped by the long-term impact of livestock grazing should be viewed as a natural component of the mountain ecosystems in Scandinavia.

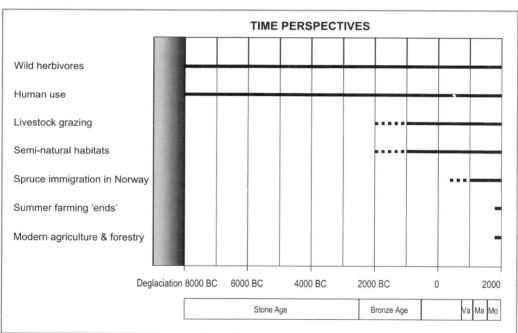

Figure 13.1 Time perspective for the human use of mountains in Scandinavia. Literature references to the different aspects are given in the text. The human use of wild herbivores through hunting was a basis for human immigration into Scandinavia along with the withdrawal of the continental ice. Domestic livestock grazing in mountains is indicated in palaeoecological studies through the existence of lowland grassland plants. The age of the mountains' semi-natural habitats is inferred through the pollen data. The immigration of spruce into Norway is documented in pollen data. The solid lines denote estimated wide distribution; the dotted lines denote estimated limited distribution. 'Va' = Viking Age; 'Ma' = Middle Age; 'Mo' = Modern Time.

The development of the various summer farming systems, which are forms of transhumance systems, is estimated to date back to the 14th century in central Norway (Engen *et al.*, 2000). Transhumance systems have been used world-wide to utilize biological resources in mountains with seasonal production (Allan *et al.*, 1988). Among the Scandinavian countries, the intensive use of mountains by livestock grazing has remained in

Norway only, although land use changes have been large-scale and significant (Olsson *et al.*, 2000). The end of traditional mountain summer farming (approximately 50-60 years ago) coincided with the application of modern agriculture and forestry (Figure 13.1).

The aim of this chapter is to show how aspects of the human use of mountains, especially through livestock grazing, have shaped landscapes and plant diversity in the Norwegian mountains, and to discuss some aspects of future use and management related to sustainable food production.

13.2 The relationship between summer farming, livestock grazing and biological diversity

13.2.1 The influence of grazing on landscapes

The main challenge and driving force within the summer farming system in the Scandinavian mountains was to collect fodder for livestock during summer and to be stored for winter. During summer, the mountains were used as grazing land where the livestock roamed over sizeable areas under the care of shepherds. Along with the development of vegetation and thus of fodder availability, animals were moved to higher altitudes over the summer, thereby effectively using the biological production (Olsson & Bele, 1996; Olsson & Dodgshon, in press). Apart from the livestock grazing, a variety of human uses of the mountain ecosystem contributed to a significant shaping of the landscape in general, implying a shift from mountain birch forest and woodlands to semi-natural heathlands, grasslands and mountain wood-pastures (Olsson *et al.*, 2000). The overriding impact of grazing by large herbivores introduced an ecological disturbance factor which changed the ecological competition (Huston, 1979). This often led to more species-rich communities compared to the non-grazed situation (Dullinger *et al.*, 2003). Along with the contemporary scientific view of ecological communities (Pickett *et al.*, 1992; White & Jentsch, 2001), it follows that ecological communities and ecosystems that are governed by grazing will change significantly when the grazing regimes change, or when grazing ends.

Today, the semi-natural habitats in Norwegian mountains, maintained by livestock grazing, are sites for many red-listed plant species that found a refuge here when similar lowland habitats had been destroyed (Austrheim *et al.*, 1999). However, during the second half of the 20th century, this land use was abandoned due to changes in agricultural practices and a demand for high yielding livestock. The mountains are no longer an essential part of the agro-ecosystem, since their function as a fodder source for livestock has been replaced by arable field production. This development has led to large-scale forest succession and ecosystem changes in the mountains (Figure 13.2a). Parallel to this development is the decline of biological diversity due to the change and destruction of habitats for many species adapted to semi-open landscapes and semi-natural habitats. Our studies in the Jotunheimen mountain range, in mid-Norway, illustrate this process.

During the 20th century there was a significant change in the composition of livestock in Norwegian agriculture (Figure 13.2b), both nationally and in the mountains. By recording changes in the patterns and size of habitats over a 35 year period, we found that the great decline in semi-natural habitats, grasslands and heathlands was correlated with the parallel changes in livestock composition (Olsson *et al.*, 2000). There was large-scale expansion of forested habitats, especially pine (*Pinus sylvestris*) and birch (*Betula pendula*)

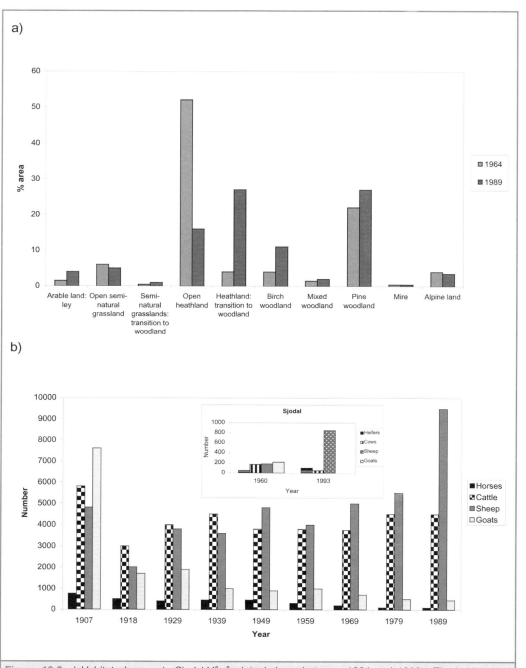

Figure 13.2 a) Habitat changes in Sjodal-Vågå, Jotunheimen between 1964 and 1989. The habitats are recorded from aerial photos and confirmed by field studies. The habitat category 'transition to woodland' means secondary forest succession on former open land. b) Livestock changes in Sjodal-Vågå, Jotunheimen. The changes in number and composition of livestock in farms in the parish of Vågå, which have summer farms in Sjodalen. The livestock changes are reflected in changes in livestock grazing in the Sjodal mountains. The insert figure denotes livestock grazing in the Sjodal mountains 1960-1993. Source: Agricultural statistics, Norway, and county statistics, Vågå, Norway and intertviews in 1994.

woodlands. The decline in the number of cattle (*Bos taurus*) was small, but there was a considerable rise in the number of sheep (Olsson *et al.*, 2000). However, what really matters is that the cattle now graze freely in the mountains only to a very limited extent. They are kept on arable fields while the sheep are let out to roam freely in the mountains. However, sheep (*Ovis aries*) prefer the grazing in the alpine part of mountains, rather than the subalpine regions which were formerly strongly influenced by grazing. The result is that these regions experience forest succession and the tree limit is rising – in some mountains by several hundred metres (Aas & Faarlund, 1996).

This development of plant communities often takes place over large areas since it derives from agricultural changes on many farms in the same region. Thus, the succession processes also have landscape effects (*c.f.* M. Edwards, this volume). The landscape changes from being a mosaic, with open semi-natural grasslands and heathlands intermingled with wetlands and woodland groves, to becoming more homogeneous, containing only wooded areas and wetlands.

13.2.2 Biodiversity responses to grazing at community and population levels

There is still grazing by mixed livestock herds in some mountain valleys at high altitude in the Jotunheimen mountain range. In the small mountain valley of Griningsdalen, 1,010 m above sea level, a grazing exclosure was established in 1986. Fifteen years later, it was completely dominated by willows (*Salix* species) and juniper (*Juniperus communis*) shrubs with 100% canopy cover, and several young birches were also found. The surrounding areas are classified as species-rich, semi-natural grasslands which have been continuously grazed long before and since 1986. Patches of sub-alpine birch (*Betula*) woodland are found within 500 m of the grazing exclosure. This experiment indicates that livestock grazing in this subalpine mountain valley, at this high altitude, of over 1,000 m, has been crucial in preventing shrubland and woodland development, and most probably in creating and maintaining subalpine grasslands. Hence grazing has significant landscape effects in this region (Olsson, 1996).

At the community level, the changes in the landscape are expressed by the change of species composition. By using the concept of 'functional groups' of vegetation, this change is demonstrated (Figure 13.3). Over 15 years without grazing, there were significant changes in the numbers of species and functional groups. The greatest changes were in the strict grassland species group, belonging to a short-statured grass-sward, which decreased significantly; and in woodland species, which increased more than four-fold over the 15 years (Endresen, 2001). An obvious shift from a grassland community to a woodland/shrub community had occurred. This was also visible in the field layer species composition (Figure 13.3).

The impact of grazing by domestic livestock and semi-domestic reindeer in the alpine zone influences the distribution of grassland patches and areas of shrubs, composed of willow and juniper. The influences of environmental factors like soil moisture, soil depth, topography, aspect and elevation are certainly crucial, but a general trend can be identified: increased grazing impacts in this zone transform the shrub patches to semi-natural grassland patches (Størseth, 2002). A study of diversity and species richness in the two alpine habitats revealed that the semi-natural grasslands had almost twice the number of plant species per square metre and 35% higher species diversity compared with the shrub patches (Figure 13.4).

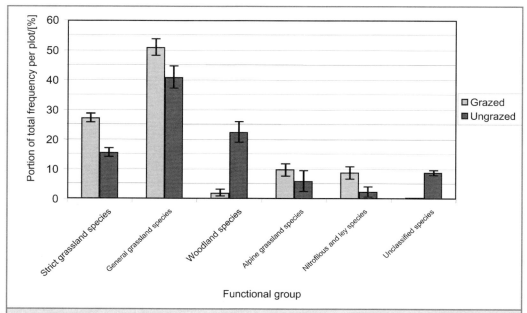

Figure 13.3 Differences in field layer plant species in sub-alpine semi-natural grassland, continuously grazed or ungrazed for 15 years. The species are grouped in different functional groups according to habitat of origin (Austrheim *et al.*, 1999). Standard errors are marked on each bar. The 'woodland species' and 'strict grassland species' groups show the largest differences – shown in a larger font. The figure is redrawn from Endresen (2001).

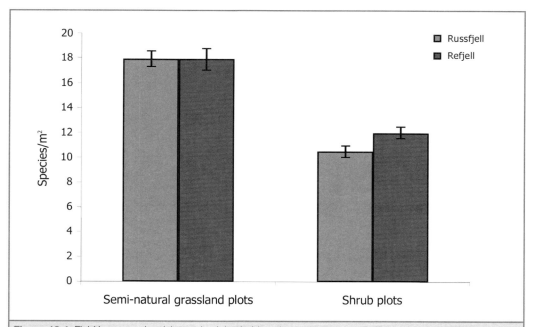

Figure 13.4 Field layer species richness in alpine habitats in two mountains, Refjell and Russfjell. The grassland habitats are derived from the shrub habitats through grazing by semi-domesticated reindeer *(Rangifer tarandus)*. and sheep. The figure shows the mean number of species per square metre. Standard errors are shown on each bar. The figure is redrawn from Størseth (2002).

The grassland species *Primula scandinavica* (Figure 13.5), endemic to the Scandinavian high mountains, provides an example of the influence of grazing on the maintenance of viable populations of grassland species. The plant is confined to montane grasslands and in patches with sparse vegetation. We studied population response to land use, expressed as different forms of ecological disturbance, including livestock grazing and trampling, tractor tracks, wind and snow erosion (Aarnes, 2003). There was a significant positive response to ecological disturbance, expressed as an increased number of established seedlings with increased exposure of bare soil. This implies that grazing and trampling resulted in increased establishment of this species. However, we do not yet know at what level this disturbance is too high and results in negative impacts on seedling establishment and population development. To be able to predict the response of species and communities, we need to investigate the historical disturbances in space, time and frequency, and to obtain landscape data (White & Jentsch, 2001).

Figure 13.5 The grassland species *Primula scandinavica*, endemic to the Scandinavian high mountains, grows in semi-natural subalpine montane grasslands grazed by livestock and semi-domestic reindeer. It also grows in alpine habitats with natural disturbance from wind, water or grazing animals. The plant is closely related to *Primula scotica*, endemic to Scotland. (Photo: E.G.A. Olsson)

For ecological communities that have been shaped by grazing, the cessation of grazing leads to

- a change in ecosystem through forest succession from a grassland to a forest ecosystem;
- a decrease in field layer vegetation diversity because field layer species-richness is higher in semi-natural grasslands than in boreal forests; and
- a decrease in the habitat area available for threatened species (50% of the red-listed vascular plant species in Norway belong to semi-natural habitats created by human use, often grasslands).

13.2.3 Consequences of biodiversity changes related to the cessation of grazing

The number of visitors to Norwegian mountains is increasing. Visitors have many different reasons for spending time there, but one common reason is the fascination of the open mountain landscape, its aesthetics and its accessibility via tracks and walking paths (c.f. the Scottish situation described by Mackay, this volume). These qualities are, to a large degree, a function of long-term livestock grazing, including the grazing of semi-domestic reindeer. As the examples have shown, the disappearance of grazing in these mountains leads to landscape, community and population changes. There will be a dominance of shrubs and trees in the subalpine parts of the mountains. The change will also lower accessibility and make mountain walking more difficult. The field layer flora would change towards fewer herbs and graminoids. For those visitors interested in the alpine flora, the vascular plant diversity will be reduced.

13.3 Agricultural development in Europe 1960-2000

A general trend in contemporary European agriculture is amalgamation into larger farming units in terms of both area and number of livestock, so as to facilitate rationalisation of management. This development has been taking place in western Europe since the 1960s. In Sweden, the number of farming units decreased by 50% in the period 1970-2002; the decrease was highest, 65%, for the small holdings of 2 to 10 ha (Agricultural Statistics Sweden, www.scb.se/eng). This development can be expressed as a polarisation of land use towards intensified use of arable land in some regions, and abandonment in others, especially grazing land. The development has been similar in Norway although at a much slower pace, partly due to a high level of government subsidies to farming in marginal and remote areas. The farming units are smaller: the mean area in 2002 was 16.7 ha (Agricultural Statistics Norway, www.ssb.no/english/subjects/). An important difference between Sweden and Norway is that the common land in the mountains is still grazed by livestock, mostly sheep in Norway, while in Sweden this land use disappeared some 100 years ago. It is forestry land now in Sweden.

This development trend in agriculture and changes in agro-ecosystems have created common ecological problems across western Europe. The spread of livestock diseases has increased and is facilitated by large-scale management, including transport between regions. Another common problem is the disappearance of semi-natural habitats shaped and maintained by livestock grazing. As in Norway, such habitats are home for many red-listed plant and animal species. Such species may approach extinction when their habitats are overgrown by forests. Several countries are investing government money to maintain a number of such habitats under different schemes of green policies (Stenseke, 2001; Department for Enviroment, Food & Rural Affairs (Defra) 'England Rural Development Plan' (www.defra.gov.uk/erdp); Dodgshon & Olsson, 2003). In Sweden, some 400,000 ha of semi-natural grasslands are maintained annually by paying farmers to use the habitats for livestock grazing (Swedish Agricultural Board, 2001). Ironically, the same area of land is cropped annually and harvested for soya in Brazil for export to Sweden for use as cattle feed. The productivity of the Swedish semi-natural grasslands that were shaped by livestock grazing is considered too low for present breeds of cattle. This is the argument for importing soya for cattle feed, which raises questions about the long-term sustainability of milk and meat production.

Within the European Union (EU) there is an ongoing discussion to reform the Common Agricultural Policy. A general problem is the huge economic support given to food production which consumes almost 50% of the EU budget. This has led to over-production of food products, creating 'mountains' of butter, meat, eggs, etc., and also generates ecological problems such as the ones mentioned above. Today, there is an intention to change the economic incentives from stimulating production to producing and maintaining ecological qualities like the biodiversity of landscapes and biological communities, and other environmental qualities in parallel with food production (Defra's 'England Rural Development Plan' www.defra.gov.uk/erdp).

13.4 The relationship between livestock grazing in the mountains and sustainable food production for the future

In such a framework, the Norwegian mountain commons become increasingly interesting, as do other marginal areas in Europe. Used for range grazing by livestock, they maintain landscapes, ecological communities and species shaped and developed within semi-natural ecosystems since prehistory. At the same time, livestock are kept and maintained at a good standard from an animal health perspective, and human food of the highest quality is produced. However, a number of arguments can be raised against these statements. One could be the risk of over-use and the irreversible exploitation of common resources in the mountains. This has happened in northern Norway. The management system of the commons, owned by the government and managed by local and regional users, has to be revised according to the complexity of the present situation. The diversity of user groups is increasing, with some based in distant urban areas (Sandberg, 1999).

In several countries, like the UK, the ecological problem is not undergrazing of mountains – it is the opposite, overgrazing. Large numbers of sheep for meat production, and of red deer (*Cervus elaphus*) and grouse (*Lagopus lagopus*) for hunting purposes, are being raised in the Highlands of Scotland. This has resulted in local and regional overgrazing problems, no regeneration of woody species and, generally, a lowered biological diversity (Gimingham, 2002; Thompson *et al.*, this volume). Another argument against re-using mountains for livestock grazing, especially for cattle, could be that the available forage is too low in proteins to allow for rapid growth development, and thus the meat production would decrease. On the other hand, the mountain vegetation is diverse and has a high content of vitamins and minerals and might contribute to good livestock health. There would be no need for the import of soya proteins over long distances, and the total cost for production might be lower. The total production would take place within a regional framework and would merit the label of 'sustainable food production'.

Decreasing profitability within agriculture has driven the large-scale rationalisation processes. These in turn have caused many environmental problems. What future can be envisioned for farming and living in rural areas? Food for humans will have to be produced for local markets, at least to some extent. Consumers are a powerful driver of production. There is rising interest and awareness of the importance of quality of food products, and the linkage to the ethics of livestock rearing and animal welfare. In western Europe, the increasing public interest in such issues led to a decrease in the consumption of commercial meat as a protest against unethical livestock husbandry and for fear of the transmission of animal diseases to humans. This is a positive sign which gives hope for the future. Here is the seed for awareness of the relationship between food production and landscape ecology, the insight that it matters how the food is produced – both for the landscape and for the consumer. Food that is produced in an environmentally beneficial way will allow high standards of animal welfare, and implies higher quality standards for human food. This will contribute to the quality of human life.

With a global perspective and within a framework of sustainable development, food has to be produced in a sustainable way in the western world. The transport of high-value protein from the south to be used for livestock feed in the north cannot be defended as sustainable. Food production has to be founded to a large degree within a region's ecosystems. Returning to the Norwegian mountains, the challenge for the near future for local farmers is to maintain agriculture, including extensive grazing, in the mountains that will foster the production of high-quality food items. The somewhat lower production level could be compensated for by the excellent quality of that food, yielding higher consumer prices. There is also a very large potential in the development of niche products like special goat *(Capra* spp.) cheeses, alpine lamb meat, etc., that could generate incomes and also be attractive for export to other parts of Europe. Multi-functionality is a concept for European agriculture that is valid also for the Norwegian mountains, with possibilities for combining agriculture with involvement in other activities like mountain ecotourism.

13.5 Conclusions

The changes in agrarian production over the last 50 years have led to major changes in landscapes and ecosystems, and have affected many ecological components. The rationalisation and industrialisation of agriculture detached the links between local and regional ecosystems and food production when inputs for high-yielding agriculture, transported over long distances, became available because of cheap fossil fuel. Sustainable development in all parts of the world demands a halt to the transfer of high-quality protein from the south to maintain high-yielding livestock production in the north. Food production has to fall back on the regional and local ecosystems to a larger extent than now. In such a framework, the mountain areas in Norway and many other countries can again be valuable resources, used for extensive grazing of livestock. This would maintain the biodiversity qualities of landscapes and ecosystems, and contribute to ethical livestock rearing and the production of human food of the highest quality, thus contributing to a sustainable food production for the future.

Acknowledgements

The research for this chapter was funded by the Norwegian Research Council, through the 'Changing Landscapes' programme, for the research project 'Customary rights, cultural practice and biological diversity in mountain agricultural landscapes'. Mona Endresen, Ketil Størseth and Espen Aarnes kindly provided data from their Masters thesis projects at the Department of Biology, Norwegian University of Science and Technology, Trondheim. All and two referees are gratefully acknowledged.

References

Aarnes, E. (2003). The influence of disturbance on *Primula scandinavica*. Masters thesis. Plant Ecology Department, Norges teknisk-naturvitenskapelige universitet (Norwegian University of Science and Technology), Trondheim.

Aas, B. & Faarlund, T. (1996). The present and the Holocene subalpine birch belt in Norway. Special issue. EFS Project. *European Palaeoclimate and Man 13*. European Science Foundation. Gustav Fischer Verlag, Stuttgart.

Austrheim, G., Olsson, E.G.A. & Grøntvedt, E. (1999). Land-use impact on plant communities in semi-natural sub-alpine grasslands of Budalen, central Norway. *Biological Conservation*, **87**, 369-379.

Allan, N.J.R., Knapp, G.W. & Stadel, C. (1988). *Human Impact on Mountains*. Rowman and Littlefield Publishers, Totowa, USA.

Barth, E. (1979). Fangstgraver for rein i Rondande og andre fjell (Reindeer pitfalls in Rondane and other mountains). In *Fortiden i søkerlyset. ^{14}C datering gjennom 25 år*, ed. by R. Nydal, S. Westin, U. Hafsten

& U. Gulliksen. Laboratoriet for radiologisk datering, Trondheim. pp. 139-148.

Dodgshon, R.A. & Olsson, E.G.A. (2003). Conserving European mountain habitats: what can their past tell us about their future? In *Ecosystems and Sustainable Development IV: Volume 1 (ECOSUD 2003)*, ed. by E. Triezzi, C.A. Brebbia & J.-L, Uso. WIT Press, Southampton. pp. 137-149.

Dullinger, S., Dirnböck, T., Greimler, J. & Grabherr, G. (2003). A resampling approach for evaluating effects of pasture abandonment on subalpine plant species diversity. *Journal of Vegetation Science*, **14**, 243-252.

Endresen, M. (2001). Effekter av husdyrbeite på vegetasjonen i subalpine enger i Griningsdalen, Jotunheimen (Influence of livestock grazing on subalpine mountain landscape and vegetation, Griningsdalen, Jotunheimen). Masters thesis. Plant Ecology Department, Norges teknisk-naturvitenskapelige universitet (Norwegian University of Science and Technology), Trondheim.

Engen, A., Lauritzen, P.R. & Øvsteng, M. (2000). *Sjodalen - fra fangst til fritid.* Bokprosjektet Sjodalen, Otta.

Gimingham, C.H. (2002). Towards an integrated management strategy. In *The Ecology, Land Use and Conservation of the Cairngorms*, ed. by C.H. Gimingham. Packard Publishing Limited, Chichester. pp. 185-199.

Hafsten, U. (1992). The immigration and spread of Norway spruce (*Picea abies* (L.) Karst.) in Norway. *Norsk Geografisk Tidskrift*, **46**, 121-158.

Huston, M. (1979). A general hypothesis on species diversity. *American Naturalist*, **113**, 81-101.

Kaland, P.E. (1992). Pollenanalytiske undersøkelser utenfor boplassen i Kotedalen. In *Kotedalen – en boplass gjemnnom 5000 år. Bind 2. Naturvitenskapelige undersøkelser*, ed by. K.L. Hjelle, A.K. Hufthammer, P.E. Kaland, A.B. Olsen & E.C. Soltvedt. Universitetet i Bergen, Bergen. pp. 65-89.

Kvamme, M., Berge, J. & Kaland, P.E. (1992). *Vegetasjonshistoriske undersøkelser i Nyset-Steggjevassdragene.* Arkeologiske rapporter. Historisk museum. Universitetet i Bergen, Bergen.

Moe, D., Indrelid, S. & Fasteland, A. (1988). The Halne area, Hardangervidda. Use of a high mountain area during 5000 years – an interdisciplinary case study. In *The Cultural Landscape – Past, Present and Future*, ed. by H.H. Birks, H.J.B. Birks, P.E. Kaland & D. Moe. Cambridge University Press, Cambridge. pp. 429-444.

Olsson, G.A. (1996). Orörd vildmark eller fjället kulturlandskap. Biodiversitetsproblematik i norska fjällområden. In *Infallsvinklar på biodiversitet*, ed. by A. Langaas. SMU-Rapport nr 6/96. Norwegian University of Science and Technology, Trondheim. pp. 12-28.

Olsson, E.G.A. & Bele, B. (1996). Historical awareness as a tool for conservation of biodiversity: examples from the Norwegian summer farming landscapes. In *Proceedings from the Permanent European Conference for the Study of the Rural Landscape, Papers from the 17th session, Trinity College, Dublin*, ed. by F.H.A. Aalen & M.A. Hennessy. Trinity College, Dublin. pp. 102-107.

Olsson, E.G.A. & Dodgshon, R.A. (in press). Seasonality in mountain areas: a study in human ecology. In *Seasonal Landscapes*, ed. by. H Palang, H. Sooval; I & A. Printsmann. Springer, Dordrecht.

Olsson, E.G.A., Austrheim, G. & Grenne, S.N. (2000). Landscape change patterns in mountains, land use and environmental diversity, Mid-Norway, 1960-1993. *Landscape Ecology*, **15**,155-170.

Paus, A. & Jevne, O.E. (1987). Innerdalens historie belyst ved den pollenanalytiske metoden. In *Kulturhistoriske undersøkelser i Innerdalen, Kvikne, Hedmark*, ed. by A. Paus, O.E. Jevne & L. Gustafson. Rapport. Arkeologisk Serie 1987-1. Vitenskapsmuseet. Universitet i Trondheim, Trondheim. pp. 1-89.

Pickett, S.T.A., Parker, V.T. & Fiedler, P.L. (1992). The new paradigm in ecology: implications for conservation biology above the species level. In *Conservation Biology*, ed. by P.L. Fiedler & S.K. Jain. Chapman and Hall, New York & London. pp. 65-90.

Sandberg, A. (1999). *Conditions for community-based governance of biodiversity.* NF-Report 11/99. Nordland

Research Institute, Bodø.

Stenseke, M. (2001). *Landskapets värden – lokala perspektiv och centrala utgångspunkter.* Kulturgeografiska institutionen, Göteborgs universitet, Gothenburg.

Størseth, K. (2002). Impact of different grazing regimes of reindeer and domestic livestock on alpine landscape and vegetation, Jotunheimen mountain area. Masters thesis in management of natural resources/Plant ecology. Plant Ecology Department, Norges teknisk-naturvitenskapelige universitet (Norwegian University of Science and Technology), Trondheim.

Swedish Agricultural Board - Jordbruksverket (2001). Progress in the program for environment and rural development. Utvecklingen inom Miljö- och landsbygdsprogrammet år 2001. Jordbruksverket, Jönköping.

White, P.S. & Jentsch, A. (2001). The search for generality in studies of disturbance and ecosystem dynamics. *Progress in Botany,* **62**, 399-449.

14 Landscape history and biodiversity conservation in the uplands of Norway and Britain: comparisons and contradictions

M.E. Edwards

Summary

1. The uplands of Britain and Norway have been shaped into cultural landscapes by interactions between people and the environment over millennia. The original shapers of the land were the local inhabitants, but at the turn of the 21st century a diverse constituency of users has both an interest in and an impact on what happens to the landscape.

2. Woodland conservation management in both regions values unmodified or barely modified habitats. However, woodland ecosystems have been affected by centuries of human use, and what is being conserved is greatly transformed from the 'natural' state. Future ecological trajectories will reflect past changes as well as present management regimes, and are likely to be hard to predict or control.

3. Particularly rapid land use change is occurring under today's social and economic pressures. In the uplands of Mid and West Norway, centuries of cultural use based on grazing stock on sub-alpine summer farms have shaped a vegetation mosaic that is valued by biologists and managers for its biodiversity; by recreational users for its scenic appeal; and by academics, and increasingly the general public, for its links to cultural history. Land abandonment is leading to the re-growth of forest and scrub, a transformation that is considered by many to be negative. Creative solutions are required to maintain this biologically rich cultural landscape.

4. In upland Britain, exemplified by Wales, centuries of intensive use have shaped a largely treeless, heavily grazed, and biologically impoverished vegetation cover. In comparison with Norway, less importance is placed on the cultural heritage of the landscape. Instead, nature conservation is focused on small, remnant woodlands, cliffs and mountain summits, which are biodiversity 'hot spots'. However, increasing land abandonment, as farm subsidies decrease, may lead to woodland re-growth which would enhance biodiversity. The lack of ecological 'templates', in the form of areas of relatively undisturbed habitat, frees managers to seek novel ways to increase upland biodiversity as opportunities allow.

Edwards, M.E. (2005). Landscape history and biodiversity conservation in the uplands of Norway and Britain: comparisons and contradictions. In *Mountains of Northern Europe: Conservation, Management, People and Nature*, ed. by D.B.A. Thompson, M.F. Price & C.A. Galbraith. TSO Scotland, Edinburgh. pp. 163-178.

14.1 Introduction

In north-west Europe, people and the land share a long history of interaction. The uplands of Scandinavia and Britain are both landscapes in the original sense of the word: land shaped by human action (Olwig, 1996, 2000). Ecologically, the status of the majority of uplands in north-west Europe reflects the interaction of the natural environment and human activities during preceding centuries and millennia, particularly woodland in terms of clearance, burning, and grazing, as revealed by historical and palaeoecological studies (e.g. Birks, 1988; Simmons, 1996, this volume). Land use practices are constrained by the natural environment and determined by the interplay of social factors that affect local populations, such as resource requirements, customary rights and traditional uses, and economic constraints. Past impacts shape the context within which strategies of conservation management are developed, as do the prevailing attitudes towards landscape, cultural history, and biodiversity held by land managers, the institutions on whose behalf they work, and society in general. Concerns about how society views and values the roles of the upland landscape and who controls its management and use underlie current debates about multiple use of the land and which strategies are best for conserving biodiversity.

A comparative approach can provide new perspectives by highlighting contrasts. The contrasted cases should share fundamental features so that meaning can be drawn from an assessment of observed differences. This seems valid in the case of Norway and Britain. Both have extensive uplands with a long history of human use, which is particularly well-documented by palaeoecological and archaeological studies, and they have shared cultural roots. Both are subject to the influence of a cool-temperate oceanic climate, and ecological processes, if not the details of ecosystems, are similar (e.g. Ratcliffe & Thompson, 1988). This chapter compares past human impact in upland Norway and Britain and its effect on vegetation and landscape, current attitudes towards biodiversity, and the resultant strategies for its conservation. A considerable literature exists about the landscape and ecological history of the different regions of upland Britain (e.g. Ratcliffe, 1977; Birks, 1988; Birks *et al.*, 1988; Ratcliffe & Thompson, 1988; Chambers, 1993). To provide details on all regions is beyond the scope of this chapter, and examples are restricted to Wales (primarily mountainous Mid and North Wales), the region with which the author is most familiar. The uplands of Mid Norway and adjoining West Norway are characterized by mountain summer-farming districts that are of considerable conservation interest and the subject of numerous studies (e.g. Birks *et al.*, 1988; Austrheim *et al.* 1999; Olsson *et al.*, 2000; Bryn & Daugstad, 2001; Vandvik, 2002; Vandvik & Birks, 2002; Olsson, this volume).

14.1.1 Study areas

The mountains of Mid and North Wales (Figure 14.2a) are for the most part <1,000 m above sea level. At c. 53°N, the potential natural vegetation is temperate forest, although there is virtually none today (Figure 14.1). Birks (1988) estimates the maximum Holocene tree line (the upper limit of continuous or discontinuous woodland with trees >2 m tall) to have reached >635 m; this is likely to be higher than the potential treeline under the present cooler climate. Edgell (1969) places the limit of tree growth as low as 350 m in highly exposed areas. Virtually all of the upland area is covered by acid grassland, bracken (*Pteridium aquilinum*), or heather (*Calluna vulgaris*), with the exception of the highest mountain tops, a few small deciduous woodlands (Ratcliffe, 1977; Edwards, 1981, 1986)

Figure 14.1 Remnant deciduous woodland, North Wales (Photo: H.J.B. Birks).

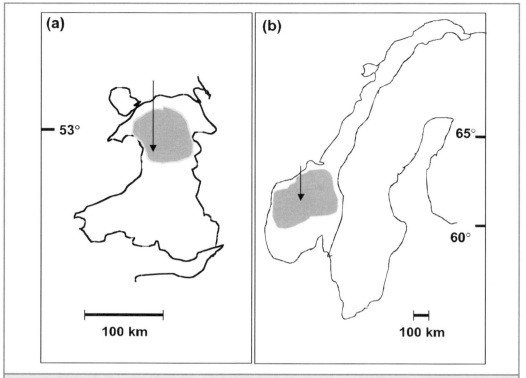

Figure 14.2 (a) Map of Wales showing the mountains of Mid and North Wales (shaded). Arrow indicates approximate location of Coed Gallywd. (b) Map of Norway showing the mountains of Mid and West Norway (shaded). Arrow indicates the approximate location of Sjodalen.

and recent plantations of exotic conifers. Hill farming for sheep is the dominant land use, with some more intensive agriculture taking place at low elevations.

The uplands of Mid and West Norway lie between c. 60° and 63°N (Figure 14.2b). The region lies within the latitudinal boreal forest zone, which is dominated by Scots pine (*Pinus sylvestris*) and Norway spruce (*Picea abies*). Alpine tundra covers extensive high-elevation plateaux and mountain peaks (many summits are >1,000 m). The tree line tends to rise in elevation along a gradient from c. 300-400 m at the coast to >1,000 m in the more continental interior (Moen, 1999). The uppermost forest zone, where many summer farms are situated (see below), is commonly dominated by mountain birch (*Betula pubescens* ssp. *czerepanovii*). Conventional agriculture is largely confined to valley floors. The forested valley sides are managed for timber, usually on a small scale, and areas at and above the tree line are used for grazing. The most common stock animals are sheep, goats and reindeer, with cattle and horses less frequent (see Moen (1999) and Olsson (this volume) for more details).

14.2 Evolving relationships between people and the land

Geographers have long recognized landscape as a useful concept with which to examine the intersection of culture and nature. By tracing the evolution of the landscape concept in north-west Europe over the last few centuries, Olwig (1996, 2000) demonstrates how changing societal attitudes are related to patterns of land ownership, rules of access, and conservation concerns. While Scandinavia and Britain share many common cultural roots, differences in their social and political development are reflected in somewhat divergent attitudes today. In Scandinavia, one meaning of the word landscape is 'land *of* the people', which implies a form of cultural identity attached to the land. This definition emphasizes human activities as part of landscape and requires that the culture and the physical manifestations of landscape be considered together (Olwig, 2000). In contrast, in Britain particularly, from the 17th century onwards, under the influence of renaissance and enlightenment thinking, the meaning of landscape evolved into 'scene', a concept that emphasizes the importance of the viewer, who is outside the scene (Olwig, 2000). This viewpoint underpins the predominant contemporary attitude: for many people, landscapes are to be viewed or experienced for pleasure rather than places to be used and lived in.

In Scandinavia, there continues to be widespread acknowledgement of the culture that is embedded in the landscape, for example, in the form of customary rights of access and traditional uses that imbue places with local meaning and still shape the land. According to Olwig (2000), the principle of *allemannsretten* (which allows people broad access to the land, with the exception of farm infields) emerges from these ancient uses. In contrast, while customary rights and traditional uses certainly existed, and still exist, in Britain, they are curtailed in many areas, and public access to large parts of the British countryside is far more restricted (though the uplands, particularly in Scotland, tend to have broader access than the lowlands).

In the post-war decades, urban populations have grown far more than rural ones in both countries. With increasing information, education, transport, and leisure time, many who do not make their living from the land nevertheless appreciate the rural landscape and wish to have a say in what happens to it. There are now numerous groups with an interest in the uplands, and many of their constituents live far from these landscapes. The contemporary land manager is faced with multiple demands on the land: for production, amenity, and conservation. Competing values are observable in conflicts such as those over forestry and recreation (Jones, 1991) and predator control in Norway, and over fox hunting and the impact of agriculture on the environment in Britain. Upland landscapes are in transition, from being (literally) shaped by those who live in them to a resource claimed by many and increasingly managed by a diverse range of institutions, rather than the local inhabitants.

14.3 Biodiversity in the uplands

'Biodiversity' is an important social and political construct and is central to current strategies for biological conservation (Gaston, 1996). It refers to the biological system in its entirety – genes, species, and ecosystem function – and as such is a concept, not a practical definition. In practice, managers decide which elements of an ecosystem are the most critical for conservation. The ecological rationale of such a decision is site-specific and related to the properties of ecosystems and the biota, but overall approaches are influenced by more generally prevailing views of what is important. On the one hand, there is acknowledgement of the pervasive role of human use in shaping present ecosystems. In many cases, an ecosystem is dependent in its present form on the continuance of certain human activities (such as grazing of domestic stock). On the other hand, the systems that are considered the least altered are often the most valued. These seemingly contradictory preferences, which can be linked to the ideas of 'cultural landscape' and 'wilderness', appear not only in ecological thinking but also more broadly in societal attitudes to biodiversity and landscape.

The wilderness concept, which originated in its current form in the early 20th century national park movement in the United States, has proved highly influential within conservation thinking (Lowenthal, 1999). The concept was taken up in Scandinavia and associated with an idyllic pastoral past in which people used – but had little influence on – the landscape (Sernander, 1920, cited by Faegri, 1988). Only later, when exclosure experiments revealed how scrub and forest re-colonized grazed land, was the extent of human influence on the landscape appreciated. In fact, the term 'cultural landscape' was not common in Scandinavia until the second half of the 20th century (Faegri, 1988). In recent decades, however, it has become widely used by researchers, planners, managers, and the media (Jones, 1991). In Britain also, the uplands were considered 'wilderness' until palaeoecology revealed the broad extent of human impact, and it is now generally considered that few, if any, ecosystems are free from cultural influence.

Over most of both Britain and Norway, then, biodiversity exists within cultural landscapes. Human use alters features of the ecosystem and modifies habitats, and as activities change over time, so do biological responses and the appearance of the land (Jones, 1995; Hallanaro & Usher, this volume). In both Norway and Britain, under the pressures of millennia of human use and long-term climate change, soils and vegetation have followed different trajectories and reached different states, varying at each locality as a result of the integrated response of the climate-vegetation-soil system to a particular disturbance regime. However, Faegri (1988, p. 1) has commented "The cultural landscape can only be understood by its antithesis: untouched, unspoiled nature". The realization of this dualism is the basis for understanding and appreciating both.

This suggests that, beyond the ideal of wilderness, there is a more pragmatic scientific reason to value 'untouched nature' – to understand the effects of human impact by reference to systems that are free of that impact. While Faegri's 'untouched nature' may be hard to find, a system unaffected by humans remains an important theoretical touchstone in this ecological complexity, and habitats that approach an unmodified state are highly valued.

A consequence of the long history of land use in both Britain and Norway is that it is difficult to assess the level of 'naturalness' of a system. Nor is it easy, given the present economic changes affecting land use, to predict the future trajectory of a particular ecosystem, once traditional practices stop and/or it is placed under conservation management. This is particularly so when the details of the history of the landscape are unknown (see Lundberg, 2002). A challenge to managers is to understand past pressures on the system and predict responses to an imposed management regime, which itself will depend on whether the guiding strategy is to protect a cultural landscape or an 'undisturbed' system. The following sections illustrate contradictions that can arise in contemporary biodiversity conservation due to these problems and to a tendency to polarize the wilderness and cultural landscape concepts, rather than acknowledge a complex gradient of human impact (Faegri, 1988).

In Norway and Britain, national legislation defines various kinds of protected area, and international obligations such as those linked to the 1992 United Nations Convention on

Biological Diversity further determine conservation activities (see Thompson *et al,*. this volume). Norwegian legislation of 1970 incorporates both cultural landscape and wilderness values. National Parks are conceived as mainly undisturbed nature and are therefore mostly in mountain areas; hiking, grazing, hunting and fishing are allowed, in recognition of traditional and more recent uses. Protected Landscapes are designated in recognition of their biological and/or cultural importance. Activities such as summer farming and forestry are allowed, although farming practices reflect the current agricultural economy and often do not mirror the traditional uses of earlier centuries. Except in a few sites specifically representing culturally shaped ecosystems, all human activities in Nature Reserves are strongly curtailed, and emphasis is placed on the undisturbed, or nearly undisturbed, nature of the habitat (Gjaerevoll, 1988; Moen, 1999).

Inherent in the British legislation of 1949 and 1981 is the assumption that virtually all landscapes have been shaped by human action and that most are occupied and productive areas. The concept of wilderness or undisturbed nature is not emphasized. National Parks are designated for nationally valued scenery and recreational opportunities but specifically recognized as living landscapes, within which most types of rural land use take place. Areas of Outstanding Natural Beauty protect scenic landscapes but allow most forms of rural land use. National Nature Reserves (NNRs) are the most strictly managed areas, representing key habitats and species assemblages; some of these areas would be considered as most closely approaching 'undisturbed nature'.

14.3.1 Norway

Landscapes and biological systems are often particularly diverse (that is, structurally heterogeneous and species-rich) under non-intensive forms of production (Hobbs & Huenneke, 1992). This is the case with summer farming in the sub-alpine zone and its associated semi-natural ecosystems (Austrheim *et al.*, 1999; Vandvik, 2002; Olsson, this volume). Traditional land use featured minimal use of pesticides and artificial fertilizers, extensive livestock grazing systems in the uplands, and communal systems of land tenure, and created a semi-natural vegetation mosaic of open and wooded habitats that is highly valued for its plant diversity. Evidence of domestic grazing dates from the mid-Holocene, and grazing that had some impact on forest structure and/or composition occurred as early as the late Bronze Age (c. 2,500 years before present (yrs BP)) (Olsson, this volume). A trend to more intensive use, with burning, grazing, and, locally, soil erosion occurred at about 1,000 yrs BP (Kvamme, 1988).

Summer farming has declined dramatically since World War II, with changes in methods of agricultural production. The last 50 years have seen a widespread shift to intensive, high-input farming, and extensive-use areas have been converted or abandoned. From about 44,000 summer farms at the beginning of the 20th century, the number had declined to about 1,900 in the early 1990s (Moen, 1999). Cultural historians who draw attention to the rapid loss of these traditional farming systems and the knowledge and customs that relate to them are also voicing a more general concern that the rural cultural heritage in Norway is being eroded. In Norway, farmers and farming play a central role in politics and society; there are strong links between the farmer and Norwegian national identity; and the tradition of the *seter*, the summer farm of the uplands (see Figure 14.3), has important cultural value (Jones & Daugstad, 1997; Daugstad, 2000).

With the decline in summer farming, birch forest and scrub are invading sub-alpine mountain pastures previously kept open by grazing (see Figure 14.4). This re-growth is both a visual transformation and a marked structural change. Biologists tend to view it as a negative development with respect to biodiversity, and recreational users of the uplands express concerns about reduced access and scenic appeal. Furthermore, there is a growing impetus to protect the threatened species and plant communities that are closely tied to human activities, because they are considered a cultural product and are imbued with cultural value (Jones, 1995; Olsson, this volume).

Figure 14.3 *Seter* buildings and grazed summer pasture, Sjodalen, Norway (Photo: M. Edwards).

In contrast, within the forest zone in Mid and West Norway, conservation management tends to focus on the preservation of localities of ancient or 'undisturbed' forest. It seems that many – including managers, farmers, and recreational users – perceive woodland as a place with the potential to be 'undisturbed nature'. The presence of trees, rather than open land, is taken as a significant indicator of naturalness. However, areas of accessible forest were often grazed and/or used for timber in previous centuries. Under rural population pressure c. 150-100 years ago, levels of deforestation were likely much higher than today, and the remaining forests may have had a more open structure. Much low- and mid-elevation forest is now largely re-grown (Moen, 1999). It is in the conservation of forest biodiversity that managers are particularly prone to misread the landscape. Modern woodlands frequently reflect past cultural activity (see Rackham (1976, 1988) for example), but wooded landscapes more readily fit the idea of 'wilderness', and it is often difficult to detect cultural influences within the forest. Two examples illustrate the point.

Figure 14.4 Scrub encroachment on abandoned summer pastures, Daugstad, Norway (Photo: M. Edwards).

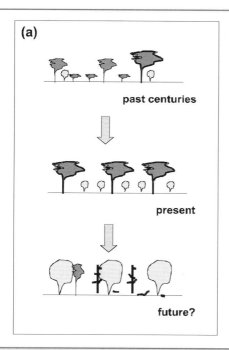

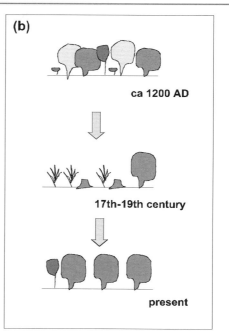

Figure 14.5 (a) Hypothesized development of upland pine forest in Sjodalen, Mid Norway (see text for details). (b) Changes in woodland structure and composition with time at Coed Ganllwyd, North Wales (see text for details).

First, in the mountain valley of Sjodalen (Figure 14.5a), a patch of 'ancient' forest containing impressive Scots pine trees has developed from a traditionally managed forest grazing system. Currently fenced off and unmanaged (perhaps more as a consequence of lack of funding than deliberate policy), its future trajectory is that of senescence without replacement of pine unless disturbances (to which the pine forest is naturally adapted) are re-introduced by active management (M. Edwards & E. Brovold, unpublished data; Figure 14.5a). Second, Lundberg (2002) describes a site in coastal West Norway where lowland moist forest, assigned high conservation value for its 'near natural' status, was a treeless outfield less than 100 years ago.

In summary, two parallel but contrasting emphases on landscape and biodiversity exist in Norway. One aims to conserve the sub-alpine *seter* landscape and its culturally related biodiversity. In this case, managers and users alike generally regard the re-colonization of abandoned land by the 'natural' forest as a negative feature. In contrast, the other strategy is the conservation of 'natural' or near natural forests in lower-elevation areas. However, because much forest land in Norway is in a process of dynamic change after a several centuries of relatively intensive exploitation, and often relatively little is known of past land use (Lundberg, 2002), what is actually being conserved and what will develop in the future are often unknown.

14.3.2 Wales

In Britain, farming does not have the same political weight or social meaning as it does in Norway, and upland farming is particularly marginalized. The open landscapes of the Welsh mountains are appreciated by many for scenic and recreational values. A 1996 survey by the Countryside Council for Wales (CCW) reported that the feature of Welsh National Parks listed most frequently by visitors was scenic appeal (Countryside Council for Wales, 1996), but they are not generally linked to an appreciation of cultural heritage.

In many parts of upland Britain, a major landscape transformation occurred in prehistoric times, apparently a result of the susceptibility of the natural system to prehistoric clearance and burning, possibly combined with climate change (Moore, 1975), and/or the intensity and duration of prehistoric land use. In Wales, vegetation clearance occurred on a larger scale than in upland Norway. Chambers *et al.* (1988) report charcoal from mid-Holocene sediments in north-west Wales that may be connected with burning to encourage game. Walker (1993) describes probable human impact for at least the past 4,000 years in the mountains of Mid Wales, and Moore (1993) and Smith & Cloutman (1988) also report early forest clearance and blanket bog development. By c. 3,500 yrs BP (early Bronze Age) there was considerable deforestation (Chambers *et al.*, 1988) and soil erosion (Walker & Taylor, 1976).

The earliest written accounts, such as that of Giraldus Cambrensis (c. 1194), suggest North Wales was little cultivated and probably still well wooded, at least from the perspective of one travelling through the inhabited valleys. However, according to Linnard (1979), at the end of the Middle Ages, the region was severely deforested, and local people experienced timber and fuel shortages. Sheep densities increased steadily from the 17th to 19th centuries (Hughes *et al.*, 1973). The transhumance system (about which relatively little is known; Chambers *et al.* (1988)) declined, and sheep were kept on the mountains all year. Only in the agricultural depression at the end of the 19th century was grazing pressure somewhat lowered (Hughes *et al.*, 1973). Wealthy landowners attempted reforestation schemes in the 18th and early 19th centuries (Linnard, 1979); these are the origin of many of the present deciduous woodlands, some of which are designated as National Nature Reserves (NNRs). In the 20th century, afforestation with exotic conifers has added to the forest estate, but has been viewed negatively with regard to landscape aesthetics and biodiversity (e.g. Ratcliffe 1977; Ratcliffe & Thompson, 1988).

In this largely treeless landscape, the vegetation over almost the whole altitudinal gradient has little resemblance to the potential natural vegetation. Extensive early clearances would have led to the loss of many elements of the woodland flora and fauna. High levels of burning and grazing until the early 20th century continued the impoverishment of soils and vegetation, which in turn has probably reduced populations of upland species that characterize open landscapes, a trend exacerbated by conifer planting (Ratcliffe & Thompson, 1988). For the moment, hill farming continues, but over the last decade, reductions in subsidies have led to some farms being given up.

In Britain, as in Norway, there is an emphasis on conserving woodlands, perhaps even more so, given the impoverishment of Britain's forest estate. In upland Wales, important biodiversity hotspots are cliffs, mountain tops, inaccessible woodland fragments that escaped the axe (though not necessarily the grazing of sheep) during previous centuries of rural impoverishment (Ratcliffe, 1977), and patches of woodland created or conserved through the management practices of large estates during the 17th and 18th centuries (Linnard, 1979; Edwards, 1981). The wooded area is small, however, and the woodlands have been modified extensively by the impact of past cultural practices. For example, at Coed Ganllwyd NNR, an enclosed oak-woodland in North Wales (Figure 14.5b), the present generation of trees regenerated after clear felling in the 19th century. Palaeoecological studies show that, since early Medieval times, there has been a loss of tree-species diversity associated with repeated disturbances – probably coppicing that favoured sessile oak (*Quercus petraea*). Recent soil acidification and poor regeneration are a consequence of the dominance of oak and lack of disturbance (Edwards, 1981, 1986; Figure 14.5b).

In summary, the extent of the transformation of the upland landscape is far greater in Britain, and this is recognized by ecologists and managers if not the public at large. While undoubtedly impoverished biologically, the uplands of Wales, along with the other British upland regions, are unique in an international context precisely because of the extreme modifications they have undergone (Ratcliffe & Thompson, 1988). Some remaining woodland areas, though greatly altered, are highly valued because they are the closest remaining examples of what the natural land cover of the uplands would be like (Figure 14.6).

Figure 14.6 Welsh upland vegetation near Coed Ganllwyd (Photo: M. Edwards).

14.4 Implications for future management

It seems, from the limited observations made, that in both countries woodland conservation goals are orientated towards creating or retaining systems that come as close as possible to the presumed natural state. However, the lasting imprint of past human activity (seen both in current structure and composition and in successional trends) may be underestimated. There is a tendency to enclose conservation woodlands, as grazing is commonly seen as a threat. In fact, active management to disturb the system and allow regeneration, and, in the case of Welsh woods, intervention to encourage increased tree species richness by controlled grazing, may be a more useful biodiversity management approach (see Mitchell & Kirby, 1990). In both countries, the status and future trajectories of the forest reflect a continuing response to past changes and management; management for the *status quo* will fail to conserve what is currently there.

Major differences between sub-alpine *seter* landscapes in Norway and the landscapes of upland Britain include levels of diversity and the degree of ecosystem transformation caused by clearing and grazing. The example of Wales shows that the uplands have undergone tremendous physical and biological change over the past several millennia, and they are undoubtedly much more impoverished biologically than the uplands of Norway, in part because the transformation has been so extensive, leaving little in the way of a mosaic of woodland and open vegetation. While *seter* farming has clear ecological effects in terms of increasing soil nutrient levels and fostering unusually diverse vascular plant assemblages in the most intensively used areas (Vandvik & Birks, 2002), the palaeoecological record suggests that the ecosystem has not been radically transformed as it has in Britain. With reduced human pressure, the *seter* landscape is reverting to something that more approaches the 'natural' state. It also remains unclear how much species diversity will be reduced by the loss of open habitats through succession, as is feared by some ecologists. Wild fauna must have grazed this ecosystem throughout its post-glacial history, and our reaction to present changes may fail to place events in a broad enough historical context (see, for example, Vera (2000)).

In Norway, traditional cultural activities are responsible for the diverse and aesthetically appealing landscape in the *seter* areas. However, strong socio-economic forces are rapidly transforming the culture that created this landscape and thus the landscape itself. The traditional shapers of the summer-farm landscape have largely abandoned their practices, and areas of the Norwegian uplands may be returning to something approaching the natural state. This is unwelcome to many who now view, use and, increasingly, manage the land (for example, for National Parks). Rather, the cultural landscape is most highly prized for a variety of reasons. There is thus a convergence of interest in maintaining, on the one hand, the subalpine vegetation mosaic and, on the other, the Norwegian rural cultural heritage – and in the future, this may be a key conservation strategy. In part, the viewers - no longer living the farming life but interested in national identity and recreation – may become the shapers. Ideas being tried out include educational projects specifically aimed at retaining knowledge of traditional practices, and cultural tourism, that is: come for a holiday – and run the summer farm!

In Britain, grazing pressure on the uplands is likely to decrease with the decline in farm subsidies. In Wales, it is probably too early to observe an ecological response, for example, in the form of natural regeneration of scrub or woodland. A process such as woody re-growth that enhances a vegetation mosaic is likely to increase habitat and species diversity, because the present open vegetation is species-poor. Given the more widespread intensive use levels, greatly transformed soils, and relative lack of local seed sources compared with upland Norway, any future response may be far slower. Ironically, the abundant re-growth of upland forest – 'unwanted' in Norway, but desirable in Britain – may be hard to achieve.

Conservation management in an upland region such as Wales will be less about retaining or replacing traditional cultural practices, and more about fostering a reversion to a more natural system after millennia of human impact in a landscape that bears little resemblance to the presumed natural state. From a biodiversity perspective, this could provide a chance for innovative forms of management, given that there is no clear ecological or cultural 'ideal' – providing that the needs of other interest groups, such as the surviving year-round sheep farms and the ever-increasing number of recreational users from urban centres, can also be accommodated. There are fates for the uplands other than afforestation with – dare one say – Norway spruce.

Acknowledgements

Thanks to Michael Jones, Gunilla Olsson, Kenneth Olwig, Gunhild Setten, and H. John B. Birks, who contributed valuable ideas in connection with this chapter. The manuscript was much improved after helpful reviews from Jeffrey Maxwell, Odd Inge Vistad, Angus MacDonald, and H. John B. Birks. This work was supported in part by the Norwegian Research Council. Welsh field data were obtained during a project funded by the Natural Environment Research Council (UK).

References

Austrheim, G., Olsson, E.G.A. & Grøntvedt, E. (1999). Land use impact on plant community patterns in semi-natural sub-alpine grasslands in Budal, central Norway. *Biological Conservation, 87*, 369-379.

Birks, H.J.B. (1988). Long-term ecological change in the British uplands. In *Ecological Change in the Uplands,* ed. by M.B. Usher & D.B.A. Thompson. Blackwell, Oxford. pp. 37-56.

Birks. H.H., Birks, H.J.B., Kaland, P.E. & Moe, D. (Ed.) (1988). *The Cultural Landscape, Past, Present, and Future.* Cambridge University Press, Cambridge.

Bryn, A. & Daugstad, K. (2001). Summer farming in the subalpine birch forest. In *Nordic Mountain Birch Ecosystems,* ed. by F.E. Wielgolaski. Man and the Biosphere Series, Volume 27. Parthenon, New York. pp. 307-315.

Cambrensis, G. (circa 1194). *The Itinerary through Wales and the Description of Wales, circa 1194.* Everyman Edition, Dent, London (1908).

Chambers, F.M. (Ed) (1993). *Climate Change and Human Impact on the Landscape.* Chapman and Hall, London.

Chambers, F.M., Kelly, R.S. & Price, S.-M. (1988). Development of the late-prehistoric cultural landscape in upland Ardudwy, north-west Wales. In *The Cultural Landscape, Past, Present, and Future,* ed. by H.H. Birks, H.J.B. Birks, P.E. Kaland & D. Moe. Cambridge University Press, Cambridge. pp. 333-348.

Countryside Council for Wales (1996). Peace in the parks - but too many cars, survey reveals. Press Release dated 21st November 1996. Countryside Council for Wales, Bangor. http://www.ccw.gov.uk/News/index.cfm?Action=Press&ID=182.

Daugstad, K. (2000). Mellom romantikk og realism: om seterlandskapet som ideal og realitet. Doctoral Dissertation. Norwegian University of Science and Technology, Trondheim.

Edgell, M.C.R. (1969). Vegetation of an upland ecosystem. *Journal of Ecology, 57*, 335-359.

Edwards, M.E. (1981). Ecology and Historical Ecology of Oakwoods in North Wales. Ph.D. dissertation, University of Cambridge, Cambridge.

Edwards, M.E. (1986). Disturbance histories of four Snowdonian woodlands and their relation to Atlantic bryophyte distributions. *Biological Conservation, 37*, 301-320.

Faegri, K. (1988). Preface. In *The Cultural Landscape, Past, Present, and Future,* ed. by H.H. Birks, H.J.B. Birks, P.E. Kaland & D. Moe. Cambridge University Press, Cambridge. pp. 1-4.

Gaston, K.J. (1996). What is biodiversity? In *Biodiversity: A Biology of Numbers and Difference,* ed. by K.J. Gaston. Blackwell, Oxford. pp. 1-9.

Gjaerevoll, O. (1988). Nature conservation in Norway. In *Ecological Change in the Uplands,* ed. by M.B. Usher & D.B.A. Thompson. Blackwell, Oxford. pp. 313-321.

Hobbs, R.J. & Huenneke, L.F. (1992). Disturbance, diversity, and invasion: implications for conservation. *Conservation Biology, 6*, 324-337.

Hughes, R.E., Dale, J., Williams, I.E. & Rees, D.J. (1973). Studies in sheep population and environment in the mountains of North Wales. *Journal of Applied Ecology, 10*, 113-132.

Jones, M. (1991). The elusive reality of landscape: problems of interpretation. *Norwegian Journal of Geography, 45*, 229-244.

Jones, M. (1995). Forvaltning av biodiversitet og kulturlandscap. Abstract for workshop: Biodiversitetsforskning (7.12.95). Centre for Environment and Development, Trondheim University, Trondheim.

Jones, M. & Daugstad, K. (1997). Usages of the 'Cultural Landscape' concept in Norwegian and Nordic landscape administration. *Landscape Research, 22*, 267-281.

Kvamme, M. (1988). Pollen analytical studies of mountain summer farming in Western Norway. In *The Cultural Landscape, Past, Present, and Future,* ed. by H.H. Birks, H.J.B. Birks, P.E. Kaland & D. Moe. Cambridge University Press, Cambridge. pp. 349-367.

Linnard, W. (1979). The history of forests and forestry in Wales up to the formation of the Forestry Commission. Ph.D. thesis. University of Bangor, Bangor.

Lowenthal, D. (1999). From landscapes of the future to landscapes of the past. *Norwegian Journal of Geography,* **53**, 139-144.

Lundberg, A. (2002). The interpretation of culture in nature: landscape transformation and vegetation change during two centuries at Hystad, Stord, SW Norway. *Norwegian Journal of Geography,* **56**, 246-256.

Mitchell, F.J.G. & Kirby, K. (1990). The impact of grazing and human disturbance on the conservation of semi-natural woods in the British uplands. *Forestry,* **63**, 334-353.

Moen, A. (1999). *National Atlas of Norway: Vegetation.* Norwegian Mapping Authority. Hønefoss.

Moore, P.D. (1975). Origin of blanket mires. *Nature,* **241**, 350-353.

Moore, P.D. (1993). The origin of blanket mire, revisited. In *Climate Change and Human Impact on the Landscape,* ed. by F.M. Chambers. Chapman and Hall, London. pp. 217-224.

Olsson, E.G.A., Austrheim, G. & Grenne, S.N. (2000). Landscape change patterns in mountains, land use and environmental diversity, Mid-Norway, 1960-1993. *Landscape Ecology,* **15**, 155-170.

Olwig, K.R. (1996). Recovering the substantive nature of landscape. *Annals of the Association of American Geographers,* **86**, 630-653.

Olwig, K.R. (2000). The place ecology of landscape. In *Planering för landskap,* ed. by E. Skärbäck. Stad & Land 166. Sveriges landbruksuniversitetet, SLU, Alnarp. pp. 37-55.

Rackham, O. (1976). *Trees and Woodlands in the British Landscape.* Dent, London.

Rackham, O. (1988). Trees and woodland in a crowded landscape – the cultural landscape of the British Isles. In *The Cultural Landscape, Past, Present, and Future,* ed. by H.H. Birks, H.J.B. Birks, P.E. Kaland & D. Moe. Cambridge University Press, Cambridge. pp. 53-77.

Ratcliffe, D.A. (Ed) (1977). *A Nature Conservation Review. Volumes I and II.* Cambridge University Press, Cambridge.

Ratcliffe, D.A. & Thompson, D.B.A. (1988). The British uplands: their ecological character and international significance. In *Ecological Change in the Uplands,* ed. by M.B. Usher & D.B.A. Thompson. Blackwell, Oxford. pp. 9-36.

Sernander, R. (1920). Den svenska hagens historia. Svenska betes – og vallföreningens årskrift, **1920**, 5-13.

Simmons, I.G. (1996). *Changing the Face of the Earth: Culture, Environment, History.* 2nd Edition. Blackwell, Oxford.

Smith, A.G. & Cloutman, E.W. (1988). Reconstruction of the Holocene vegetation history in three dimensions at Waun Fignen Felen, and upland site in South Wales. *Philosophical Transactions of the Royal Society (B),* **322**, 159-219.

Vandvik, V. (2002). Pattern and Process in Norwegian Upland Grasslands: an Integrated Ecological Approach. Dr. Scient. thesis. University of Bergen, Bergen.

Vandvik, V. & Birks, H.J.B. (2002). Pattern and process in Norwegian upland grasslands: a functional analysis. *Journal of Vegetation Science,* **13**, 123-124.

Vera, F.W.M. (2000). *Grazing and Forest History.* CABI Publishing, New York.

Walker, M.J.C. (1993). Holocene (Flandrian) vegetation change and human activity in the Carneddau area of upland mid-Wales. In *Climate Change and Human Impact on the Landscape,* ed. by F.M. Chambers. Chapman and Hall, London. pp. 169-183.

Walker, M.F. & Taylor, J.A. (1976). Post-neolithic vegetation changes in the western Rhinogau, Gwynedd, North-West Wales. *Transactions of the Institute of British Geographers, New Series,* **1**, 323-345.

15 Transhumance and vegetation degradation in Iceland before 1900: a historical grazing model

Amanda Thomson

Summary

1. Extensive land degradation in Iceland has been attributed to overgrazing by livestock introduced by the Norse settlers.

2. An environmental simulation model (Búmodel) was developed to look at the influence of livestock management on land degradation in Vestur Eyjafjallahreppur in southern Iceland before 1900.

3. The model shows that inflexibility in livestock management greatly increased the risk of over-grazing under colder climate conditions. The model can be used to explore alternative management strategies.

15.1 Introduction

The country of Iceland has suffered extensive land degradation over the past 1,100 years: up to 40% of the vegetation and soil cover has been totally removed, and much of the remaining vegetated area is suffering from ongoing erosion and depleted productivity (Arnalds *et al.*, 2001). This degradation has been attributed to livestock grazing, following the introduction of large herbivores to the highly dynamic ecosystem by the first Norse settlers (see Figure 15.1).

Vegetation that is overgrazed is less resilient and more sensitive to adverse climatic fluctuations and other natural shocks. Grazing above a threshold percentage of the annual production has a negative impact upon further vegetation growth, and may increase vulnerability to vegetation degradation and soil erosion (Thorsteinsson, 1980; Evans, 1998). Therefore the ability to adjust livestock numbers and management practices in response to increasing environmental stress was crucial to farm survival. A spatial computer simulation model, Búmodel, has been developed to explore the relationship between vegetation degradation and seasonal management practices in Iceland before the introduction of modern farming practices (c. 1900).

15.2 The Icelandic pre-modern agricultural system

Pre-1900 Icelandic agriculture was based upon extensive livestock production, with a mixture of sheep, dairy cattle and horses being kept. Hay production, from manured hay meadows and uncultivated pastures, was essential for supporting the cows and some sheep during the winter months; the other livestock were grazed outside in all but the harshest weather. Upland

Thomson, A. (2005). Transhumance and vegetation degradation in Iceland before 1900: a historical grazing model. In *Mountains of Northern Europe: Conservation, Management, People and Nature*, ed. by D.B.A. Thompson, M.F. Price & C.A. Galbraith. TSO Scotland, Edinburgh. pp. 179-184.

Figure 15.1 Glaciation, volcanism, and a sub-arctic maritime climate have combined to create a highly dynamic landscape, illustrated by this river valley in the Þórsmörk National Park, north-east of Vestur Eyjafjallahreppur (Photo: Amanda Thomson).

rangelands were owned communally by the local agricultural community and were exploited for summer grazing. (This is a simplified version of the agricultural system: there were differences between regions and between centuries before 1900.)

15.3 Búmodel: a grazing simulation model

Búmodel is a spatially-explicit mathematical simulation model (Figure 15.2), producing results which can be analysed in GIS (Thomson & Simpson, in press a). Model input data come

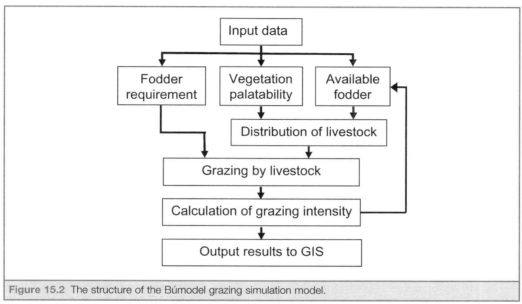

Figure 15.2 The structure of the Búmodel grazing simulation model.

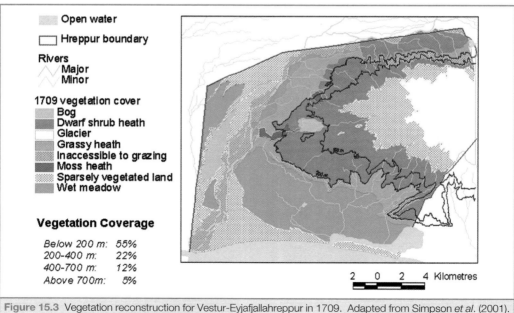

Figure 15.3 Vegetation reconstruction for Vestur-Eyjafjallahreppur in 1709. Adapted from Simpson *et al*. (2001).

from contemporary Icelandic agricultural research, historical records and palaeo-environmental evidence. The environmental inputs concern the distribution, growth parameters and palatability of vegetation in the landscape, the location of different land use zones (hayfield, outfield, rangeland) and climate. The management inputs concern the numbers and size of livestock (Aðalsteinsson, 1990), the distribution of livestock between land use zones and the management of the hayfield (not discussed in this chapter).

Búmodel was applied to the district of Vestur Eyjafjallahreppur (in southern Iceland, 20°W, 63.6°N) for 1709, when the climate entered a prolonged cool period. A farm census taken at that time (Magnússon & Vídalín, 1913-1990) lists livestock numbers on each farm. Thirty-nine farms shared a common rangeland of 139,000 ha: 2,818 lambs, yearling sheep and wethers grazed this area during the summer. The historical landscape was reconstructed (Figure 15.3) from palaeo-environmental vegetation data (Hallsdóttir, 1987), together with known altitudinal and vegetation relationships (Simpson et al., 2001).

15.4 Grazing simulation results

Búmodel was run for the rangeland zone, using constant sheep numbers grazing from June to September, under three climate scenarios with a range of mean annual temperatures derived from instrumental records: baseline (4.6°C), extreme cold (3.5°C), and warm (5.8°C).

The vegetation cover and climate scenario had the greatest impact upon grazing utilisation. The quantity of utilisable biomass (grazeable vegetation) is much reduced under extremely cold conditions, to 60% of a cell average of 390±115 kg ha^{-1} under baseline conditions. In contrast, under the warm climate scenario the amount of utilisable biomass only increased by 20% relative to the baseline amount. Grazing has a greater impact under colder conditions, as the growing season and vegetation production is reduced, and

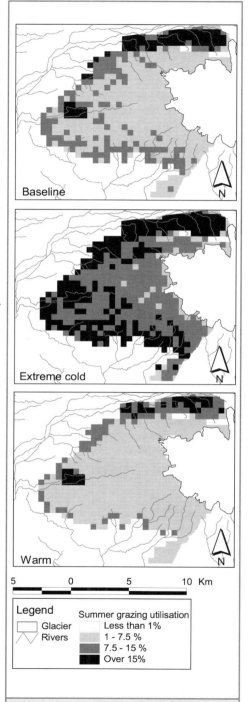

Figure 15.4 Búmodel-predicted summer utilisation of rangeland vegetation under three climate scenarios.

Figure 15.5 Heavily degraded uplands in Vestur Eyjafjallahreppur showing the effects of aeolian deposition (the orange soil 'islands') and erosion, which strips the vegetation cover and exposes the gravel beneath (Photo: Amanda Thomson).

Figure 15.6 Even heavily eroded pastures are still sometimes used for sheep grazing today (Vestur Eyjafjallahreppur), although efforts are now being made to restrict this practice (Photo: Amanda Thomson).

livestock fodder requirements increase. Much of the rangeland is covered by dwarf shrub heath, which is highly sensitive to grazing at levels greater than 15% of the annual production (Thorsteinsson, 1980). In the extreme cold scenario, 40% of cells have >15% utilisation, compared to 11% in the baseline scenario and 6% in the warm climate scenario (Figure 15.4).

15.5 Conclusions

This work demonstrates the use of modelling to investigate the interactions between vegetation, climate and management in a simple pastoral system. These results demonstrate that the Icelandic transhumance system was delicately balanced. If herd sizes were inflexible then large areas of rangeland were open to over-utilisation in cold years, and hence increasingly vulnerable to degradation and soil erosion (Figures 15.5 and 15.6). Búmodel can be used to explore the implications of different management strategies (Thomson & Simpson, in press b). It can provide a spatial and temporal perspective to the issue of overgrazing in the past, and can be used to analyze current grazing practices that can still cause problems today in Iceland and other parts of Northern Europe.

References

Aðalsteinsson, S. (1990). Importance of sheep in early Icelandic agriculture. *Acta Archaeologica,* **61**, 285-291.

Arnalds, O., Thorarinsdóttir, E.F., Metusalemsson, S., Jonsson, A., Gretarsson, E. & Arnason, A. (2001). *Soil Erosion in Iceland.* Soil Conservation Service, Agricultural Research Institute, Reykjavik.

Evans, R. (1998). The erosional impacts of grazing animals. *Progress in Physical Geography,* **22**, 251-268.

Hallsdóttir, M. (1987). *Pollen analytical studies of human influence on vegetation in relation to the Landnám tephra layer in southwestern Iceland.* Ph.D. thesis, Lund University.

Magnússon, Á. & Vídalín, P. (1913-1990). *Jarðabók. Vol. 1-13.* Hið Íslenzka Fræðafélag, Copenhagen.

Simpson, I.A., Dugmore, A.J., Thomson, A. & Vésteinsson, O. (2001). Crossing the thresholds: human ecology and historical patterns of landscape degradation. *Catena,* **42**, 175-192.

Thomson, A.M. & Simpson, I.A. (in press a). Modelling the impact of historical land management decision in sensitive landscapes. The design and validation of a grazing model for Iceland. *Environmental Modelling and Software.*

Thomson, A.M. & Simpson, I.A. (in press b). Modelling historical rangeland management and grazing pressures, Mývatnssveit Iceland. *Human Ecology.*

Thorsteinsson, I. (1980). Grazing intensity – proper use of rangelands. *Íslenzkar Landbúnaðarransóknir,* **12**, 113-122.

16 Terrain sensitivity on high mountain plateaux in the Scottish Highlands: new techniques

Stefan M. Morrocco

Summary

1. High mountain plateaux in the Scottish Highlands are widely regarded as being sensitive to disturbance, but assessment of their sensitivity has hitherto been exploratory and qualitative.
2. Measuring terrain performance under applied stress is an efficient and accurate method of quantifying terrain sensitivity.
3. The degree of sensitivity between mountain geo-complexes varies considerably, with moss-dominated communities being at least four times more sensitive to disturbance than grass-dominated communities.
4. This research points the way to broader work to assess the sensitivity of mountain landscapes to natural and human influences.

16.1 Introduction

High mountain plateaux in the Scottish Highlands exhibit a wide range of responses to both natural stresses and human impacts. Assessment of terrain sensitivity has, however, to date remained exploratory (Gordon *et al.*, 1998; Haynes *et al.*, 2001; Thompson *et al.*, 2001), potentially limiting the effectiveness of conservation management and remedial strategies for high plateau environments. In Scotland, the high mountain plateaux are areas of moderately undulating to flat land above approximately 800 m above sea level (asl) in the Eastern Highlands and generally above 600 m asl in the Western Highlands. These areas are characterised by a mosaic of plant communities, substrate types and topographical features that form small scale geocomplexes of widely differing sensitivity to stress. The term substrate refers to any non-living minerogenic and organic material overlying bedrock, such as *in situ* material derived from the weathering of bedrock, aeolisols (aeolian transported and deposited material), soils such as podzols and brown soils, and organic soils such as blanket peat. Terrain sensitivity is a measure of the resistance and resilience of such geo-complexes to stress, in terms of both geomorphological and ecological sensitivity (Brunsden, 2001; Miles *et al.*, 2001; Thompson *et al.*, 2001; Welch *et al.*, 2005).

This chapter introduces new techniques for quantifying terrain sensitivity on high plateaux in the Scottish Highlands, presents some preliminary results associated with these techniques, and discusses the possible significance of these results. A selection of 12 high

Morrocco, S.M. (2005). Terrain sensitivity on high mountain plateaux in the Scottish Highlands: new techniques. In *Mountains of Northern Europe: Conservation, Management, People and Nature*, ed. by D.B.A. Thompson, M.F. Price & C.A. Galbraith. TSO Scotland, Edinburgh. pp. 185-190.

plateaux (Figure 16.1) was used to assess terrain sensitivity on as broad a range of plateau types as possible.

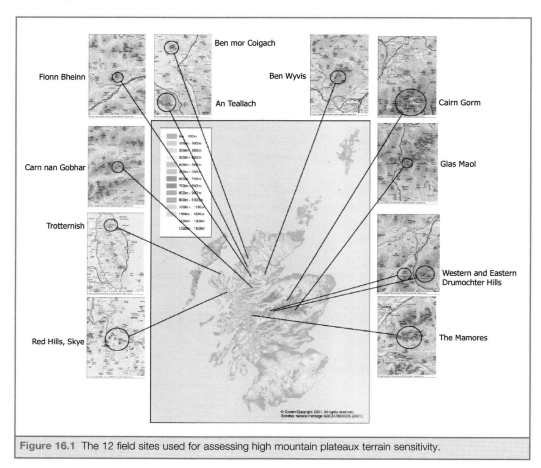

Figure 16.1 The 12 field sites used for assessing high mountain plateaux terrain sensitivity.

16.2 New techniques for assessing terrain sensitivity

The response of terrain to stress falls into two categories: a mechanical response and a physiological response. My work focuses on the response of the terrain to applied physical stress. Repeated measurements of terrain resistance (response to stress) were made using three instruments: a) a shear rake measured the tensile strength of the vegetation mat, b) a shear vane measured the shear strength of the underlying substrate, and c) a hand-held penetrometer measured the compressional strengths of the substrate.

16.3 Preliminary results

Preliminary analyses of the work show that significant variation in vegetation mat resistance exists between the different vegetation communities on all plateaux. Figure 16.2 illustrates this variation for the Cairn Gorm/Ben Macdui plateau. The two grass heath communities (National Vegetation Classification types U7 and U7a; see Averis *et al.* (2004)) have the strongest vegetation mats while the rush heath (U9) has the weakest vegetation mat. The

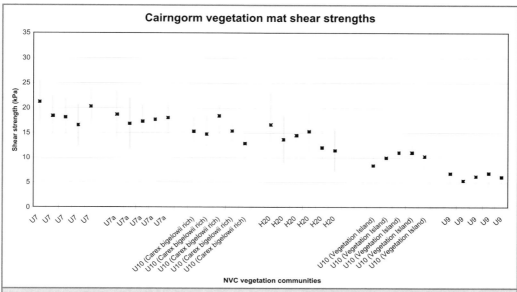

Figure 16.2 Shear strength values for six plant communities on the Cairn Gorm-Ben Macdui plateau. The central points represent the mean of a data set of 30 measurements; each error bar represents one standard deviation from the mean. The plant communities have been given their National Vegetation Classification codes (Rodwell, 1992; Averis *et al.*, 2004).

dominant species of the six vegetation communities are summarised in Table 16.1. The other plant communities have resistance characteristics which lie somewhere between these two extremes. These results show that variation in the shear strength of the vegetation mat also exist *within* individual communities.

Plateau substrate shows a general increase of strength with depth at all field sites for both shear and compressional strength, especially where the regolith is matrix-supported. Figure 16.3 shows shear and compressional strengths for three pits on Ben Wyvis. Figures 16.3a and 16.3b show that the U7 grass-heath and healthy U10 moss-heath vegetation mats have a higher shear strength than the underlying substrate, which then gradually increases in strength with depth. Figure 16.3c shows that damaged (poached and wind torn) U10 moss-heath has a lower shear strength than the underlying substrate and that the strength of the regolith underlying this association only marginally increases with depth.

16.4 Discussion

The variations in shear strength between the different vegetation communities, depicted by Figure 16.2, means it is possible to identify which areas are most likely to undergo change or damage wherever there is an increase in extrinsic stress. Figure 16.2 shows that the plant communities dominated by moss species, such as *Racomitrium lanuginosum,* have the lowest vegetation mat strengths and are thus most susceptible to physical disturbance under stress. The variation in shear strength is partly a reflection of the root structure of different species. The moss carpet, which lacks penetrating roots (Tallis, 1959; Averis *et al.*, 2004), is easily exfoliated or eroded, while those plants with rhizomes and roots that bind with the underlying soil are much stronger (see also Welch *et al.*, 2005). This contrast also explains

Table 16.1 The dominant species of the six NVC plant communities investigated on the Cairn Gorm-Ben Macdui plateau (Rodwell, 1992; Averis *et al.*, 2004).

NVC vegetation communities	Dominant species
U7	*Nardus stricta* *Carex bigelowii* *Diphasiastrum alpinum* *Polytrichum alpinum* *Cetraria islandica*
H20	*Vaccinium myrtillus* *Empetrum nigrum spp. hermaphroditum* *Racomitrium lanuginosum* *Rhytidiadelphus loreus* *Hylocomium splendens* *Pleurozium schreberi* *Hypnum jutlandicum*
U7a	*Empetrum nigrum spp. hermaphroditum* *Cetraria islandica* *Trichophorum cespitosum* *Juncus squarrosus* *Vaccinium uliginosum* and the typical U7 species
U10 (*Carex bigelowii* rich)	*Carex bigelowii*
U9	*Juncus trifidus* *Carex bigelowii* *Racomitrium lanuginosum*
U10 (vegetation islands)	*Racomitrium lanuginosum* *Carex bigelowii* *Vaccinium myrtillus* *Deschampsia flexuosa* *Festuca vivipara* *Galium saxatile* *Salix herbacea*

variations in shear strength *within* some communities. For example, the U10 community may be moss-dominated in some areas but sedge-dominated in others, causing wide variations in the shear strength of the vegetation mat.

Once the protective vegetation cover is removed, the underlying substrate is exposed. Figure 16.3 shows that the most sensitive part of most substrate profiles is the upper 10 cm. Therefore, any vegetation disturbance is likely to trigger continued erosion of the underlying substrate, the degree of erosion being dependent on a number of factors such as substrate characteristics, location and the magnitude and scale of the stress being applied. Figure 16.3

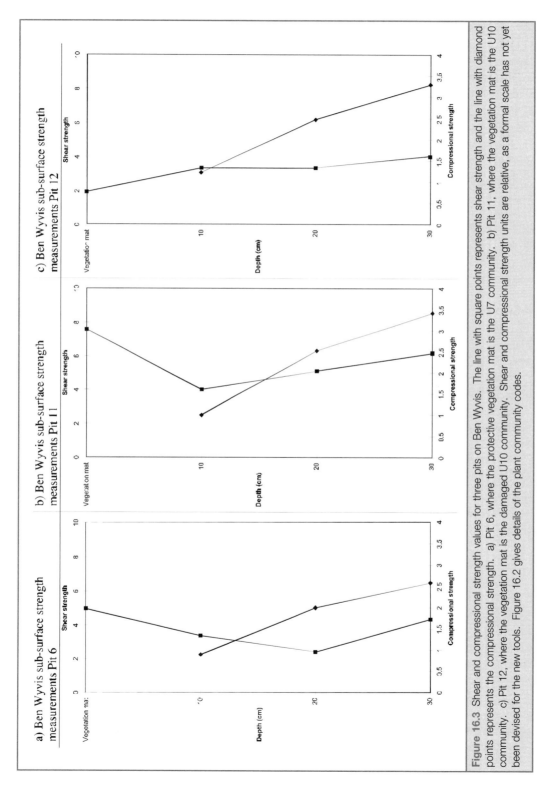

a) Ben Wyvis sub-surface strength measurements Pit 6

b) Ben Wyvis sub-surface strength measurements Pit 11

c) Ben Wyvis sub-surface strength measurements Pit 12

Figure 16.3 Shear and compressional strength values for three pits on Ben Wyvis. The line with square points represents shear strength and the line with diamond points represents the compressional strength. a) Pit 6, where the protective vegetation mat is the U7 community. b) Pit 11, where the vegetation mat is the U10 community. c) Pit 12, where the vegetation mat is the damaged U10 community. Shear and compressional strength units are relative, as a formal scale has not yet been devised for the new tools. Figure 16.2 gives details of the plant community codes.

also shows that weak vegetation tends to overlie weak substrate (Figure 16.3c). As such, this vegetation cover is highly susceptible to erosion, and therefore such terrain is vulnerable to erosion to depths of several centimetres before a more resistant substrate is exposed.

The significance of these preliminary results, in terms of conservation management, is that they highlight those geo-complexes that are most sensitive to mechanical stress. The potential impact of activities such as walking, footpath development, other recreational pursuits and grazing on different terrain types can therefore be assessed.

Measuring terrain performance under stress has proved to be an efficient and accurate method of assessing terrain sensitivity on high plateaux in the Scottish Highlands. The preliminary results indicate that significant variation exists in the sensitivity to disturbance of the different geo-complexes throughout the high mountain plateaux of the Scottish Highlands. Variation in shear strength within vegetation communities also exists and may be explained by changes in the abundance and composition of particular species over an area of plateau covered by the community. Vegetation with high shear strength values appears to provide a protective cover to the weaker underlying substrate, which would be very susceptible to erosion if the vegetation mat were removed.

Acknowledgements

This work is part of a Ph.D. study funded by the University of St Andrews and Scottish Natural Heritage with financial assistance for field work provided by the Bill Bishop Memorial Trust, BGRG, Carnegie Trust, QRA, RSGS and the Russell Trust (University of St Andrews). I thank Colin Ballantyne, John Gordon and Des Thompson for comments on a draft of this manuscript.

References

Averis, A.M., Averis, A.B.G., Birks, H.J.B., Horsfield, D., Thompson, D.B.A. & Yeo, M.J.M. (2004). *An Illustrated Guide to British Upland Vegetation.* Joint Nature Conservation Committee, Peterborough.

Brunsden, D. (2001). A critical assessment of the terrain sensitivity concept in geomorphology. *Catena,* **42**, 99-123.

Gordon, J.E., Macdonald, R., Thompson, D.B.A., Haynes, V.M. & Brazier, V. (1998). Environmental sensitivity and conservation management in the Cairngorm Mountains, Scotland. *Ambio,* **27**, 335-344.

Haynes, V.M., Grieve, I.C., Gordon, J.E., Price-Thomas, P. & Salt, K. (2001). Assessing geomorphological sensitivity of the Cairngorm high plateaux for conservation purposes. In *Earth Science and the Natural Heritage, Interactions and Integrated Management,* ed. by J.E. Gordon & K.F. Leys. The Stationery Office, Edinburgh. pp. 120-123.

Miles, L., Cummins, R.P., French, D.D., Gardner, S., Orr, J.L. & Shewry, M.C. (2001). Landscape sensitivity: an ecological view. *Catena,* **42**, 125-141.

Rodwell, J.S. (Ed) (1992). *British Plant Communities vol. 1-4.* Cambridge University Press, Cambridge.

Tallis, J.H. (1959). Studies in the biology and ecology of *Racomitrium lanuginosum* Brid. II. Growth, reproduction and physiology. *Journal of Ecology,* **47**, 325-350.

Thompson, D.B.A., Gordon, J.E. & Horsfield, D. (2001). Montane landscapes in Scotland: are these natural artefacts or complex relics. In *Earth Science and the Natural Heritage, Interactions and Integrated Management,* ed. by J.E. Gordon & K.F. Leys. The Stationery Office, Edinburgh. pp. 105-119.

Welch, D., Scott, D. & Thompson, D.B.A. (2005). Changes in the composition of *Carex bigelowii-Racomitrium lanuginosum* moss heath in Glas Maol, Scotland, in response to sheep grazing and snow fencing. *Biological Conservation,* **122**, 621-631.

17 Mapping the inherent erosion risk due to overland flow: a tool to guide land management in the Scottish uplands

Allan Lilly, John Gordon, Monica Petri & Paula Horne

Summary

1. There is some evidence that erosion rates in the Scottish uplands are increasing, and organic soils appear to be particularly vulnerable.
2. Upland organic soils are an important carbon store and the loss of this carbon has consequences for global climate change.
3. A GIS model based on soil texture, soil hydrology, slope angle, type of organic layer and land use has been developed to predict the spatial distribution of the risk of erosion with the aim of guiding the management of priority peatland habitats.

17.1 Introduction

Soil erosion in the Scottish uplands is a natural phenomenon, reflecting the over-steepened, glaciated slopes; a climate dominated by low winter temperatures, high rainfall and high wind speeds, and often fragile vegetation. However, land uses such as grazing and forestry, as well as human trampling, can accelerate the natural erosion rate and it is often difficult to distinguish between natural and anthropogenically induced erosion (e.g. Thompson *et al.*, this volume). In order to inform appropriate management responses to the effects of soil erosion on the natural heritage, there is a need to develop a twin-track approach, involving erosion risk modelling as well as experimental work on terrain sensitivity and the role of factors such as regolith and vegetation strength. This chapter addresses the former; the latter is addressed by Morrocco (this volume).

17.2 Geographical extent of soil erosion in Scotland

Soil erosion in the uplands is evident in a variety of forms, including debris flows, deflation surfaces, peat haggs and footpath widening, leading to habitat loss and increased sediment loads in streams. A study of the spatial extent of soil erosion in upland Scotland using aerial photographs and based on a sample of grid squares covering 20% of the upland area (defined as land in Land Capability for Agriculture classes 5, 6 and 7, after Bibby *et al.*, 1982) revealed that 12% of the sampled area was affected by soil erosion (Grieve *et al.*, 1994, 1995). Peat erosion was the most extensive type, covering 6% of the area sampled. The most severely eroded areas are in the peats of the Monadhliath Mountains, the eastern Southern Uplands and the eastern Grampians. Peat erosion is also extensive in Shetland (Birnie & Hulme, 1990; Birnie, 1993).

Lilly, A., Gordon, J., Petri, M. & Horne, P. (2005). Mapping the inherent erosion risk due to overland flow: a tool to guide land management in the Scottish uplands. In *Mountains of Northern Europe: Conservation, Management, People and Nature,* ed. by D.B.A. Thompson, M.F. Price & C.A. Galbraith. TSO Scotland, Edinburgh. pp. 191-196.

17.3 Trends in soil erosion in Scotland

Ballantyne (2004) showed that debris flow events have occurred over most of the last 7,000 years, although the frequency and extent appear to have increased in the last few centuries. Possible causes include more intense rainstorm events, woodland clearance, heather burning and increased stocking densities (Evans, 1998). Davidson & Grieve (2004) summarised a number of individual case studies of upland soil erosion, but noted that assessment of any wider trends was generally hampered by a lack of monitoring data at a national scale. McHugh *et al.* (2002) suggested a 5-yearly monitoring period to assess long-term erosion rates in the uplands of England and Wales.

17.4 Erosion of organic and organo-mineral soils

Erosion of soils with organic (or peaty) surface layers is a particular concern as these soils are an important carbon store. Changes in climate and land use can reduce this store through losses as dissolved organic carbon (DOC) and by oxidation of the peat, leading to increased emissions of carbon dioxide, a greenhouse gas. Over 50% of Scotland is covered by soils that have an organic surface layer. There are many processes that initiate erosion in these soils and the response to these different pressures varies with soil type. Grazing by domestic and wild animals may play a key role in either the initiation of peat erosion or in its exacerbation (Evans, 1998). Removal of the vegetation cover by trampling or muirburn may also increase the susceptibility of upland peats to erosion, while the use of gullies as shelter or pathways may enhance existing erosion. Many areas of blanket bog in the Grampians are severely eroded, particularly where deer densities have been very high in the past. Clearly historical land use, stocking densities and past muirburn are crucial to understanding present-day patterns of erosion.

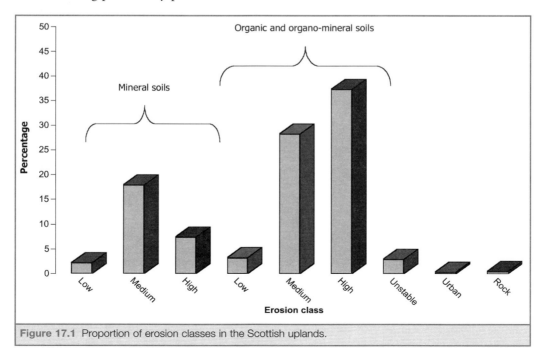

Figure 17.1 Proportion of erosion classes in the Scottish uplands.

There is an incomplete understanding of the dynamics of erosion in these soils once initiated (e.g. the role of direct oxidation of bare surfaces, desiccation and subsequent removal by wind and erosion by overland flow), and of the potential for recovery. Further work on the causes and dynamics of peat erosion is necessary for the development and implementation of management strategies for peatland priority habitats.

In order to understand the nature of soil erosion in the Scottish uplands as well as elsewhere, there is a need to identify both the inherent risk of erosion as well as those management factors that can increase the risk of erosion.

17.5 Mapping of the inherent soil erosion risk due to overland flow

Large areas of the Scottish uplands are potentially at risk from soil erosion by overland flow, particularly if the vegetation cover is damaged. A rule-based GIS model has been developed to map the inherent risk of soil erosion by overland flow in Scotland (Lilly *et al.*, 2002). The model combined data on slope angle, soil texture and soil hydrology to give a baseline estimate of the inherent sensitivity of soil to erosion. This shows that over 44% of the Scottish uplands, as defined in SNH's Natural Heritage Futures framework (Scottish Natural Heritage, 2002), have soils in the high erosion risk category (Figure 17.1). However, many upland areas have a vegetation cover that is likely to be permanent, while others are subject to land use practices, such as muirburn, that will periodically expose the soil surface.

Land cover data (Macaulay Land Use Research Institute, 1993) were categorised into the likely periodicity of vegetation disturbance or removal, and subsequently used to modify the inherent erosion risk in order to identify those areas most likely to experience anthropogenically enhanced rates of erosion (Figure 17.2). This soil erosion risk map provides a useful tool to guide upland land management and to help to reduce the risk of accelerated erosion through human activities. Future development of the model should concentrate on improving the predictions of the susceptibility to erosion of organic surface layers through the study of the causes and dynamics of peat erosion and the inclusion of stocking densities.

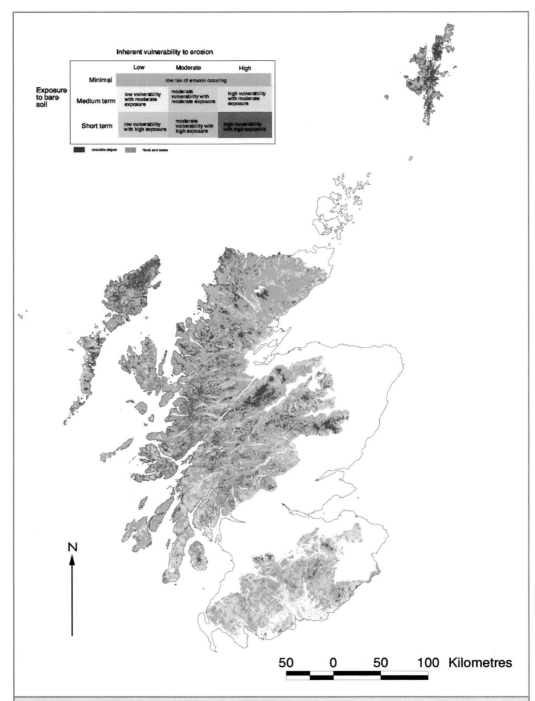

Figure 17.2 Soil erosion risk due to overland flow in the Scottish uplands. Inherent vulnerability to erosion is based on slope angle, runoff potential and soil texture (Lilly *et al.*, 2002). Potential for exposure to disturbance is derived from a reclassification of the *Land Cover of Scotland* dataset, based on the potential for disturbance to the vegetation cover (Macaulay Land Use Research Institute, 1993).

Acknowledgement

We acknowledge comments from Dr Bob Evans.

References

Ballantyne, C.K. (2004). Geomorphological changes and trends in slope processes: debris-flows. Scottish Natural Heritage Commissioned Report No. 052 (F00AC107A).

Bibby, J.S., Douglas, H.A., Thomasson, A.J. & Robertson, J.S. (1982). *Land Capability Classification for Agriculture*. Soil Survey of Scotland Monograph. The Macaulay Institute for Soil Research. Aberdeen.

Birnie, R.V. (1993). Erosion rates on bare peat surfaces in Shetland. *Scottish Geographical Magazine*, **109**, 12-17.

Birnie, R.V. & Hulme, P.D. (1990). Overgrazing of peatland vegetation in Shetland. *Scottish Geographical Magazine*, **106**, 28-36.

Davidson, D.A. & Grieve, I.C. (2004). Trends in soil erosion. Scottish Natural Heritage Commissioned Report No. 054 (F00AC106).

Evans, R. (1998). The erosional impacts of grazing animals. *Progress in Physical Geography*, **22**, 251-268.

Grieve, I.C., Hipkin, J.A. & Davidson, D.A. (1994). Soil erosion sensitivity in upland Scotland. Scottish Natural Heritage Research, Survey and Monitoring Report No. 24.

Grieve, I.C., Davidson, D.A. & Gordon, J.E. (1995). Nature, extent and severity of soil erosion in upland Scotland. *Land Degradation and Rehabilitation*, **6**, 41-55.

Lilly, A., Hudson, G., Birnie, R.V. & Horne, P.L. (2002). The inherent geomorphological risk of soil erosion by overland flow in Scotland. Scottish Natural Heritage Research, Survey and Monitoring Report, No. 183.

Macaulay Land Use Research Institute (1993). *The Land Cover of Scotland 1988*. Macaulay Land Use Research Institute, Aberdeen.

McHugh, M., Harrod, T. & Morgan, R. (2002). The extent of soil erosion in upland England and Wales. *Earth Surface Processes and Landforms*, **27**, 99-107.

Scottish Natural Heritage (2002). *Natural Heritage Futures: an Overview*. Scottish Natural Heritage, Perth.

18 Mapping downhill skiing's environmental impact: evaluating the potential of remote sensing in Scotland

Will Cadell & K.B. Matthews

Summary

1. For remote areas with heterogeneous land cover, it is frequently more efficient to use remotely sensed imagery to identify areas that have been, or are at risk of being, damaged.

2. Image analysis reduces the need for expensive ground-based survey and provides an explicitly spatial data source that may be used as a benchmark for monitoring and for management support. This chapter presents the integration of GPS survey, image analysis and Geographic Information Systems (GIS) to identify areas of vegetation damage in the Glenshee ski centre.

3. The chapter concludes that the tools and methods have significant potential benefits for natural heritage management applications.

18.1 Introduction

The use of Scottish mountain areas for recreation is increasingly popular (Mackay, this volume). This popularity can mean that fragile ecosystems may be damaged, particularly when visitor numbers are concentrated close to facilities such as ski centres (Bayfield, 1980). Field-based studies have been conducted on the soil erosion and vegetation damage near ski lifts, which have shown that the damage can be significant and extend beyond demarcated ski runs (Watson, 1985; Welch *et al.*, 2005).

Remote sensing techniques enable the analysis of large areas with a greatly reduced need for labour-intensive and expensive fieldwork. The development of software which allows the creation of high resolution maps from aerial photography (orthorectification), combined with developments in image analysis methods, provides sources of information for natural heritage managers at greatly reduced cost (Wright *et al.*, 2003). Imagery obtained from light aircraft platforms is particularly useful in mountainous regions where satellite imagery is frequently obscured by cloud.

Conventional image analysis methods classify each pixel of an image into a single land cover class. This process is prone to error, particularly when significant classes are intimately mixed (Richards & Jia, 1999). An alternative is image segmentation, where the image is divided into polygons using spectral, morphological and contextual information (Willhauck, 2000). This is a particularly appropriate methodology for identifying fragmentation of land covers that may indicate damage.

Cadell, W. & Matthews, K.B. (2005). Mapping downhill skiing's environmental impact: evaluating the potential of remote sensing in Scotland In *Mountains of Northern Europe: Conservation, Management, People and Nature,* ed. by D.B.A. Thompson, M.F. Price & C.A. Galbraith. TSO Scotland, Edinburgh. pp. 197-202.

18.2 Materials and methods

The imagery used was for the Glenshee ski centre (Figure 18.1a), taken in June 1996 at approximately 1:10,000 scale. This was captured using a vertical, medium-format, uncalibrated camera with 40 mm wide-angle lens, mounted on light aircraft platform. The imagery was orthorectified using *Erdas OrthoBASE* and analysed using *eCognition*. The orthorectification process creates a photomap using ground control and digital elevation model (DEM) data. Ground control was obtained from O.S. Landline, a Profile based DEM and a differential-GPS (dGPS) survey. The dGPS survey was principally of benefit in determining the elevation (to an accuracy of ±2 m) of features visible on the aerial photography but not mapped on the O.S. Landline coverage. The photomap created has a ground resolution of 1 m and a location precision of ±1.5 m (Figure 18.1b).

The eCognition analysis has two stages: segmentation and classification. Segmentation breaks the image down into a series of land cover polygons. The segmentation algorithm has three parameters: scale, colour and shape. Scale controls the average size of polygons and thus their heterogeneity. Colour and shape determine the relative importance of reflectance or morphology in defining polygons. eCognition was set up to derive small polygons (average size 500 m²), with a bias in favour of colour rather than shape for the segmentation (0.6/0.4). This gave a segmentation that reflects the fairly high degree of fragmentation of the land covers present and emphasises the differences in colour between the heather and grass classes.

The polygons derived from the segmentation are interpreted using information from a dGPS site survey of land cover types. The site survey locates examples of cover types and these are then linked to the polygons defined in the segmentation phase. The minimum dimension for a sample site to ensure that it can be identified in the segmented imagery is either the accuracy of the orthophotograph (±1.5 m) or the field navigation (dGPS ±1 m), whichever is larger. The sites chosen should thus be at least 3 x 3 m and for safety a minimum size of 5 x 5 m was used, equating to 5 by 5 pixels in the original orthophotograph. Since the land cover survey was conducted in May 2002 there is a significant risk that there will have been land cover change in the years since the photography was obtained. In order to minimise the possible misclassification impacts, the survey sites were located as far from boundaries as possible and avoided any areas where disturbance appeared to be recent.

The polygons of the segmented image are classified using a fuzzy classifier that determines, for each polygon, the membership probability for each of the land cover classes in the classification scheme. This analysis explicitly represents the uncertainty about which class(es) are present in any polygon. The resulting classification allows the mapping of the classes with the highest probability for each polygon (a conventional classification map) and probability maps for the individual classes.

18.3 Results

Defining the segmentation parameters and the classification scheme is an iterative process and relies heavily on expert interpretation of both the features of interest and the underlying causes. In the Glenshee study, four principal classes were used: exposed-peat, heather, grass and mineral (screes). Five mixture classes were also identified: heather-peat, grass-peat, grass-heather, heather-mineral and grass-mineral. Open water, snow and mineral modified

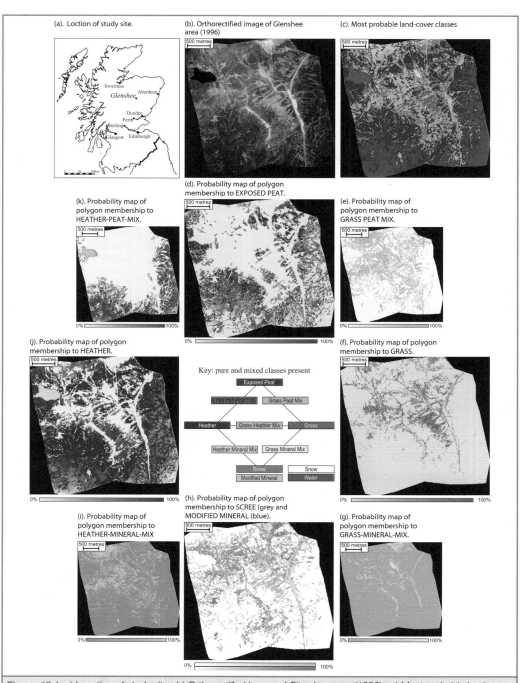

Figure 18.1 a) Location of study site. b) Orthorectified image of Glenshee area (1996). c) Most probable land cover classes. d) Probability map of polygon membership to exposed peat. e) Probability map of polygon membership to grass-peat mix. f) Probability map of polygon membership to grass. g) Probability map of polygon membership to grass-mineral mix. h) Probability map of polygon membership to mineral (grey) and modified mineral (blue). i) Probability map of polygon membership to heather-mineral mix. j) Probability map of polygon membership to heather. k) Probability map of polygon membership to heather-peat mix. The key shows pure and mixed classes present.

by construction classes completed the scheme. These classes and their inter-relationships are set out in the legend to Figure 18.1c-k.

Figure 18.1c presents a map of the highest probability class for each polygon. From this map it is possible to contrast the non-skied and skied areas. The non-skied areas have either a heather matrix with compact grass and exposed peat patches or sub-alpine land cover with mineral elements dominant. The skied areas are associated with a grass or grass mixture matrix and have strong linear features associated with the ski infrastructure.

Figure 18.1d-k shows the probability maps for principal and mixture classes. From these maps it can be seen that there is a considerable degree of complexity to the pattern of land cover within the study area. The maps are particularly helpful in highlighting the importance of soils in directly influencing land cover classifications since there are significant areas where the exposed-peat or mineral classes and their mixtures occur with high probability. While such classes may indicate damaged vegetation, there are examples of each mixture occurring naturally within the study area, for example the heather-mineral mix associated with wind-clipped heather on the summits.

18.4 Interpretation

The key to interpretation of damage classes from the land cover maps is contextual information. The GIS provides an analytical framework within which that interpretation can take place. Figure 18.2 presents three graphs that breakdown the percentages of land cover within all areas of the study site, the skied areas and the pisted areas. Two maps showing the pattern of land cover for the skied and pisted areas are also presented. From the graphs it can be seen that there is an increased proportion of grass and grass-peat-mix at the expense of heather for both the skied and pisted areas. For the pisted areas the grass classes dominate.

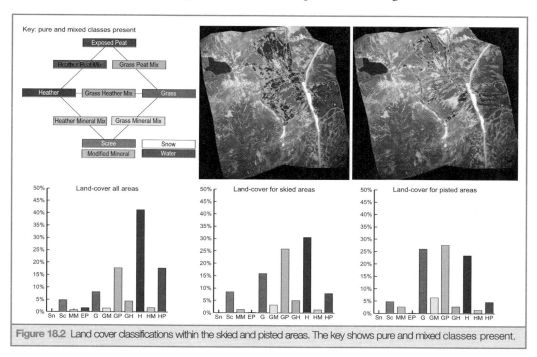

Figure 18.2 Land cover classifications within the skied and pisted areas. The key shows pure and mixed classes present.

While the increase in the grass classes probably indicates that change has occurred, without multi-temporal imagery it is difficult to make definitive statements on the dynamics of change. If the grass areas are reseeded areas that have been stabilized then, while damage has occurred, it is not ongoing and the trade-off of economic benefits from the ski centre against environmental degradation can be judged.

The grass-peat areas also present difficulties of interpretation since these may indicate areas in which skiing and piste management are thinning the vegetation so that the underlying peat is exposed and is then eroding. Alternatively these may be areas that have recently been reseeded and are undergoing regeneration. Of potentially greater concern would be the smaller but significant areas of grass mineral-mix. While these do occur naturally in some of the higher altitudes (Figure 18.1g), the areas associated with the pistes may indicate significant erosion. Since the peat is relatively thin in several areas of the site (<30 cm) it is possible that this has been stripped away to leave the mineral subsoil that is supporting thin grass.

Visualisation may also have role to play when there are complex spatial relationships between phenomena. Figure 18.3 shows a 3-D viewshed with the non-skied areas masked and the grass-mineral mix class overlaid. From this visual it is possible to see that, while the grass-mineral mix occurs naturally as small polygons within a larger matrix, the larger area at the confluence of several pistes probably indicates ongoing damage caused either by shaving or piste grooming.

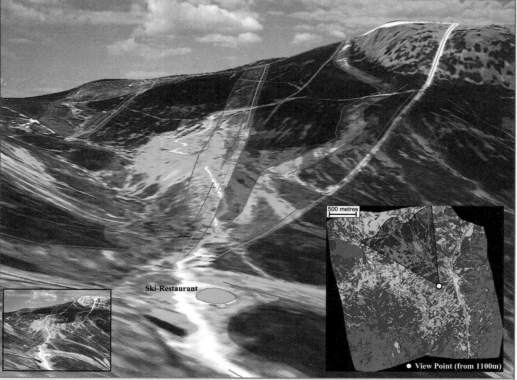

Figure 18.3 3-D viewshed showing grass-mineral mix class with non-skied areas masked, snow fences in red and uplifts in blue.

18.5 Conclusions

Remote sensing has been used previously as a tool in natural heritage management (Lillesand & Kiefer, 2000) but the combination of aerial imagery, dGPS, orthorectification, image segmentation and fuzzy classification presents significant opportunities for natural heritage management. The proposed methodology could be significantly improved, however, if the mixture classes did not have to be predefined or identified in the field. While it is relatively easy to identify areas of homogeneous vegetation that can serve as signatures for a land cover class, obtaining signatures for the mixture classes, with their heterogeneous composition, is time consuming and prone to error.

With sub-pixel classification of the image segmentation, it would be possible not only to identify changes in polygon boundaries but also to receive early warning about reductions of proportions of sensitive land cover types within polygons. The measurement of the dynamics of impact or the testing of hypotheses about the classes of land cover most susceptible to damage does, however, depend on the availability of multi-temporal photographic coverage.

While the applications to date have focused on downhill skiing's impact on the mountain environment, it is possible to envisage a much wider range of applications, such as path erosion, depending on the scale and quality of imagery available. The methods developed would be of particular benefit if used proactively as part of the process of evaluating proposals for future developments and management.

Acknowledgements

The research presented here is funded by grant-in-aid from SEERAD, and the aerial photography was supplied by SNH.

References

Bayfield, N. (1980). Replacement of vegetation on disturbed ground near ski lifts in the Cairngorm Mountains, Scotland. *Journal of Biogeography*, 7, 249-260.

Lillesand, T.M. & Kiefer, R.W. (2000). *Remote Sensing and Image Interpretation.* Fourth edition. Wiley, New York.

Richards, J.A. & Jia, X. (1999). *Remote Sensing Digital Image Analysis.* Third edition. Springer, Berlin.

Watson, A. (1985). Soil erosion and vegetation damage near ski lifts at Cairn Gorm, Scotland. *Biological Conservation*, **33**, 363-381.

Welch, D., Scott, D. & Thompson, D.B.A. (2005). Changes in the composition of *Carex bigelowii-Racomitrium lanuginosum* moss heath in Glas Maol, Scotland in response to sheep grazing and snow fencing. *Biological Conservation*, **122**, 621-631

Willhauck, G. (2000). Comparison of object classification techniques and standard image analysis for the use of change detection between SPOT multi-spectral satellite images and aerial photos. In *International Society of Photogrammetry and Remote Sensing, Vol. XXXIII, Part B3.* International Society of Photogrammetry and Remote Sensing, Amsterdam. pp. 35-42.

Wright, G.G., Matthews, K.B., Cadell, W.M. & Milne, R. (2003). Reducing the cost of multi-spectral remote sensing: combining near-infrared video imagery with colour aerial photography. *Computers and Electronics in Agriculture*, **38**, 175-189.

19 Mountain tourism in Northern Europe: current patterns and recent trends

Peter Fredman & Thomas A. Heberlein

Summary

1. In many parts of the Northern European mountain region economies have been heavily dependent on extractive industries. But with logging, farming, mining and fishing declines in labour needs, tourism often takes on a new meaning for such communities.

2. Today, tourism has become an important means of potential economic and social development, while there are also concerns about the environmental impacts of tourism.

3. To understand the complex social and natural systems that many mountain regions feature, one must first describe their components, but there is often a lack of comprehensive data about tourism and recreation. This chapter focuses on recreation patterns and trends in mountain tourism in Sweden for the last 20 years, and makes comparisons with other mountain regions of Northern Europe.

4. In Sweden, mountain tourism has both increased and moved south, closer to the large urban regions. Much of the increase is ascribed to a growth in mechanized winter activities such as snowmobiling and downhill skiing. Traditional mountain outdoor activities like hiking and cross-country skiing have been stable or declining over time. Similar patterns of change are found in Finland, Norway and Iceland.

5. In the analysis, both societal changes and activity-specific changes are used to explain these changes in mountain tourism.

19.1 Introduction

Tourism is not a static phenomenon. It changes as society changes. During the last century, mountain tourism has developed to be a powerful economic force in some parts of the world, but expectations about tourism development have sometimes exceeded the real benefits (Weiermair *et al.*, 1996; Godde *et al.*, 2000). Except for remoteness and natural beauty, mountain areas have little economic potential after the physical and biological resources are extracted, and it is common to turn to tourism in an effort to maintain human communities. It is sometimes argued that the quality of the natural landscape is an essential part of a community's economic well-being (Power, 1996). Sweden is no exception. Only

Fredman, P. & Heberlein, T.A. (2005). Mountain tourism in Northern Europe: current patterns and recent trends. In *Mountains of Northern Europe: Conservation, Management, People and Nature*, ed. by D.B.A. Thompson, M.F. Price & C.A. Galbraith. TSO Scotland, Edinburgh. pp. 203-212.

about 150,000 people live in the 15 mountain municipalities – which cover approximately one-third of the country – and this population is declining. The region features large employment declines in forestry, mining and other extractive industries during the last 30–40 years, while current regional policies emphasise tourism (Statistics Sweden, 2000). Other comparable mountain regions in Northern Europe characterized by similar conditions are found in Norway, Finland, Iceland and Scotland.

Just as mountain communities change over time, so does tourism. While much of the early mountain tourism was a summer phenomenon, there was a strong increase in winter tourism starting in the 1950s. Since the end of the 1980s, the international demand for winter holidays in mountain areas has been relatively stable, while the demand for summer holidays has decreased (Freitag, 1996). How do we explain these changes in tourism patterns? Most of the data available are cross-section surveys that measure participation, and a main problem in assessing change and the potential causes of change is the lack of comparable long-term data. But long-term recreation data are essential for assessing visitor impacts, facilities planning, and estimating the economic contribution that tourism provides, as well as the value of the recreation experience to the visitors themselves (Loomis, 2000).

In this chapter we go beyond the current set of one shot-case studies in tourism research and to explore the changes in tourism patterns in the Swedish mountains for the last 20 years. This is possible by replicating, 16 years later, a national survey of Swedish mountain tourism conducted in 1984. Comparing the two national surveys will allow us to determine if tourism is increasing or decreasing and how activities and destinations are changing. We also contrast the changes identified in the Swedish mountains with some comparable regions in Finland, Norway and Iceland.

Documenting change is a start, but it is important to understand what accounts for the change. To that end we examine two factors that can be associated with change: societal changes and activity-specific changes. Mountain tourism often involves strenuous activities and one would anticipate declines in participation as the population ages. For some outdoor recreation activities, a positive correlation between income and participation is expected. There is evidence that people's time demands are increasing (Schor, 1992) and that people are taking shorter vacations. We would expect this to be manifest in mountain tourism. Independent of these broader influences, there are changes in the activities themselves. Outdoor equipment has improved in performance and diversity and has become more reliable. Specific activity-focused social programs and government subsidies might be initiated which promote certain activities in certain regions, etc. These factors could affect both use levels and use patterns.

19.2 Study method and region

Data for this study primarily come from two sources. The first study, by the Swedish Environmental Protection Agency in 1984 (Naturvårdsverket, 1985), surveyed 2,850 randomly chosen Swedish citizens with a response rate of 75%. The second survey, by the European Tourism Research Institute (ETOUR), in 2000 surveyed 2,145 Swedish citizens, with a response rate of 64%. The latter sample came from 3,500 randomly chosen Swedish individuals who had previously participated in a telephone survey about mountain tourism (Heberlein *et al.*, 2002).

Both surveys were conducted as postal surveys and sent to individuals aged between 15 and 70 years. Questions were asked about the number of visits to the mountain region, choice of activities and accommodation for the last 5 years. The surveys measured mountain visits during the 5-year periods 1980–1984, and 1995 (May)–2000 (April) respectively. For the purpose of the survey, the mountain region was defined by the Swedish Environmental Protection Agency (Naturvårdsverket, 1985) as "areas with mountains reaching above the tree line, including the valleys and places with mountains reaching above the tree line within sight. Densely populated areas are not included". Figure 19.1 provides a map of the 15 mountain municipalities of Sweden, situated in the western parts of the counties of Dalarna, Jämtland, Västerbotten and Norrbotten. The study area – as defined above – is located within these 15 municipalities.

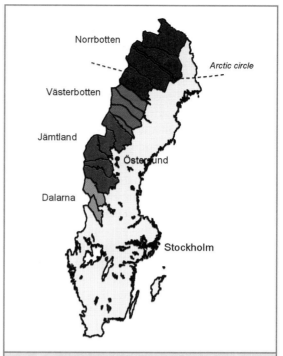

Figure 19.1 Map of Sweden and the 15 mountain municipalities in the counties of Dalarna, Jämtland, Västerbotten and Norrbotten.

19.3 Changes in Swedish mountain tourism

The main results from the two surveys are presented in Tables 19.1–19.3 below. For both surveys, figures are given as estimated proportions of the population studied. Statistical inferences for population proportions are based on Mann-Whitney U tests. The data show that statistically significant changes in travel pattern to the mountain region have occurred during the 20-year period (Table 19.1). The mountains of Dalarna, the southern-most county of the mountain region, hsd become the most popular location by the late 1990s. Thirty-four per cent of the Swedish adult population (2,117,000 individuals) visited this part of the Swedish mountain region during the 5 years at the end of the century. Thirty-two per cent had visited the mountains of Jämtland (1,980,000 individuals). Levels of visits to the mountains of Västerbotten and Norrbotten are much lower; only 7.4% of the population visited Västerbotten, and 10.3% made a visit to Norrbotten (461,000 and 641,000 individuals

Table 19.1 Changes in visitation to different parts of the Swedish mountain region. Bold is significant at 5% level or above.

	1980-1984 (%)	1995-2000 (%)	Unit change (%)	Change (No. of individuals)	Sig.
Norrbotten	11.2	10.3	-0.9	-22,000	n.s.
Västerbotten	**9.7**	**7.4**	**-2.3**	**-112,000**	**p<0.05**
Jämtland	29.7	31.8	2.1	229,000	n.s.
Dalarna	**22.2**	**34.0**	**11.8**	**806,000**	**p<0.001**

respectively). The estimated 11.8% increase in Dalarna (806,000 individuals) is more than the total visits to Västerbotten or Norrbotten in the late 1990s. In the northern parts of the mountain region, we identify a drop in the proportion of the Swedish population that visited Västerbotten (2.3% down), while no significant changes are observed for Jämtland and Norrbotten.

Not only is the geography of mountain visits changing, but activities are changing as well (Table 19.2). In the early 1980s, 22% of the Swedish population made at least one trip to the mountains to downhill ski. In the late 1990s this had increased to 36% of the population, an increase of 970,000 individuals. We find no significant change in the proportion of Swedes who went cross-country skiing on a trip to the mountains – it is stable at about 22% (approximately 1.4 million individuals). Only about 2% of the population made even one trip to the mountains in a 5-year period where they cross-country skied, including at least one overnight stay. This showed no significant change between the periods. There was also no significant change in hiking: about 21% of the population visited the mountains for a day hike during each of the 5-year periods (almost 1.3 million individuals). About 6% of the Swedish population did a hike on trails (including at least one overnight stay) during the two periods. These numbers were both stable over the 20 year period. Snowmobiling, like downhill skiing in the mountains, showed large increases between the early 1980s and the late 1990s. In the early 1980s, 9% of the Swedish population (550,000 individuals) visited the mountains at least once to snowmobile during a 5-year period, but by the end of the 1990s this had increased to 16% (1.0 million individuals).

Table 19.2 Changes in activity participation in Swedish mountain tourism. Bold is significant at 5% level or above.

	1980-1984 (%)	1995-2000 (%)	Unit change (%)	Change (No. of individuals)	Sig.
Downhill skiing	**22.0**	**36.4**	**14.4**	**970,000**	**p<0.001**
XC-ski day-trip	22.9	22.4	-0.5	45,000	n.s.
XC-ski overnight	2.2	1.8	-0.4	-15,000	n.s.
Hike day-trip	21.6	20.8	-0.8	20,000	n.s.
Hike trail overnight	5.7	6.8	1.1	87,000	n.s.
Snowmobile	**9.4**	**16.1**	**6.7**	**450,000**	**p<0.001**

Changes in accommodation (Table 19.3) are probably related to changes in activity patterns. In the late 1990s, 46% of the Swedish population (2.9 million individuals) rented a cabin or stayed in a mountain lodge at least once in the past 5 years. This was up from 37% in the early 1980s. Second-home owners show a similar increase between the two periods, but levels in this category change from 6 up to 9%. The percentage staying at least once in a tent or a recreational vehicle (RV) along the road dropped from 10.2% in the 1980s to 6.5% in the 1990s. Also, people staying in a tent in the backcountry dropped from 7.6% to 5.6%. The numbers staying in cabins at least once along the extensive Swedish mountain trail system were stable at slightly less than 5% of the population.

Table 19.3 Changes in accommodation in Swedish mountain tourism. Bold is significant at 5% level or above.

	1980-1984 (%)	1995-2000 (%)	Unit change (%)	Change (No. of individuals)	Sig.
Rented cabin, lodge	**37.3**	**46.3**	**9.0**	**685,000**	**p<0.001**
Second home	**6.1**	**8.9**	**2.8**	**192,000**	**p<0.01**
RV or tent at road	**10.2**	**6.5**	**-3.7**	**-197,000**	**p<0.001**
Cabin along trails	4.4	4.8	0.4	40,000	n.s.
Tent in backcountry	**7.6**	**5.6**	**-2.0**	**-103,000**	**p<0.05**

19.4 Why did the changes occur?

The Swedish population both increased and aged between the two periods studied. In 1984 there were 5,900,000 Swedes between the ages of 15 and 70. In 2000, this number had increased to 6,226,000 (Statistics Sweden, 1985, 2000). With a slight increase in the number of Swedes, we would expect to see more people visiting the mountain region, which is what the data show. It may also be that more mechanized activities, such as downhill skiing and snowmobiling, attract an ageing population, rather than more strenuous backpacking and cross-country skiing.

We also see the percentage of mountain visits to Dalarna in the south increase faster than the Swedish population. Shorter vacations, decreases in free time and an increased population concentration to urban areas in the south of Sweden are consistent with the increasing number of visits to Dalarna, because it is closer to the major population centres. In Sweden, the average disposable income increased by 11% between 1981 and 1997. This is consistent with the growth in the high cost activities of downhill skiing and snowmobiling, rather than the relatively low cost hiking and cross-country skiing. As income levels increase, hikers and cross-country skiers can afford to look for more exotic experiences in other countries, which might substitute for a vacation in the Swedish mountains. This is consistent with the results of Heberlein & Fredman (2002), who found income to be a constraint to visiting the Swedish mountains among downhill skiers, but not among backpackers.

How then do you account for the stimulus given by a national hero? When Ingemar Stenmark – a quiet young man from the north of Sweden – won his first slalom World Cup in 1976, and then dominated the sport for almost a decade, he started a national downhill ski boom. Having seen the runs on television in the morning, young children ran out to the local slopes full of enthusiasm. Today these youngsters are in their thirties, bringing their own families to the ski-slopes. Similar influences are reported from Austria where successful downhill skiers have increased the popularity of the sport (McGibbon, 2000). Hence, downhill skiing has found fertile ground in Sweden because it attracts multiple age groups who can participate in various social contexts and during different life-phases – some looking for relaxing and sociable activities, others for more adventurous experiences.

There is a concentration of large ski areas in the southern mountains, partly as a consequence of recent structural changes in the industry. Downhill skiing – or rather hauling skiers up the mountains – is an economic activity which generates money. There were also expectations of positive external effects to society from tourism, which led to extensive subsidies by the national government for ski development in the 1970s. Increased resources for marketing, increased capacity and development of new infrastructure have all promoted downhill skiing participation in Sweden.

Snowmobiles allow one to travel in the winter country in a way that motor boats cruise the waters during the summer. Technical development and design have made the snowmobile machines more reliable and fun to ride. It is an activity that is at the same time comfortable, adventurous and physically demanding, which attracts people. An increased number of snowmobile rentals allow people who travel to the mountains primarily to participate in other activities to try out this activity.

Both hiking and cross-country skiing are neither increasing nor decreasing in the Swedish mountains. Why? As hiking and cross-country skiing have matured as recreational activities, there has been increased competition from other activities offering faster and more exotic experiences. Hiking and cross-country skiing – particularly when overnight stops are involved – require more time and commitment compared to downhill skiing and snowmobiling. The latter activities offer more action and involve more intensive experiences in short time periods, which fits with general trends in outdoor recreation and tourism (Urry, 1995; Sandell, 2000). Compared to downhill skiing and snowmobiling, there is less marketing for hiking and cross-country skiing because the commercial opportunities are more limited. It is also difficult to buy package tours with all services included. Nor has new infrastructure been created for these traditional activities – the extensive system of trails and mountain hotels and huts was in place long before 1980.

19.5 Some comparisons with Finland, Norway and Iceland

To what extent do the changes in tourism patterns in the Swedish mountains compare to other countries in Northern Europe? Is tourism moving closer to urbanized areas in other countries as well? Is there an increase in mechanized winter activities? Are more traditional activities like backpacking and cross-country skiing declining?

To answer some of these questions, data were collected about changes in mountain tourism in Finland, Norway and Iceland (see Tables 19.4-19.6 respectively; see also Mackay, this volume). A lack of comparable studies in these countries does, however, limit the potential for comparisons. Yet another complication is the definition of mountain tourism in contrast to non-mountain tourism and mountain recreation. To Swedes a 'real' mountain reaches above the timber line (which is at about 700 m) and the mountain region is easy to define and obvious to most people. In Finland, Norway and Iceland the situation is different. In Finland, only a few mountains in the very north reach above the timber line while many regions are characterized by forested hills. In Norway and Iceland, bare mountains dominate the landscape in many regions, making most of the country a mountain tourism destination. Consequently, the comparisons made here are limited to typical mountain activities and, in several studies, no distinction is made between recreation by local residents and tourists.

Data from Finland show a decline in berry picking, while hunting and fishing have been stable over time (Table 19.4). Increases for the last 5-10 years are found in boating, the number of registered snowmobiles and downhill ski-ticket sales. In Norway, day hiking in the mountains has increased between 1986 and 1996 while cross-country skiing has been

Table 19.4 Changes in mountain tourism (or related activities) in Finland. Data source: Sievänen (2001).

	1991	1995	2000	2001
Fishing (% of population)	45		46	
Hunting (% of population)	9		8	
Picking berries (% of population)	65		56	
Boating (% of population)	34		47	
Snowmobiles (registrations)		70,374		93,486
Downhill ski (ticket sales, million Finland marks)		150		205

Table 19.5 Changes in mountain tourism (or related activities) in Norway. Data sources: Vorkinn *et al.* (1997); Teigland (2000).

	1986	1989	1993	1996	1999
Day hiking (% of population)	33		53	57	
XC-skiing (% of population)	40		45	41	
Biking in nature (% of population)		18	35	42	
Downhill skiing (% of population)		21		24	
Snowmobiling (% of population)		5			10-15

Table 19.6 Changes in mountain tourism (or related activities) in Iceland. Data sources: Statistics Iceland (2002); Icelandic Tourist Board (2002).

	1990	1999	2000	2001
Foreign visitors arrivals	141,700		302,900	
Nature observation (% of summer visitors)		74	69	79
Hiking (% of summer visitors)		55	48	54
Glacier/snowmobile trip (% of summer visitors)		23	19	20
Mountaineering (% of summer visitors)		13	10	12
Hunting/fishing (% of summer visitors)		5	4	3

quite stable (Table 19.5). Increases are also observed in biking, downhill skiing and snowmobiling. In Iceland, tourism is one of the fastest growing sectors of the economy and the number of international arrivals has more than doubled in the last 10 years (Table 19.6). Foreign visitors are mainly interested in recreational activities connected with nature, and there has been a large increase in the supply of such activities in recent years (Icelandic Tourist Board, 2002). Data on trends in tourism activities are not available, and for the last 3 years only minor changes can be recognized in the activities studied (Table 19.6). In the summer season, 'nature observation' is the most commonly reported activity, followed by visits to the Blue Lagoon (an outdoor swimming area) and backpacking. Looking at overnight stays in different parts of Iceland between 1998 and 2000, we find increases in all regions characterized by a mountainous landscape – with the highest growth in the Western Fjords, one of the more peripheral parts of Iceland (Statistics Iceland, 2002).

19.6 Discussion

Our comparison of two national surveys shows that mountain visitors in Sweden have increasingly become downhill skiers and snowmobilers. There is an increase in motorized recreation, while tenting, camping and non-motorized recreation are stable or have declined. There seems to be a shift from the self-reliant individual Swede to one who goes to ride machines up and over the mountains. Today, downhill skiing is the most powerful industry in Swedish mountain tourism, but much of the growth has been concentrated in a few major ski areas (Bodén & Rosenberg, 2004). A similar consolidation of the ski industry is also observed internationally for the last few decades (Hudson, 2000). Our data suggest that Swedes may, more than 20 years ago, be purchasing their nature experiences – staying in hotels, buying ski lift tickets and renting snowmobiles. We believe that both subsidies to the industry and commercial opportunities have been driving this development.

There are a number of activity-specific events that continue to play a role in changes in participation. The heavy impact of the local hero Ingemar Stenmark cannot be discounted, and the focus on downhill skiing as a family activity fits with the Swedish culture. As income levels increase, people can afford more expensive activities, while hikers and cross-country skiers can meet the expense of more exotic experiences in other countries, which might substitute for a trip to the Swedish mountains. The advent of new technology and international trends are surely driving the snowmobile market. Snowmobile rentals can be part of a week-long winter mountain vacation, where downhill skiing is mixed with cross-country skiing and a day on a snowmobile. Heberlein *et al.* (2002) found that only 7% of the mountain snowmobilers participated solely in snowmobiling – 74% also went downhill skiing, 32% did cross-country skiing and 23% participated in fishing.

Findings from other countries confirm several of the changes observed in Sweden. An increased number of people participate in mechanized activities, while participation in more traditional outdoor recreation activities like fishing, hunting, cross-country skiing and hiking is stable over time. Norway appears to show a slightly different pattern, as there are significant increases in day hiking while downhill skiing was stable between 1989 and 1996. In part, this is a matter of which time period one focuses on, and looking another 10 years back in time we will find significant increases in downhill skiing in Norway as well (Teigland, 2000).

The changes in mountain tourism identified in this chapter have both environmental and socio-economic implications for the future development of Northern Europe. While downhill skiing is concentrated in certain areas where development and environmental impact can be controlled, snowmobiling, hiking and cross-country skiing demand larger backcountry areas, which increases the possibilities of user conflicts. The economic development potential of both downhill skiing and snowmobiling is, however, probably far beyond that of hiking and cross-country skiing. This points towards the importance of research on outdoor recreation activities and monitoring changes in participation over time.

Such research will aid a better understanding of the complex social and natural patterns of mountain areas as new economies arise based on services and the exploitation of non-extractive values.

Acknowledgements

This research was financed by the European Tourism Research Institute (ETOUR) and the research program Fjällmistra. We would like to thank Ebbe Adolfsson, Dr. Goran Bostedt, Professor Lars Emmelin, Professor Kreg Lindberg, Mr. Stig Lundkvist and an anonymous referee for valuable comments on earlier drafts of this paper.

References

Bodén, B. & Rosenberg, L. (2004). *Kommersiell turism och lokal samhällsutveckling. En studie av sex svenska fjälldestinationer.* ETOUR, rapportserien, R2004:115. European Tourism Research Institute, Östersund.

Freitag, R.D. (1996). International Mountain Holidays – A pan-European overview based on results of the European Travel Monitor. In *Alpine Tourism. Sustainability: Reconsidered and Redesigned*, ed. by K. Weiermair, M. Peters & M. Schipflinger. Proceedings of the International Conference at the University of Innsbruck, May 1996. ITD-Edition, Innsbruck. pp. 12-41.

Godde, P.M., Price, M.F. & Zimmermann, F.M. (Eds) (2000). *Tourism and Development in Mountain Regions.* CABI Publishing, Wallingford.

Heberlein, T. & Fredman, P. (2002). *Motivation, Constraints and Visits to the Swedish Mountains.* European Tourism Research Institute, working paper 2002:2. European Tourism Research Institute, Östersund.

Heberlein, T., Fredman, P. & Vuorio, T. (2002). Current tourism patterns in the Swedish mountain region. *Mountain Research and Development*, **22**, 142-149.

Hudson, S. (2000). *Snow Business. A Study of the International Ski Industry.* Cassell, London.

Icelandic Tourist Board (2002). *Iceland 2002. Handbook.* Icelandic Tourist Board, Reykjavik.

Loomis, J.B. (2000). Counting on recreation use data: a call for long-term monitoring. *Journal of Leisure Research*, **32**, 93-96.

McGibbon, J. (2000). Tourism pioneers and racing heroes: the influence of ski tourism and consumer culture on local life in the Tirolean alps. In *Reflections on International Tourism: Expressions of Culture, Identity and Meaning in Tourism*, ed. by M. Robinson, P. Long, N. Evans, R. Sharpley & J. Swarbrooke. Business Education Publishers, Sunderland. pp. 151-166.

Naturvårdsverket (1985). Svenskarnas fjällvanor: en undersökning om vilka fjäll svenskarna besöker, hur de bor och vad de gör. Rapport 3019. Naturvårdsverket (Swedish Environmental Protection Agency), Stockholm.

Power, T.M. (1996). *Lost Landscapes and Failed Economies. The Search for a Value of Place.* Island Press. Washington, D.C.

Sandell, K. (2000). Fritidskultur i natur. In *Fritidskulturer*, ed. by L. Berggren. Lund, Studentlitteratur. pp. 213-239.

Schor, J. (1992). *The Overworked American.* Basic Books. New York.

Sievänen, T. (2001). *Luonnon virkistyskäyttö 2000* (Outdoor recreation 2000). Metsäntutkimuslaitoksen Tiedonantoja 802. Metsäntutkimuslaitos, Vantaa,

Statistics Iceland (2002). *Iceland in Figures 2000–2001.* Volume 6. Statistics Iceland. Reykjavik.

Statistics Sweden (1985). *Statistical Abstract of Sweden 1985.* Statistics Sweden, Stockholm.

Statistics Sweden (2000). *Statistical Yearbook of Sweden 2000.* Statistics Sweden, Stockholm.

Teigland, J. (2000). *Nordmenns friluftsliv og naturopplevelser. Et faktagrunnlag fra en panelstudie av langtidsendringer 1986-1999*. VF-rapport 7/2000. Western Norway Research Institute, Sogndal.

Urry, J. (1995). *Consuming Places*. Routledge, Abingdon.

Vorkinn, M., Aas, Ö. & Kleiven, J. (1997). *Friluftslivsutövelse blant den voksne befolkningen – utviklingstrekk og status i 1996*. Östlandsforskning, rapport nr. 07/1997. Östlandsforskning, Lillehammer.

Weiermair, K., Peters, M. & Schipflinger, M. (1996). *Alpine Tourism. Sustainability: Reconsidered and Redesigned*. Proceedings of the International Conference at the University of Innsbruck, May 1996. ITD-Edition, Innsbruck.

Neil Bayfield, Rob Brooker & Linda Turner

Summary

1. Conservation and management activities in mountain areas depend upon integrated monitoring, especially with respect to environmental change at the national and international levels.

2. Procedures for organising multi-site monitoring networks are still being developed.

3. An examination of some existing, successful networks (ECN, GLORIA and SCANNET) provides useful information on the strengths and weaknesses of current network organisation.

4. The future development of large-scale integrated monitoring networks is considered, indicating the need for linking networks and developing multi-disciplinary approaches.

20.1 Introduction

Conservation and management of any environment require accurate monitoring of the state of that environment. Without such monitoring it is impossible to assess whether the aims of management or conservation policies are being met.

In addition, the design and implementation of management scenarios are critically dependent upon future changes in the environment. Such environmental changes are commonly driven by factors external to the particular area of interest with respect to management, for example the impacts of anthropogenic drivers such as pollution or climate change. Therefore most land management policy in remote areas such as mountain regions, whilst not being able to influence directly the drivers themselves, must account for the potential impact of such drivers when setting management or conservation targets. Monitoring is a crucial part of providing predictions of the future impacts of environmental change. It provides important data for the testing and refinement of predictive models, and, when conducted in sufficient detail, it can enhance understanding of the relationships between the different components of ecosystems, such as the vegetation, hydrology and soils.

Bayfield, N., Brooker, R. & Turner, L. (2005). Some lessons from the ECN, GLORIA and SCANNET networks for international environmental monitoring. In *Mountains of Northern Europe: Conservation, Management, People and Nature*, ed. by D.B.A. Thompson, M.F. Price & C.A. Galbraith. TSO Scotland, Edinburgh. pp. 213-222.

Perhaps because of an increasing awareness of its benefits, long-term monitoring is again, after a period of neglect (Heal, in Sykes & Lane, 1996), becoming a cornerstone of the global environmental research effort. However, because so many of the external drivers influencing our environments are at the national, continental or even global scale, environmental monitoring efforts also need to be co-ordinated across large scales. This is perhaps a more recent trend in environmental monitoring, being aided by developments in information technology and the increasing ease with which we can exchange and collate information.

The techniques and protocols for the co-ordination of such large-scale monitoring programmes are still being developed. In this chapter we discuss the co-ordinated national-scale monitoring conducted by the UK Environmental Change Network (ECN). We outline the approach of the ECN and, in particular, the example of the Cairngorms ECN site, which is also a member of the arctic SCANNET and alpine GLORIA networks. By describing the three networks and examining their respective strengths and weaknesses, we provide some pointers for the improved design and operation of new or realigned international networks for recording environmental condition and environmental change.

20.2 The UK Environmental Change Network (ECN)

The ECN is the UK's long-term environmental monitoring programme. It is a multi-agency programme sponsored by a consortium of 14 UK Government departments and agencies and co-ordinated by the Centre for Ecology and Hydrology.

20.2.1 Objectives

- To establish and maintain a selected network of sites within the UK from which to obtain comparable long-term datasets through the monitoring of a range of variables identified as being of major environmental importance.
- To provide for the integration and analysis of these data, so as to identify natural and man-induced environmental changes and improve understanding of the causes of change.
- To distinguish short-term fluctuations from long-term trends, and predict future changes.
- To provide, for research purposes, a range of representative sites with good instrumentation and reliable environmental information.

20.2.2 Sites

The ECN programme operates a network of 12 terrestrial and 42 freshwater sites throughout the UK (Figure 20.1). Sites range from upland to lowland, moorland to chalk grassland, small ponds and streams to large rivers and lakes. Each of the sponsoring organisations provides one or more sites and covers the costs of ECN measurements at those sites, as well as contributing to the cost of network co-ordination activities.

20.2.3 Protocols

The ECN has been operating across the UK since 1993, with monitoring starting formally in 1993 at terrestrial sites and 1994 at freshwater sites. The ECN network records a wide

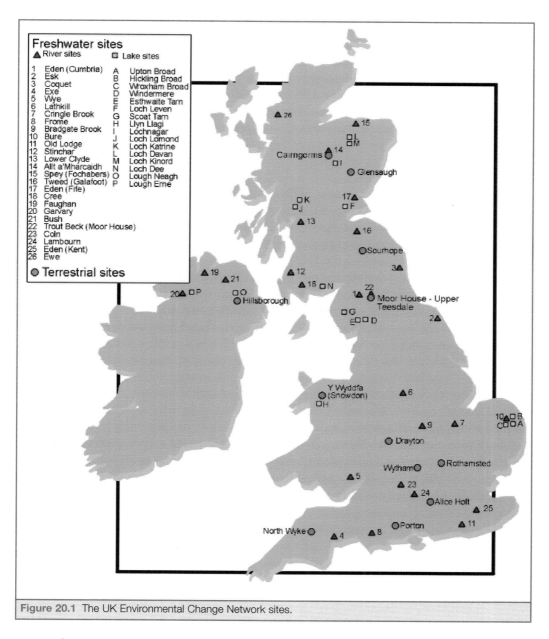

Figure 20.1 The UK Environmental Change Network sites.

range of environmental change indicators, based on a set of key physical, chemical and biological variables which drive and respond to environmental change. A set of standard protocols is used for all data collection, and all data are stored on a central database (Sykes & Lane, 1996). The protocols for the collection of data are designed to enable the easy analysis and interpretation of these long-term data sets, i.e. providing consistency both across years and across sites, thereby forming part of the quality assurance procedures in ECN. All terrestrial and freshwater protocols are available as books for purchase or citation, and can also be found on the ECN website (www.ecn.ac.uk/protocols/index.asp; 14 July 2005).

20.2.4 Database and data users

The ECN database stores all data and meta-data from ECN's core measurements collected at its network of sites. These data, along with other historic data from ECN sites, are held in standardised structures in order to support the cross-disciplinary analyses necessary for environmental change research. The database uses the Oracle 8i RDBMS with links to Arc GIS for spatial data handling and is maintained and developed by the ECN Central Co-ordination Unit. Data are regularly sent in from sites and are quality assured before being lodged in the database; a system of quality flags and codes ensures that the data are properly qualified in the meta-database.

Wide availability of data is central to ECN's success as a resource for environmental change research, policy purposes and public information (Rennie *et al.*, 2000; Rennie & Jackson, 2001). ECN promotes the use of its data through a number of different data access methods, including targeted data products, web developments and events to suit a range of users from scientists and policy-makers to students and the general public.

20.2.5 Strengths and weaknesses of the ECN approach

Major strengths of the network are: the use of standard published protocols; the central database; the established network of sites, most of which have been operating for at least 9 years; and the wide availability of data to other users. There are few comparable networks elsewhere in the world with this combination of features. Although there are many networks in existence, comparatively few use uniform protocols and a central database. The web page is very informative, with live data links to weather stations, data summaries, downloadable protocols and separate sections for educational users. The rapid rate of website update, and the immediacy of data availability, may make it of more interest and use to a wider end-user community.

Weaknesses are that site coverage is not comprehensive (several substantial tracts of the UK, such as the western highlands of Scotland, are so far unrepresented), and the large number of variables recorded mean that operating costs are comparatively high. Furthermore, the indicators chosen are not necessarily the most relevant or cost-effective that could be used, but those selected by a group of experts when the scheme was set up. It would now be pertinent to examine the runs of data collected to assess the statistical power of the indicators, and their cost-effectiveness and relevance to current perceived environmental threats and impacts. This would permit the selection of revised indicators or protocols for the future. This highlights the need for continual review of protocols, even in highly standardised monitoring networks such as the ECN.

20.3 The Cairngorms ECN site: a link to the GLORIA and SCANNET networks

The Cairngorms terrestrial site joined the ECN network in summer 1999. It is sponsored by a consortium of Scottish Natural Heritage, the Centre for Ecology and Hydrology (CEH), the Scottish Environment Protection Agency (SEPA) and the Macaulay Institute (MI), and is managed by CEH.

The site is situated near Aviemore in Strathspey. The 10 km^2 site is part of the Cairngorms National Nature Reserve and well within the new Cairngorms National Park. It has been used as an intensive research site since the 1970s. Therefore, monitoring at this site is being underpinned by an ever-improving understanding of the mechanisms that drive ecosystem variability and the interactions between ecosystem components. Such manipulative research in conjunction with long-term monitoring is essential to understand fully the long-term drivers of environmental change (Heal, 1991).

The site is ideally placed to monitor changes in:

- tree colonization: a reduction in deer grazing began 15 years ago and colonization is now widespread;
- climate change: the site straddles the zones of increasing winter precipitation and decreasing summer precipitation. There is also evidence of increasing windiness;
- hydrology: it is one of the longest recorded snow sites in the UK and is also a gauged catchment;
- pollution: it lies in a relatively clean part of the UK and provides a good control area; and
- vegetation change: there is an excellent altitudinal sequence of communities from Caledonian pine woodland at 300 m up to subnival fellfield at 1,100 m.

The site has had an automatic weather station since 1984 and has been used for long-term hydrological and snow studies by the Institute of Hydrology and the Macaulay Institute for about 15 years. From 1997-1999 it was one of the ECOMONT (land use

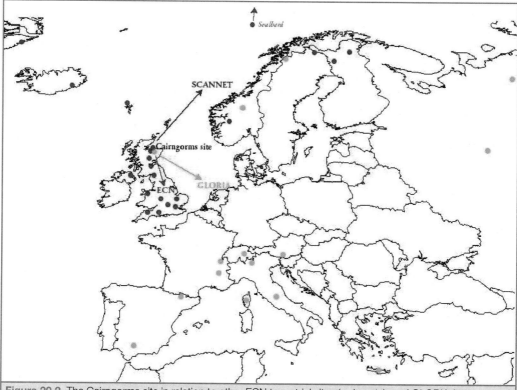

Figure 20.2 The Cairngorms site in relation to other ECN terrestrial sites (red spots), and GLORIA (green spots) and SCANNET sites (blue spots).

change in mountain areas of Europe) sites (Cernusca *et al.*, 1999). It has been a freshwater ECN site for several years and continues to be used for vegetation, soils and nutrient cycling studies by several universities (e.g. Callaway *et al.*, 2002; Perring, 2002). The site forms part of the larger Feshie catchment in the NICHE programme (National Infrastructure for Catchment Hydrology Experiments).

This is the first arctic alpine ECN site in the UK, and an important link not only to other upland ECN sites, but also to the mountain GLORIA network and arctic SCANNET network (Figure 20.2). It provides an excellent demonstration of how one site can play a pivotal role in linking several national and international monitoring networks, thereby providing added value to the sponsoring consortia of each.

20.4 The SCANNET network

SCANNET (Scandinavian/North European Network of Terrestrial Field Bases (www.envicat.com/scannet/Scannet; 14 July 2005) is a network of field site leaders, research station managers and user groups in northern Scandinavia and Europe who are collaborating to improve comparative observations and access to information on environmental change in the North. There are bases in Scotland, Finland, Sweden, Norway, Svalbard, Faroes, Greenland and Iceland.

SCANNET focuses particularly on northern landscapes in some of Europe's last wilderness areas containing specialised and diverse plants and animals and large stores of soil carbon. These regions are experiencing rapid environmental and social changes and are particularly vulnerable to predicted climatic changes. By linking researchers and databases, and by driving toward standardised protocols across sites (as with the ECN), SCANNET seeks to facilitate research into these changes and their implications for both northern and lower latitudes.

In addition, SCANNET aims to provide better access to improved information on environmental change to a wide range of users. An interesting aspect of the approach of SCANNET is the involvement of local stakeholders, as well as research scientists, in the development of monitoring protocols. The variables immediately considered for monitoring by scientists may not be those that local stakeholders consider to be important indicators of the quality of their environments. Part of the SCANNET work programme therefore uses decision modelling workshops to gauge the opinion of all stakeholder groups, including scientists, as to what they consider to be the important characteristics of environmental quality in their region. At present this process is only partially completed. However, preliminary results indicate that, across the SCANNET region at least, there is wide variation in what are considered to be key indicators. For example, in the Faroes, the concerns of local stakeholders are closely related to marine productivity, whilst in Iceland the retreat of glaciers is considered important.

This highlights a potential difficulty for monitoring networks. It is unrealistic to try to be all things to all people. For instance, a standardised international monitoring programme across a wide range of environments, that includes variables of interest to all potential stakeholder groups, may be impossible. However, in areas with more restrained geographic and stakeholder coverage, for example in mountain systems, the potential for achieving consensus in the development of fully standardised monitoring programmes may be greater.

As well as trying to make data on climate variability more accessible, helping to standardise monitoring protocols and fill gaps in network observations, SCANNET also aims to provide regional climate change scenarios, and to examine variations in trends of biodiversity, species performance and phenology across sites. Therefore monitoring is again closely linked to other types of scientific research.

Here, synergy between the ECN and SCANNET networks comes from testing of new protocols for snow and phenological recording, and holding stakeholder workshops to determine perceived priorities for environmental recording across the SCANNET area. In addition, ECN can provide SCANNET with established protocols for monitoring a wide range of variables, and now has nearly 10 years of experience in the pitfalls of organising and running a large-scale integrated monitoring programme.

20.4.1 Strengths and weaknesses

The network has a central secretariat, web pages and a database based in Abisko, Sweden. It also has the added advantage of having had a programme of stakeholder consultation to prioritise the key variables to be considered, so providing an audit trail to justify the direction of its work.

The main weakness is that, because the network is based upon established sites with existing programmes of monitoring, the protocols vary from site to site. Therefore, SCANNET is faced with the task of retrospectively combining datasets. However, the aim is to get more common protocols and to increase the compatibility of the data collected. New protocols for snow and phenological recording are, for example, being developed. There has so far been no attempt to calculate statistical power values for the protocols, partly because there are currently few protocols in common use across the network.

20.5 The GLORIA network

GLORIA (Global Observation Research Initiative in Alpine Environments) is a long-term observation network for detecting climate change effects on mountain biota on a global scale (www.gloria.ac.at/res/gloria_home/; 14July 2005). The current network includes 18 partners from Russia to Spain, Crete and Scotland, but the network is potentially worldwide.

The alpine zone is the only terrestrial biogeographic unit with a world-wide distribution. High mountain ecosystems, therefore, provide a unique opportunity for comparative climate impact research at the global scale. By concentrating on one particular type of environment, and one target ecosystem component (i.e. mountain biota rather than also including hydrology, glaciology, etc.) it may be far simpler for GLORIA, than for either SCANNET or the ECN, to operate as a closely integrated monitoring network for a wide range of variables.

Because GLORIA aims to provide a network of standardised monitoring sites in all major mountain systems on Earth, its goals need to be met by applying a cost-effective and relatively simple approach, practicable even under severe conditions with limited logistical support. The Multi-Summit approach – GLORIA's basic strategy – provides this method, and is designed to compare biodiversity patterns along the fundamental climatic gradients, vertically as well as horizontally. The GLORIA recording procedure involves very detailed recording of vegetation in permanent plots on mountain summits. Plots are recorded on each aspect (N, S, E, and W) of each summit, and temperature regimes are recorded with temperature loggers. The data are held in a central ACCESS database. Every aspect of the protocol has been carefully documented and tested, and the field manual is available to download (www.gloria.ac.at/res/gloria_home/; 14 July 2005).

Four summits fringing the Cairngorms ECN site have been recorded using these GLORIA protocols. In addition, because of the detailed monitoring already occurring at the ECN site, it is proposed as one of the so-called 'master station sites' where more detailed recording will be undertaken. At least part of this additional recording could be based on the ECN protocols, so there are strong synergistic links between the networks.

20.5.1 Strengths and weaknesses

As with SCANNET and ECN, there is a central secretariat and web-based information source, in this case based in Vienna. The GLORIA protocols are outstandingly well documented and tested. The database is also very well structured. The protocols can probably be easily adapted to almost any part of the world.

Weaknesses are that the focus was initially quite narrowly on temperature as the main driver of change. As the programme developed, however, it became apparent that precipitation, windiness, recreation, grazing, nitrogen inputs and other drivers were potentially or actually important at many of the sites. Attempts have since been made to include these variables in the programme, but they are not as well considered or as comprehensively covered as the temperature recording. Although the network has been operational for two years, there are still no data on the statistical power of the techniques used, although this is currently under investigation.

20.6 Discussion

With increasing emphasis being placed upon high quality, integrated monitoring of the state of the environment, it is necessary and productive to examine the current state-of-play of large-scale environmental monitoring.

The ECN is one of the first, and is still amongst the most comprehensive, environmental change networks in the world. The newer GLORIA and SCANNET networks have several features in common with the ECN (Table 20.1). In particular a central secretariat, well-organised web pages, and central database are probably desirable for any larger scale international network. Common protocols, fully described and published, are also very desirable to ensure that data are compatible. Common weaknesses lie in the selection of the protocols, and the demonstration of cost effectiveness, relevance and statistical power to

Table 20.1 Comparison of features of the ECN, GLORIA and SCANNET networks.

	ECN	GLORIA	SCANNET
Central database	✓	✓	✓
Synergy with other networks	✓	✓	✓
Statistical power values	✗	In preparation	✗
Published protocols	✓	✓	✗
Stakeholder consultation	✗	✗	✓
Central secretariat	✓	✓	✓
Web page	✓	✓	✓

predict change. Only the SCANNET network is so far attempting to incorporate stakeholder perceptions into its selection of protocols, and only GLORIA is attempting to assess statistical power values. Neither network has completed these tasks yet, but they are important to establish both the relevance and reliability of ongoing recording.

The Cairngorms ECN site provides an example of how a carefully chosen central site can act as a link between multiple networks. However, perhaps future monitoring programmes will not be dependent upon one or two sites for synergistic linkages. An idealised international network would tackle these issues at inception; it would have all the advantages listed above, and would be well funded for the foreseeable future. As this book goes to press, we can report that both GLORIA and SCANNET have now finished their initial phases, and final reports and papers from these projects are available (e.g. Callaghan *et al.*, 2004); both are still working as active groupings. SCANNET is developing stronger high latitude links to initiatives such as CEON (the Circumarctic Environmental Observatories Network; www.ceoninfo.org/: 20 July 2005), and GLORIA is extending its activities with the continued initiation of new sites. The ECN project is now becoming a major focus for work through the ALTER-Net project to develop a long-term ecosystem research network for Europe (A Long-Term Biodiversity, Ecosystem and Awareness Research Network; www.alter-net.info/default.asp: 20 July 2005).

One of the most important developments is the extension of these monitoring efforts to encompass what can be described as the 'human dimension'. This involves an interdisciplinary approach to monitoring, by integrating work on the natural environment with the social and economic drivers and pressures that exist within human populations (local or dispersed). Within ALTER-Net there is currently much discussion of the concept of 'LTER Platforms' where such integrated monitoring and research might take place. However, as we cautioned at the outset, and as has since been highlighted in the context of developing a monitoring programme for Mountain Biosphere Reserves (Brooker & Turner, 2004), developing a single, unified, 'one size fits all' monitoring system to encompass the multiple factors involved in these complex relationships could be extremely difficult. Irrespective, it is clearly an exciting time for environmental monitoring networks within Europe.

References

Brooker, R.W. & Turner, L. (2004). Designing global change monitoring networks for terrestrial ecosystems: lessons from the GLORIA and UK Environmental Change Network (UK ECN). In *Global Environmental and Social Monitoring*, ed. by C. Lee & T. Schaaf. UNESCO, Paris. pp. 78-85.

Callaway, R.M., Brooker, R.W., Choler, P., Kikvidze, Z., Lortie, C.J., Michalet, R., Paolini, L., Pugnaire, F.I., Newingham, E., Aschehoug, E.T., Armas, C., Kikodze, D. & Cook, B.J. (2002). Positive interactions among alpine plants increase with stress. *Nature*, **417**, 844-848.

Callaghan, T.V., Johansson, M., Heal, O.W., Saelthun, N.R., Barkved, L.J., Bayfield, N., Brandt, O., Brooker, R., Christiansen, H.H., Forchhammer, M., Høye, T.T., Humlum, O., Järvinen, A., Jonasson, C., Kohler, J., Magnusson, B., Meltofte, H., Mortensen, L., Neuvonen, S., Pearce, I., Rasch, M., Turner, L., Hasholt, B., Huhta, E., Leskinen, E., Nielsen, N. & Siikamäki, P. (2004). Environmental changes in the North Atlantic Region: SCANNET as a collaborative approach for documenting, understanding and predicting changes. *Ambio*, Special Report **13**, 39-50.

Cernusca, A., Tappeiner U. & Bayfield N (Eds.) (1999). *Land use Change in European Mountain Ecosystems. ECOMONT – Concept and Results.* Blackwell Wisssenschafts Verlag, Berlin.

Heal, O.W. (1991). The role of study sites in long-term ecological research: a UK experience. In *Long-term Ecological Research*, ed. by P.G. Risser. Wiley, Chichester. pp. 23-44.

Perring, M. (2002). What is controlling the pattern of regeneration of native Scots pine now that deer grazing is declining? A study of the Allt á Mharcaidh, Upper Speyside. MSc thesis, University of Aberdeen, Aberdeen.

Rennie, S.C. & Jackson, D. (2001). Using environmental data on the Internet. *Teaching Geography*, **26**, 33-35.

Rennie, S.C., Lane, A.M.J. & Wilson, M. (2000). Web access to environmental databases: a database query and presentation system for the UK Environmental Change Network. *Proceedings of the 2000 ACM Symposium on Applied Computing*, **2**, 894-897. DOI: http://doi.acm.org/10.1145/338407.338678.

Sykes, J.M. & Lane, A.M.J. (Eds) (1996). *The United Kingdom Environmental Change Network: Protocols for Standard Measurements at Terrestrial Sites*. The Stationery Office, London.

PART 4:

Management Influences, Practices and Conflicts

Fire management trial in 2004 within the Abernethy Forest National Nature Reserve, Cairngorms, Scotland, UK (Photo: Neil Cowie, Royal Society for the Protection of Birds).

Having explored the many human influences on mountain areas, the book moves on to consider land management issues. Maxwell & Birnie (Chapter 21) introduce us to the changing regimes of land management. They analyse the range of land ownership regimes and the environmental, access, recreational, sporting, tourism, agricultural and forestry interests. They conclude that an important means of achieving sustainable development in mountain regions will be to accommodate changing property rights and responsibilities. A clear example of an unambiguous approach to land management is given by Beaumont *et al.* (Chapter 22). They describe the management objectives and practices applied to Abernethy Forest Nature Reserve, which is part of the Cairngorms National Park. Some novel and contentious management trials are described here, such as burning within Scots pine (*Pinus sylvestris*) woodland to stimulate regeneration and benefit habitat conditions and food supplies for woodland grouse, notably the capercaillie (*Tetrao urogallus*) and the black grouse (*T. tetrix*). Beaumont *et al.* present a clear account of how one nature conservation charity is managing a substantial nature reserve to conserve an important assemblage of birds and habitats.

Sandberg explores the influences of ownership on the management of mountains in Northern Europe (Chapter 23). This study presents a penetrating account of how fundamental changes in ecological and social conditions influence the management of predators, in particular. Gunslay (Chapter 24) develops this theme in considering the economic and socio-cultural influences on the management of reindeer (*Rangifer tarandus*) by Swedish Saami reindeer herders. This study shows the very delicate balance between environmental, societal and cultural influences in ensuring that Saami herding is sustained into the future.

Increasingly, public participation techniques are being developed to manage environmental conflict. Wallsten (Chapter 25) describes the highly successful public participation process used to establish the Fulufjället National Park in Sweden. The 'inside-out process', as Wallsten describes it, focused on local perspectives and considered socio-economic opportunities; this approach was borne out of the evident failure of the earlier public consultation processes to designate this, Sweden's twenty eighth, and most recent, National Park. This is an uplifting chapter which reveals the success of an imaginatively planned process of involving local people in a site designation process. Ferguson & Forster (Chapter 26) describe the processes and techniques involved in designating the Cairngorms National Park, the United Kingdom's largest National Park, which was established in 2003. As with Wallsten's experience in Sweden, the main challenge was achieving the right level of public consultation during the designation process, again undertaken with success. What has been achieved in these two National Parks offers important, constructive lessons for people involved in the designation of further National Parks. Dougherty (Chapter 27) provides an interesting consideration of some of the broader philosophical issues surrounding the development of management policies for National Parks. He notes that whilst in North America the wilderness debate is prominent, in the United Kingdom this is not so.

The final two chapters in this part of the book deal with two very different community interests. Pillai (Chapter 28) examines how community land ownership in Scotland might promote sustainable development in rural communities. She explores the difficult balance between the economic development of these communities and the protection of environmental interests. Tourism offers one important means of economic regeneration, but as we have already seen, this can bring with it some potential conflicts with environmental interests. Pettersson & Vuorio (Chapter 29) consider this challenge in relation to Northern Sweden where the Saami people have begun to engage in tourism. A lot has been learned through the assessment of visitors' attitudes towards the Saami tourism establishments. Like so many of the other studies reported in this part of the book, there is a sense of optimism because the research highlights conservation and management measures which will reduce the potential for conflict between visitors and local or indigenous people.

21 Multi-purpose management in the mountains of Northern Europe – policies and perspectives

Jeff Maxwell & Richard Birnie

Summary

1. There has been a movement away from a dominant social paradigm where the countryside was regarded as a place of 'production' by European societies to one where it is equally regarded as a place of 'consumption'. This perspective and European Union policies, especially through the Rural Development Regulation, will influence greatly the multi-purpose management of mountains in Northern Europe.

2. There is the potential for even greater specialisation of the countryside, and in the mountain regions, managing land for environmental benefits could become the imperative. Sustainable mountain development is a regionally specific process of sustainable development that concerns both mountain regions and populations living downstream or otherwise dependent on these regions in various ways.

3. Accommodating changing property rights and responsibilities within a range of land ownership systems towards long-term multi-purpose management regimes will be important in achieving overall sustainable development in mountain regions: but within such a context a greater emphasis on policy integration, allowing for local interpretation and implementation, will be essential in designing relevant and multi-purpose systems of land management.

21.1 Introduction

In the final paragraph of his book 'Contested Mountains', which focuses on the 'Nature, Development and Environment in the Cairngorms Region of Scotland 1880-1980', Robert Lambert says:

"The last century has pointed out the importance of nature conservation and recreation as landuses in the Cairngorms area, but time has not solved the question of the balance between them, nor how to achieve this balance along with a strengthening of the economic prosperity of the region, without which any future designation would not be sustainable in the long run.' (Lambert, 2001, p. 271)

Maxwell, J. & Birnie, R. (2005). Multi-purpose management in the mountains of Northern Europe – policies and perspectives In *Mountains of Northern Europe: Conservation, Management, People and Nature,* ed. by D.B.A. Thompson, M.F. Price & C.A. Galbraith. TSO Scotland, Edinburgh. pp. 227-238.

This succinct summary also encapsulates many of the challenges that face the mountain areas of Europe, as we enter this 21st century. We intend to illustrate how the management issues requiring resolution in the mountain region of the Cairngorms represent a set of issues in a context of historical, cultural, environmental and social dynamic that is not untypical of the rest of the mountain regions of Northern Europe.

21.2 Changing perspectives and policies

As a part of the wider rural domain, the mountain regions reflect, in a highly focused manner, the way in which European societies have come to view the countryside and the activities that take place there. Progressively, and more rapidly in recent time, there has been a movement away from a dominant social paradigm where the countryside was regarded as a place for the 'production' of food, fibre and timber (Dunlap & Van Liere, 1978, 1984). It is now regarded as equally a place of 'consumption' for recreation and amenity: a place where habitats, wildlife, water and landscape have intrinsic worth and value. Many of these 'resources' are intangible or moving but are regarded as 'public goods' of significant value.

21.2.1 The mountain areas

Mountain areas are often remote, fragile and testing environments. For humankind they provide a sense of wildness, solitude, and mental and physical challenge. For some people this makes them a place of refuge, peace, reinvigoration and spiritual renewal – and very special. It is for these reasons that they act as a lightning rod for the contested nature of the arguments that have to do with how rural land is used, who owns it, and who has rights of access to it. But in countryside terms, they are nevertheless a part, albeit at one end of a spectrum of rural interests that are subject, in a European Union (EU) context, to policies that pertain to the whole of the countryside. It is these environmental, agricultural, forestry and rural development policies that will influence greatly the multi-purpose management of the mountain areas of Northern Europe and the extent to which sustainable development can be achieved within them.

21.2.2 European policies

The impact of European rural policies, particularly the Common Agricultural Policy (CAP), on the countryside has been well documented. A Ten Nation Study (Baldock *et al.*, 2002) on the nature of rural development in Europe reports on these. It is generally accepted that there has been a degradation in semi-natural and farmed habitats, driven by a combination of intensification and abandonment; degradation of soils and water associated with intensive farming, including excessive water abstraction and pollution associated with irrigation in Southern Europe; and inappropriate afforestation, insensitive commercial management and excessive felling, in different regions.

It is now also clear that while agricultural prosperity may still be a feature on good quality land, agricultural decline or land abandonment has become a common feature on poorer or marginal land. There has been a general relative decline in the economic importance of agriculture in the industrialised, densely populated EU states such as the UK, Germany and France. More generally, there is a marked tendency in the more advanced economic regions of the EU that rural areas have become places to live in and sites for leisure rather than places predominantly concerned with the production of food. However, in more sparsely populated and more 'rural', or less developed states, for example Sweden and Spain, depopulation of remote rural areas is still a major problem. In several countries and regionally within others, tourism has become a major feature and of greater economic importance than agriculture or forestry. This is the case in the environmentally sensitive parts of Austria and Germany and is certainly true of much of Scotland and probably much of the mountainous areas of Northern Europe.

21.2.3 Changing perspectives – changing policies

An important observation from the Ten Nation Scoping Study to which we have referred states that "from the ten nation reports, it is clear that an important axis is an agrarian versus rural perspective. Beyond confusion over terms is a more fundamental debate about the continuing centrality of farming, socially, culturally and environmentally, to the future of rural areas". This reiterates and provides further emphasis on the debate as to whether rural space is mainly for 'consumption' or 'production' activities.

Progressively since the publication of the EC's 1988 policy document 'The Future of Rural Society', European Community policy has evolved to reflect the changing nature and outcomes of this debate (Commission of the European Communities, 1988). Gradually the agricultural structures measures of the CAP are being combined with the partnership approaches to rural development employed in the Structural Funds, though many would argue that progress is much too slow. Nevertheless, Rural Development Plans have become embedded in the CAP. While each nation retains its own perspective on what rural development means, and takes varying approaches to implementation, some prevalent themes underpin national approaches. These are:

- agricultural modernisation;
- infrastructure development;
- maintaining regional populations;
- landscape protection and management;
- rural economic diversification;
- social and economic cohesion; and
- relieving rural disadvantage and deprivation.

Broadly, the intention is that rural development should build upon and conserve the varied, intrinsic assets of each area.

The Ten Nation Scoping Study reports that although there is general agreement that many of the issues that are important to the future of rural areas are not agricultural, the dominance of an agrarian perspective is still apparent in Europe. Policies continue to be developed separately within agricultural and environmental 'boxes', creating inconsistency and incoherence. While being recognised by most nations as being crucial, the lack of integration and joined-up thinking at all levels of governance in developing and delivering locally relevant, coherent rural policies is all too apparent.

The general thrust of EU policy and of proposed continuing CAP reform is about removing production subsidies towards area-based and environmental payments with an emphasis on a sustainable and multi-functional agriculture. In parallel, there is a continuing emphasis on environmental protection and conservation and the implementation of the Habitats and Birds Directives. The Water Framework Directive and the Pollution Prevention and Control legislation are progressively being incorporated into the legislation of the nation states. At the same time, the rural industries continue to be influenced by conditions of international trade, currency fluctuations and domestic consumer preferences and demand. Together, these influences, measures and instruments of change will affect the management of land and rural resources, significantly within the next five to ten years.

Interpreting how these proposed and inevitable changes will role out for all the nation states is not easy and yet it is important to have some feel as to how the different geographical entities that constitute the countryside might interact. This is vital if we are to have any hope of sensibly planning and achieving a coherent framework for rural sustainability.

21.2.4 Future perspectives

Birnie *et al.* (2002) have carried out such an appraisal for Scotland and have identified some of the issues that emerge from such an analysis. We believe developing its generality in a European context is relevant and helpful in gaining an insight as to the place of mountain areas within an overall rural perspective.

21.2.5 Where there are good soils and a favourable climate

First, in areas with production advantage, with good soils and favourable climate, dominated by arable agriculture and/or intensive livestock production, smaller units are likely to become uneconomic. Generally, the outcome will be farm amalgamations, resulting in fewer and larger agricultural units operating on a capital-intensive basis. However, there will be notable exceptions. Where there is a tradition of local production and local markets and a diversity of employment opportunity, some farmers may adopt a pluriactive strategy and continue farming on a part-time basis. However, without a change in agrarian culture, and the adoption of appropriate controls, incentives and legislation, the environmental consequences are likely to be negative. In these potentially highly productive areas, there could be a continuing loss of landscape, habitat diversity and structure, and an increased risk of soil erosion and nutrient enrichment.

21.2.6 Areas with environmental advantage

Second, in areas with environmental advantage that may be within the metropolitan travel-to-work areas that have ready access to services, or in areas with high scenic amenity, of which the mountain areas are undoubtedly a part, there is likely to be less pressure for structural change. This will be particularly likely if agri-environment support payments are area-based, accompanied by incentives and opportunities for on-farm diversification and off-farm employment. Other potential outcomes for these areas are a buoyant market for small farms in response to in-migrant lifestyle choices; a market for high conservation land through purchases by conservation and amenity trusts; and a trend for 'new' estate owners to commit to environmental objectives. Land designation, with accompanying funding, such as National Parks, will also provide a stabilising influence on structural change and environmental coherence. Taken together, these potential developments could reinforce the existing trend for these areas to be managed primarily for environment and natural heritage objectives.

21.2.7 The 'middle countryside'

Third, the most problematic areas are those with neither production nor environmental advantage, the so-called 'middle countryside'. Depending on the extent of the definition of mountain areas, this type of countryside will either be included or in such close proximity to them that what goes on there will be of inevitable significance to them. They are areas with relatively poor access to services and usually have a high dependency on primary industries for employment. In some of these areas, there may exist already a tradition of mixed agriculture and forestry. In some parts of Europe, where this is not so, agricultural land is likely to be converted to commercial forestry, particularly in countries where such land qualifies for forestry or woodland grants. In areas where this is not appropriate, it is possible that surplus land will result in the adoption of more extensive forms of land management, providing a considerable opportunity for environmental gain.

21.2.8 Further specialisation in the countryside?

As far as anything is predictable, what these scenarios suggest is that there is the potential for an even greater specialisation of the countryside than we have at present. Arable and intensive livestock agriculture is likely to be the imperative on good quality land at relatively low altitudes whereas managing land for the environment and natural heritage could become the imperative elsewhere. The consequent negative and positive impacts associated with these changes will also be geographically separated.

Birnie *et al.* (2002) point out that for Scotland such a forecast does not provide a vision of Scotland that is environmentally sustainable everywhere. There are implicit trade-offs between different areas. This will also extend to the countries of the EU as a whole, particularly if a special case is to be made for the mountain areas. However, it is clear that in the context of policy development on a European basis, in which funds are used as incentives and regulation is used to control, mountains are very much part of the whole.

If, as we postulate, the imperative for land management in the mountain areas is towards the environment, real progress can only be made if an increasing proportion of the CAP budget is so directed. So far this has been extremely limited in terms of agri-environment funding, and in any event, affects only those areas of land that are in receipt of CAP funds, that is land involved in some form of agriculture. Though there has been progressive development of rural development policy through the CAP, the continued dominance of funding towards agricultural activities leaves significant areas of land in some of the nation states with very limited access to European funds to bring about positive environmental management. Unless a higher proportion of European funding is directed towards specific environmental measures, many of the priorities for action that have been identified by the Mountain Forum (2002), will not be delivered.

This apparent stalemate in terms of policy development in Europe, and the deployment of funds for rural development and environmental programmes, arises because of a reluctance to meet 'head on' the challenge that sustainable development demands. The Ten Nation Study reports that sustainability is still not a key objective of rural development policies nor of the institutions that administer them: therein lies part of the problem. Driving towards a sustainable future requires a set of processes that have the potential to deliver consistent, coherent, locally relevant policies and multi-purpose management. We know what needs to be done, we know about the processes needed to do it, but we fail to apply them.

21.3 What sustainable development implies for rural development in mountain areas

While there is no universally accepted definition of sustainable development, there is general agreement that it is a process that aims at ensuring that current needs are satisfied while maintaining long-term perspectives regarding the use and availability of natural and other resources into the future. It also stresses the need to achieve equity between the well being of present and future generations. It has been suggested that sustainable mountain development is "a regionally specific process of sustainable development that concerns both mountain regions and populations living downstream or otherwise dependent on these regions in various ways" (Price & Kim, 1999). This is an important addendum. Mountain areas may be distinct geographically but their resources in the sense of 'public goods' have been regarded by some for generations, as 'belonging' to a wider community than those residing or having 'ownership' in the mountain area.

That this has become for some European societies a *cause célèbre* is a reflection of a greater number of people having a greater amount of leisure time and disposable income to use in areas that are of great beauty and attraction to them. The implications from a sustainability perspective are in finding practical and equitable ways of sharing and managing these public goods for both present and future generations, and ways in which property rights are distributed equitably and responsibly within a range of private to public land ownership systems (Table 21.1).

21.3.1 Strategic concepts and implications

Fundamental strategic concepts and processes lie at the heart of developing sustainable systems (O'Riordan, 2002) whether they are for mountains or any other regions. An ecosystem approach is pivotal to developing sustainable systems: it recognises the interconnectedness of the natural world and the place of humankind within it. This confers clear responsibilities upon those using land or benefiting from its resources. For those who use land for productive purposes, it requires that natural resources be managed to maintain and enhance life support systems, with all that this implies. It also carries with it an obligation to protect, conserve and enable access to those consumptive, often intangible resources that are deemed public goods. Equally, those accessing public goods have also an obligation to minimise their impact on resources that contribute to sustainable production, as well as on those resources that are inherent in sustaining the integrity and value of the ecosystem. This quite utilitarian view needs also to be mediated alongside a belief that all living things have a right of existence: a belief held by those at the far right of the spectrum of sustainability.

The implication of these apparently simple requirements is that there need to be processes that enable communities to work out what should be done and how to do it. The social dimension of sustainable development requires shared responsibility in decision making within communities; and governance through inclusive participation, respect, and trust in power sharing (O'Riordan, 2002). The reasons for this are fundamental: the need to secure both intra- and inter-generational equity and to agree the responsibilities and obligations to be acted upon. Planning needs to be visualised so stakeholders can respond meaningfully to possible sustainable futures. Whatever is agreed requires to be tested against the principles of sustainability. An explicit recognition of the trade-offs between economic, social and environmental criteria both locally and globally needs to be made. Outcomes require auditing, and evaluating in terms of sustainable production on the one hand, and consumption, on the other.

21.4 Implications for distributing property rights

Through this approach it becomes clearer as to why and how the movement away from a dominant social paradigm of the countryside being a place of 'production' to one of 'consumption' has implications for distributing property rights differently. The challenge is to accommodate changing property rights and responsibilities within a range of land ownership systems towards long-term multi-purpose management regimes (Table 21.1).

21.4.1 State and charitable ownership

If our vision of the mountain areas is one where the imperatives for management are predominantly in the public interest, state ownership can readily meet such an objective provided there is commitment and funding is made available. It is a model used in Europe to a varying extent in different countries. Arrangements for access and the kind of institution that is responsible for management also vary but all have a set of objectives that include conservation and protection. An alternative system that in Scotland has become increasingly prevalent in recent times is charitable ownership. Trusts or non-governmental charitable bodies purchase land to satisfy the public interests in landscape conservation, amenity, and the protection of habitats and wildlife. Objectives for land management are usually more specific to reflect the purpose of the charitable body. Activities are dependent upon public funding and voluntary support.

The relative simplicity of these models enables an exclusive focus on environmental and conservation management objectives that is difficult to achieve using other models. But do they really satisfy the public interest?

They are models that are predominantly managed in 'top down' mode with the assumption that the managers 'know best'. Genuine community involvement is frequently missing: where it occurs there is often an imbalance in representation as between those 'on site' communities and those living and working well outside the immediate area. Trusts and non-governmental organisations (NGOs) do have to take due notice of their members' views, but even in these organisations it is often surprising how 'top down' management dominates and it is often the local communities that have difficulty in getting their views across.

Satisfying the conservation (and protection) interest is a key objective of charitable and state ownership. However, charitable and state organisations are increasingly experiencing the tension that exists in reconciling this interest with that of public access and providing enjoyment of the countryside – often a key component of the mission statements of these organisations. This is an on-going challenge, a challenge that will be more easily overcome if the public itself is involved in determining solutions and sharing responsibilities. Satisfying the public interest and the social imperative of sustainable development requires a greater recognition of the requirement for inclusivity in decision-making.

The charitable ownership model relies less on the exchequer purse and in one sense provides a market for 'purchasing' environmental goods in the public interest that other models do not: but there is a tendency to focus on the 'charismatic' environments. This is

Table 21.1 Land ownership regimes and distribution of interests.

General Nature of Interest	CONSUMPTION		
Specific Interest Ownership Regimes	ENVIRONMENT – CONSERVATION & PROTECTION	ACCESS, RECREATION & SPORT	TOURISM
STATE	Predominant interest often satisfying specific conservation and protection objectives related to international legislation	Facilitated and managed to achieve sustainable use	Important contribution to local economy
CHARITY/NGO	Aims to conserve specific habitats and wildlife in relation to main purpose of charity/NGO	Facilitated and managed in relation to specific conservation objectives	Important contribution to local economy
PRIVATE	Variable and highly dependent upon the landowner's specific interests except where there is legislative intervention	Increasingly determined by legislation and guidance on 'responsible' access. Sporting interest may be an important income generator	Contribution to countryside amenity and landscape character
COMMUNITY	Variable throughout Europe but traditional, low intensity farming methods have generally been consistent with achieving environmental benefits but as production interests have become less viable, positive benefits to conservation, access and tourism, and the local economy have become increasingly dependent on public funding		
NATIONAL PARK	Usually the imperative of management is unequivocally towards conservation and protection of areas of high natural and cultural heritage	Facilitates and controls responsible access – it focuses on the public interest and its rights	Aims to achieve sustainable local development to benefit the local economy as well as the specific conservation interest

why, for environments for which there is market failure in attracting a charitable interest, state ownership is an option that cannot be ruled out.

21.4.2 Private ownership

Such market failure recognises that delivering the public interest involves a financial cost. The public can meet this cost through taxation or levy and rely on the state to purchase the interest on its behalf. On the other hand, it can rely on private individuals or bodies (some being charitable) to make the necessary investment and thereafter appropriate the public

PRODUCTION		FUNDING	CONTINUITY OF MANAGEMENT
AGRICULTURE	FORESTRY		
Incidental to main objective	State forests are increasingly managed as multifunctional areas of land use	Predominantly Government and Local Authority	Potentially high level of continuity of management objectives although subject to changes in political priorities (e.g. shift in UK forestry policy post-1988)
Secondary to main objective and income applied to offset costs of achieving main objective		Voluntary public contributions and Government grants	High level of continuity and consistency in management objectives
Primary purpose but increasingly managed within a set of environmental regulations and incentives. In marginal areas frequently requires inward investment and provides cash flow into local economy		Predominantly private funding with Government support in some areas (e.g. through the CAP)	Frequently a very high level of continuity and consistency in management objectives but may be disrupted by variations in intergenerational management competence
Primary Purposes - Arrangements that determine management and decision-making are community-based and have historically been successful in achieving production objectives. Under present-day conditions many of the traditional systems of communal land resource management are breaking down. New models like community businesses aim to widen options to include non-production activities such as tourism		Private but increasingly reliant on Government funding aimed at maintaining or promoting social cohesion.	Experience is variable but finding a 'good' manager(s) within a community to give the lead is key to performance
Works with local landowners and develops solutions for productive land management that are consistent with achieving its primary objective through systems of grants, incentives and development support		A mixture of Local Authority, national Government and private funding	Highly dependent upon a fully integrated partnership approach that involves the private sector, public investment and local community participation.

interest through designation/regulation or incentive. Throughout the European mountain areas, there are those who have a private interest which may be agricultural, forestry, sporting or minerals based. They have purchased land to pursue production or use objectives and believe they have certain inalienable rights to do so. As far as their activities remain compatible with the public interest and the bundle of rights to which the public believes it is entitled there is not a problem. Such inward investment is often vital to the local economy of mountain areas in Europe where the marginal economics of landowning businesses is often reliant on earnings made elsewhere.

Much of the concern about private ownership centres on the belief that too often owners have operated 'extractive regimes, benefiting the owner, failing the environment, doing little for the local community' (Cramb, 1996, quoted by Warren, 2002). Designing a set of policies and management regimes to correct these failings without disenchanting or compromising the economic viability of the private interest is the challenge. As Charles Warren (2002) argues in his book 'Managing Scotland's Environment', "at least one advantage of private ownership is long-term stewardship". He points out that land management is a long-term business for which continuity of purpose is vital. He quotes Callander (1998), who says "Stewardship – an earlier generation's word for sustainability – sums up the sense of responsibility that has long been regarded as part and parcel of owning land in Scotland". On the other hand, stability of interest across generations may be difficult to achieve and many of the problems associated with estates in Highland Scotland have been due to lack of continuity in ownership (MacGregor, 1994). Private ownership carries with it a huge commitment when the role of stewardship, or sustainability as we define it, is taken seriously. This is why it is argued in Norway (Berge, 2002) where there is community ownership (with common property rights) and where several farms are commoners and owners, that the probability of finding a good manager from among the community is better than from a single household.

21.4.3 Community ownership – common property rights

Within Europe, community ownership, collective ownership or community rights of use are all models that in some form or other provide private individuals with rights to land for 'productive' purposes. Systems of collective ownership or common property rights to pursue agriculture and forestry in the alpine and northern mountain areas of Europe, for example, have developed over centuries. Arrangements that determine management and decision-making have become sophisticated, community-based and invariably successful in meeting the rights of private individuals in the production of private goods. It is also argued that at least in the past they have produced public goods such as biodiversity, and protection against natural forces in forests (e.g. erosion), without the state intervention that has been necessary for those having exclusive private property rights (Glück, 2002). However, in practice, common property rights in Switzerland and Austria have become very dependent upon public support, and the management structures themselves have become fragile (Glück, 2000). There are clearly reservations about the ability of such regimes to deliver multi-purpose management and public goods on a long-term basis without public support. Warren (2002) expresses similar reservations about community ownership in Scotland, a recent form of tenure which has yet to prove that it will raise standards of stewardship. Observations on the design principles of Norwegian Commons (Berge, 2002) also suggest that recent ideas about resource management have not been integrated with legislation but there are clearly significant political and social reasons why forms of collective ownership and common property use will continue. The challenge is that policies will require tailoring and integrating more specifically in relation to these systems. They will also require public support to facilitate multi-purpose management to satisfy both the private and the public interest.

21.4.4 Areas of high natural and cultural heritage value – National Parks?

Finding ways to accommodate the public, private and charitable interests against this diverse pattern of ownership in extensive areas of high natural and cultural heritage value has led some countries to create National Parks. The imperative of management is unequivocally for the environment. It is crucial therefore that the area designated is that which provides maximum coherent benefit to the environment. Management is for the conservation and protection of landscape, habitat and wildlife and aims also to facilitate controlled and responsible access and an enjoyment of the countryside – it focuses on the public interest and its rights. A partnership approach that involves the private sector, public investment and the local communities is fundamental to its success. Increasingly, national park authorities recognise that continued economic development is vital to the long-term sustainability of such parks. This requires that national parks have overall powers to control the planning and development process and the funding to achieve their objectives. The challenge is that the developments do not compromise the management imperative of the environment but rather support it.

21.5 Designing policies for sustainable development in mountain areas

It is clear that from our brief analysis that there is, and there will continue to be, a mixed pattern of ownership and distribution of rights in the mountain areas of Europe. It is important to see this as being useful and necessary in achieving overall sustainable development in the mountain regions where seeking an appropriate overall balance between 'production' and 'consumptive' rights should be at the heart of policies and land management. It will also be important that the evolution of European policies, particularly those that focus on rural development, take into account the diverse economic conditions, geographical circumstances, socio-cultural factors, and traditions to be found there. The 'one size fits all' approach will simply not work. Policies must allow for local interpretation and implementation.

The Ten Nation Study concludes that within the EU, the Rural Development Regulation is emerging as the most significant future policy instrument for rural areas. Currently its budget is too highly constrained and its focus largely agricultural – this must change. The study also directs our attention to the need for a greater emphasis on policy integration. This will be essential in designing relevant and coherent multi-purpose systems of land management that satisfy sustainability criteria for the mountains of Northern Europe.

References

Baldock, D., Dwyer, J., Lowe. P., Petersen, J.-E. & Ward, N. (2002). *The Nature of Rural Development: Towards a Sustainable Integrated Rural Policy in Europe. A Ten Nation Scoping Study Synthesis Report.* Institute for European Environmental Policy, London.

Berge, E. (2002). *Design Principles of Norwegian Commons.* Proceedings of the Workshop on Future Directions for Common Property Theory and Research, Rutgers University, New Brunswick, NJ 28 February, 1997. http://www.mtnforum.org/resources/library/mccax97a.htm.

Birnie, R.V., Curran, J., MacDonald, J.A., Mackey, E.C., Campbell, C.D., McGowan, G., Palmer, S.C.F., Paterson, E., Shaw, P. & Shewry, M.C. (2002). The land resources of Scotland: trends and prospects for the environment and natural heritage. In *The State of Scotland's Environment and Natural Heritage*, ed. by M.B. Usher, E.C. Mackey & J.C. Curran. The Stationery Office, Edinburgh. pp. 41-82.

Callander, R. (1998). *How Scotland is Owned.* Canongate Books, Edinburgh.

Commission of the European Communities (1988). *The Future of Rural Society*, COM (88)371 final. Commission of the European Communities, Brussels.

Dunlap, R.E. & Van Liere, K.D. (1978). The new environmental paradigm. *Journal of Environmental Education*, **9**, 10-19.

Dunlap, R.E. & Van Liere, K.D. (1984). Commitment to the dominant social paradigm and concern for environmental quality. *Social Science Quarterly*, **65**, 1023-1028.

Glück, P. (2000). Policy means for ensuring the full value of forests to society. *Land Use Policy*, **17**, 177-185.

Glück, P. (2002). Property rights and multipurpose mountain forest management. *Forest Policy and Economics*, **4**, 125-134.

Lambert, R. (2001). *Contested Mountains - Nature, Development and Environment in the Cairngorms Region of Scotland 1880-1980*. White Horse Press, Cambridge.

MacGregor, B.D. (1994). Owner motivation and land use change in north west Scotland. Unpublished report. WWF Scotland, Aberfeldy.

Mountain Forum (2002). *Why mountains? Priorities for action*. www.mtnforum.org/members/action.htm.

O'Riordan, T. (2002). A Scottish perspective on the transition to sustainability. From The Macaulay Lecture, 8 October 2002. Unpublished. The Macaulay Institute, Aberdeen.

Price, M.F. & Kim, E.G. (1999). Priorities for sustainable mountain development in Europe. *International Jornal of Sustainable Development and World Ecology*, **6**, 203-219.

Warren, C. (2002). *Managing Scotland's Environment*. Edinburgh University Press, Edinburgh.

22 Abernethy Forest RSPB Nature Reserve: managing for birds, biodiversity and people

D.J. Beaumont, A. Amphlett & S.D. Housden

Summary

1. The area of Abernethy Forest National Nature Reserve is renowned for its outstanding and unique wildlife and conservation importance.
2. The RSPB owns and manages nearly 14,000 ha of land including high mountain plateaux, moorlands and heaths, Caledonian pine woodlands, lochs and wild rivers.
3. Management has been driven by the need for habitat enhancement and expansion of the woodland ecosystem that supports significant populations of rare and threatened species. Habitat and species management techniques continue to be developed and implemented, underpinned by sound ecological research.
4. The nature reserve is integral to the Cairngorms National Park, with its stunning scenery and wildlife interest an obvious attraction to people who wish to experience this environment.
5. The economic activity generated and supported by the RSPB nature reserve is of major local significance.

22.1 Introduction

The Royal Society for the Protection of Birds (RSPB) purchased 615 ha around Loch Garten in 1975 after being involved in this site since 1957, thanks to it being the chosen site of the first pair of ospreys to successfully re-colonise Britain after a 40 year absence. Given the outstanding wildlife interest and conservation importance of this area, other purchases of adjacent land followed, the most recent being in 2002, with the addition of 834 ha of Scots pine woodland at Craigmore. The RSPB currently owns and manages 13,715 ha at Abernethy, land that stretches from around Loch Garten at 200 m above sea level, to the top of Cairn Gorm itself at 1,309 m above sea level, as shown in Figure 22.1. The RSPB's involvement in this area now follows a policy to conserve important bird species and assemblages that occur in scarce or threatened ecosystems, which themselves are in urgent need of conservation action (Housden *et al.*, 1991). The reserve includes large areas of Caledonian pine, bog woodland, dry and wet heath, steep cliffs and montane habitats.

The reserve sits within the Cairngorm Massif and the Cairngorm National Park. The introduction to the Local Prospectus for SNH's Natural Heritage Zone provides the setting in which Abernethy is placed:

Beaumont, D.J., Amphlett, A. & Housden, S.D. (2005). Abernethy Forest RSPB Nature Reserve: managing for birds, biodiversity and people. In *Mountains of Northern Europe: Conservation, Management, People and Nature*, ed. by D.B.A. Thompson, M.F. Price & C.A. Galbraith. TSO Scotland, Edinburgh. pp. 239-250.

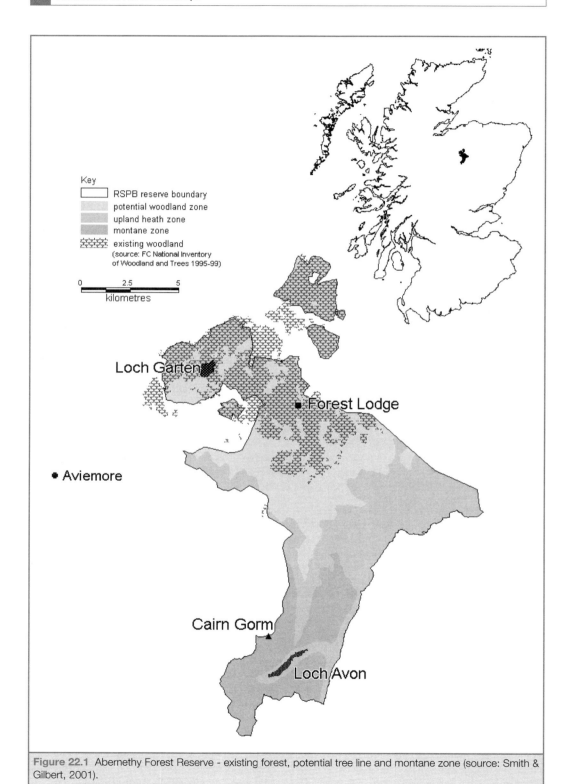

Figure 22.1 Abernethy Forest Reserve - existing forest, potential tree line and montane zone (source: Smith & Gilbert, 2001).

"The Cairngorm Massif is dominated by vast, rolling, boulder-strewn plateaux, ringed by precipitous cliffs which plunge to dark, wild corries. The highest plateaux reach over 1,200 m in altitude, supporting immense tracts of alpine heath and blanket bog which are home to dotterel, ptarmigan and snow bunting. … The area contains the largest remaining fragments of ancient Caledonian pine forest, with characteristic birds such as capercaillie, Scottish crossbill and crested tit. The nature and extent of these landscapes and habitats are exceptional within Scotland" (Scottish Natural Heritage, 2001, p. 5)

22.2 The aims of conservation management at Abernethy

The aim of habitat management at Abernethy is to restore 'present natural' ecosystems, where possible, to their former extent. Present natural conditions being defined as those which would prevail now if human intervention had not become a significant ecological factor (Peterken, 1996). Pollen evidence from peat cores at Abernethy suggests that the effects of human intervention began to change the vegetation composition around 3,000 to 4,000 years ago (O'Sullivan, 1977). Currently the reserve holds the largest remaining tract of Caledonian pinewood and the intention is to extend this up to its natural altitudinal limits, a timber line at c. 560 m above sea level, with scrub pine extending up to c. 650 m, and the pine juniper (*Juniperus communis*) scrub above to c. 800 m. The total area of woodland within the reserve is c. 4,150 ha, of which 1,935 ha are native pinewood, 13.7% of the global resource of remaining native Caledonian pinewood (Forestry Authority, 1994). If the pine forest expansion is successful the area will, over time, increase to about 6,850 ha below the timberline, with sub-alpine scrub above, i.e. an increase in the woodland area of at least 65%. The preferred method for achieving habitat restoration is by natural regeneration, thereby including the restoration of natural transitions from pine forest, to scrub, to montane zones (RSPB, 2001). This management will benefit the important bird populations at Abernethy. At present the RSPB reserve supports c. 80 regular breeding bird species, including 12 red listed and 23 amber listed species. These include nationally significant (above 1% of the UK total) populations of Scottish and parrot crossbills (*Loxia scotica* and *L. pytyopstittacus*), goldeneye (*Fuligula clangula*), capercaillie (*Tetrao urogallus*), black grouse (*T. tetrix*) and crested tit (*Parus cristatus*).

Figure 22.1 shows the present and potential extent of woodland and upland heath and the montane area. The current pine forest merges with a large area of upland heath and blanket mire, across which there with are approximately 500 ha of regenerating woodland, mostly close to existing seed sources. Upland heath and blanket mire are themselves of conservation interest, and some areas will be retained within the restored forest as natural gaps. Further up the hillside, recent work has recorded a significant amount of alpine scrub, mainly juniper, on the high slopes around Loch Avon (Amphlett *et al.*, 2001) and this habitat will continue to benefit from reduced numbers of deer.

Above the alpine scrub is the truly montane zone (land over 800 m above sea level). The montane area within RSPB management extends to 2,351 ha and is as important as the native pinewood, having its own UK Government Biodiversity Action Plan (Anon., 1999, p. 16): 1,154 ha of this area lie above 1,000 m. Here the habitat appears to be a relic from the last Ice Age and nowadays many of the species only occur in higher latitudes, close to or within the Arctic Circle. This montane area probably represents one of the least modified habitats in the UK. Even here, however, there are anthropogenic impacts on this fragile ecosystem; for example, nitrogen enrichment, together with grazing pressure, has been implicated in declines in the extent and quality of montane heaths dominated by the moss *Racomitrium lanuginosum* (Baddeley *et al.*, 1994). Increased tissue nitrogen, associated with air pollution during the 20th century, also poses a threat to snowbed bryophyte communities (Woolgrove & Woodin, 1996a,b). Recreational pressures and climate change could also be having an impact on this habitat. Snow beds that were permanent year-round features of the montane plateau are shrinking and melting earlier during most summers. This could impact upon the fragile bryophyte and lichen communities restricted to these long snow-lie areas (Averis *et al.*, 2004).

22.3 Caledonian pinewoods

22.3.1 Definition and former extent

Native pine woodlands are relic indigenous forests, dominated by self-sown Scots pine (*Pinus sylvestris*), which occur throughout the central and north-eastern Grampians and in the northern and western Highlands of Scotland. They do not support a large diversity of plants and animals compared with some more fertile habitats. However, there is a characteristic plant and animal community which includes many rare and uncommon species. The pine-birch forests of the Scottish Highlands, dominated by Scots pine, are internationally recognised as being of prime conservation importance in the EC Habitats Directive (European Community Directive 92/43/EEC on the Conservation of Natural Habitats and of Wild Fauna and Flora). Caledonian pinewoods were once a major habitat in Scotland, distributed over three-quarters of the Scottish Highlands and covering an estimated 1.5 million ha (McVean & Ratcliffe, 1962). Over a period of 4,000 years, man has contributed to the

Figure 22.2 At Abernethy the RSPB has blocked artificial drains from previous forestry operations in order to re-create naturally high water tables and areas of bog woodland. The work was funded by the European Union's LIFE - Nature Wet Woods Restoration Project. (Photo: Andy Hay – rspb-images.com)

destruction of pinewoods in many ways: fire, felling for timber, clearing for farming, grazing by stock, and by encouraging high densities of red deer (*Cervus elaphus*) and roe deer (*Capreolus capreolus*) for sport hunting. High numbers of deer result in levels of browsing that remove seedling and sapling trees and prevent forest regeneration. By the 1950s the semi-natural pinewoods had been reduced to scattered remnants covering approximately 1% of their former extent (Steven & Carlisle, 1959) and even in recent times, between 1957 and 1987, this area has been further reduced (Forestry Authority, 1994; Mackey *et al.*, 2001). At present there remain some 16,000 ha of native pinewood spread over 77 separate areas across the Highlands (Bain, 1987). Many of the remnant pinewoods, including Abernethy, were under-planted with non-native softwoods or Scots pine, as recently as the 1980s. At Abernethy, a policy of removal of non-native and recently planted trees, including Scots pine, was followed in the first years. These recent plantation areas included plantation infrastructure such as deer fences, drains and planting ridges, all of which were considered to be detrimental to the semi-natural characteristics of the forest. All deer fences have been removed and many of the drains and ditches blocked so as to restore a more natural soil water table (see Figure 22.2).

22.3.2 Forest regeneration and deer management

Expansion of the native woodland resource by natural regeneration (the preference within the Cairngorms Management Strategy) will only be achieved if the population of red deer is reduced. The Deer Commission for Scotland and Deer Management Groups have important roles to play in order to achieve this. Populations of roe and Sika deer (*Cervus nipon*), will also have to be managed (Forestry Commission, 2000).

Darling (1947) believed that red deer density should be kept below 6 per km^2 over the whole red deer range to allow woodland to regenerate while Holloway (1967), working in Deeside, considered 4 deer per km^2 would allow some regeneration. High densities of red deer were present on the high ground and in the forests of Abernethy in the late 1980s (between 10.7 and 11.7 deer per km^2) (Beaumont *et al.*. 1994), restricting the potential for successful natural regeneration, especially as 82% of the diet of woodland red deer is tree and shrub shoots (NCC, 1986).

Fencing deer out of conservation areas may seem to be an obvious solution; however, research has indicated that black grouse and capercaillie collide with deer fences, often with fatal consequences. Collisions occurred at a rate of 0.03 and 0.25 per km of fence per month for black grouse and capercaillie respectively (Baines & Summers, 1997). These two species of woodland grouse have undergone large population declines in recent years and both are on the red list of Birds of Conservation Concern (Gregory *et al.*, 2002) and are UK Biodiversity Action Plan Species (Anon., 1995, p. 106). As a result, around 60 km of deer fences have been removed from within Abernethy Forest Reserve. Removal of fences also improves the natural landscape qualities of the site, and has opened up an additional 470 ha of plantation as shelter formerly unavailable to deer, thereby contributing to a reduction in overall deer density. Without fencing as a deer management tool, the only method of control is by a regulated cull, whereby targets are set, dependent on deer numbers and deer pressure on regeneration. Monitoring the deer population and the impact on woodland regeneration is therefore an important task for RSPB Scotland staff.

Trained and qualified RSPB Scotland staff have reduced deer numbers humanely and efficiently. Being a nature reserve, there is no need to provide animals for sporting clients, although the previous owners of the estate carry out a small part of the cull (an agreement of the sale of the major part of the estate in 1988). Culled deer are sold to two local game dealers. The results of deer management have been monitored and Figure 22.3 shows the numbers of red deer and the median heights of pine seedlings over several years. Once deer numbers were reduced to a density of around 5 per km^2, the field layer, including any

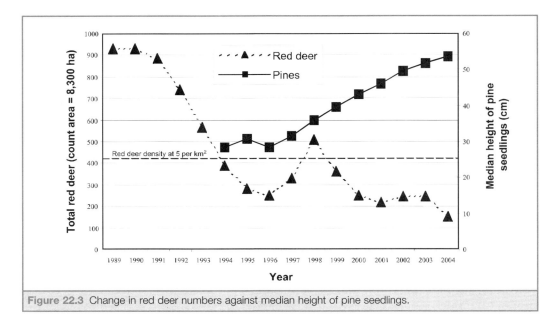

Figure 22.3 Change in red deer numbers against median height of pine seedlings.

seedlings and saplings held below the 'browse-line', began to develop. This produced an initial pulse of growth and regeneration of woodland, without fence-lines, and new woodland is appearing in patches over widely dispersed areas. The reduction of deer grazing pressure has also allowed the field layer shrubs – heather (*Calluna vulgaris*) and blaeberry (*Vaccinium myrtilis*) – to develop and grow and this lush growth is now the target of some new habitat management trials.

22.4 Innovative management

A boreal pinewood in a natural state would be quite a dynamic habitat, constantly shaped and altered by perturbations such as fire, storm damage, insect attack, fungal attack, large herbivores, floods, etc. Scots pine itself is adapted to cope with fire events and has the capacity to rapidly colonise newly burnt ground (Sykes & Horrill, 1981). However, experimental investigation of prescribed fire for pine regeneration, particularly under Scottish conditions, has been limited (see Figure 22.4).

Over the centuries of forestry management, many of these agents of disturbance have been actively discouraged and management has aimed to preclude them in order to produce a healthy crop of timber. A wide variety of highly specialised organisms usually occupies the niches created by natural perturbations. Saproxylic invertebrates for example, often require very particular types of wood rot for their larval stages (McComb & Lindenmayer, 1999). Conservation management therefore has a role in replacing or replicating this dynamism within a forest in order to increase the abundance and diversity of these habitats and to maintain populations that depend on them.

To replicate some aspects of disturbance, several trial management projects have begun at Abernethy. In 2002 and 2003, thirty 35 m x 20 m experimental sites were established in old open woodland, each with a cut, burnt and control plot. A further 10 plots were burnt at the edge of the forest. These will be monitored for the rates of regeneration, the

Figure 22.4 In 2004 the RSPB undertook trial management of the field layer vegetation. Burning is a natural element of pine forest ecosystems and much of the pinewood biodiversity is adapted to fire. Controlled burning was used to assess the impact on tree regeneration and field layer biodiversity. (Photo: Neil Cowie, RSPB).

composition of the field layer, invertebrate colonisation and use by woodland grouse. Early indications from this work are that burnt areas had 3.5 times (95% confidence limits: 1.5-8.1) the number of seedlings as areas established as controls (Hancock *et al.*, 2005). These findings are being used to develop management approaches to stimulate regeneration in artificial gaps within the woodland cover that have no intrinsic conservation interest in their current state.

The Caledonian pine forest ecosystem has also been developed and shaped by the presence of large mammals. Since the last Ice Age, wild cattle (auroch, *Bos primigenius*), beaver (*Castor fiber*), brown bear (*Ursus arctos*), moose (*Alces alces*), lynx (*Lynx lynx*), reindeer (*Rangifer tarandus*), wild boar (*Sus scrofa*) and wolf (*Canis lupus*) were exterminated in Scotland (Dennis, 1995). These animals will all have had their own particular influence in adding to the forest dynamics. The very small patches of remaining forest have been subject to artificially high densities of just one or two large mammals and this has resulted in degradation of forest habitats. As described above, deer numbers at Abernethy have been reduced to levels where tree regeneration is now taking place. The actions and dynamics of other large herbivores can now be tested within the forest. It is impractical to think about restoring auroch or moose to the forest, however cattle could be used as a trial surrogate. Trials are now underway with Highland and Luing cattle on plots in and adjacent to the pine woodland to investigate the impact of large herbivores on these habitats. These trials will continue for 5 years or more, and the results will be used to inform and advise on grazing management of native pinewoods in Scotland.

At present there is no silvicultural management within the semi-natural part of the woodland: this is Britain's largest area of non-intervention woodland. Around 140 ha of Scots pine plantation in the Mondhuie section of the reserve are being managed as continuous cover forestry. This is being done as a demonstration of forest management that allows small-scale continuous timber production, without resorting to clear-felling and planting, as restocking is by natural regeneration. Local contractors carry out the forestry operations and it is hoped that the timber produced at Mondhuie will provide a sustainable and local economic activity. In recognition of the high standards of woodland management and sustainable timber production, Abernethy Forest has been given UK Woodland Assurance Scheme certification (Forestry Commission, 2003). One of the key constituents of a natural woodland is the presence of dead and decaying wood. This is a very important habitat in its own right, being utilised by many species of invertebrate, fungi and lower plant, and it is usually scarce in managed woodlands. A large proportion of the invertebrate species associated with dead wood are local and rare, and many are declining both in Britain and throughout Europe (Kirby, 1992). At Abernethy, the RSPB is investigating ways of creating various dead-wood habitats within old plantation and semi-natural stands (RSPB, 2001). Methods to be used will be designed to mimic semi-natural processes and results. For standing 'poles', all branches will be removed from the trunk of a live tree of above 20 cm diameter at breast height (DBH), leaving the trunk standing, which will be killed by having a ring of bark removed at c.1 m above ground. To mimic the diverse effects of wind snap and wind blow, several methods will be used: tree trunks will be partially cut through at >1 m and pulled over; high stumps will be created by partially cutting through at c. 1 m above ground; some trees will be partially cut through at the base then winched over (with or without loosening around base). To mimic the effects of wind blow, whole trees will be felled and trunks damaged, or branches removed to facilitate fungal infection and the creation of rot holes. In the long term, we envisage that the forest will have 40–100 m^3 ha^{-1} (or locally more) of dead wood >20 cm diameter, with regular input of fresh dead wood in each hectare of the forest. Input rates will be approximately 1 m^3 ha^{-1} yr^{-1}. These values are derived from studies in Scandinavia and north west Russia, in near-natural pinewoods (Soderstrom, 1988; Karjalainen & Kuuluvainen, 2002; Rouvinen et al., 2002a,b; Siitonen et al., 2000).

The importance of this habitat is illustrated by the discovery of the mason bee (*Osmia uncinata*) using holes made in dead wood by the longhorn beetle (*Rhagium inquisitor*) (Edwards, 2001). Both of these species are rare, the beetle is Nationally Scarce and the bee is a UK Biodiversity Action Plan (BAP) and RDB2 species (RDB is short for Red Data Book: RDB2 is vulnerable, and RDB3 is rare). Full definitions are given in Shirt (1987). The bee is also host to the parasitic cuckoo wasp (*Chrysura hirsute*), also a BAP and RDB3 species.

22.5 Biodiversity within Abernethy RSPB Nature Reserve

The pinewoods, bryophyte-dominated snow beds, montane heaths, Loch Avon and its associated catchment are all of outstanding biological and geomorphological importance.

The reserve contains the following EC Habitats Directive Priority and Annex 1 Habitats:

- Caledonian Forest,
- bog woodland,
- blanket mire,
- active raised bogs,
- wet heath,
- dry heath, and
- oligotrophic standing waters.

The numbers of species of conservation importance are given in Table 22.1.

Table 22.1 Summary of the numbers of species from each of the major taxa present at Abernethy.

Taxon	Numbers of species of conservation importance		
	Red Data Book	Nationally rare or scarce	Biodiversity Action Plan species
Fungi		98	13
Lichens	12	104	3
Bryophytes	8	80	2
Flora (higher plants)	3	43	2
Coleoptera (beetles)	45	159	1
Diptera (true flies)	58	54	2
Odonata (dragonflies)	2	1	
Lepidoptera (butterflies and moths)	9	28	4
Hemiptera (bugs)	0	5	0
Hymenoptera			
Wasps	2	1	1
Bees	1	0	1
Ants	1	2	3
Sawflies	1	0	0
Arachnida (spiders, mites and ticks)	5	7	1
Mammals	?	10	5
Birds	13 (red listed)	33 (amber listed)	10
Totals	207	625	48

Of the 519 species for which their major habitat associations are known, 73% are associated with woodland, 14% are restricted to successional habitats (grassland, heaths and especially open disturbed areas, e.g. by running water), and 19% are associated with montane habitats. For some of these species, Abernethy is the only known location in the UK and several were new to science when discovered at Abernethy. For example, *Megaselia abernethae* and *M. gartensis* are two scuttle flies named after their place of discovery (Disney, 1988), and *Agabus wasastjernae* is a water beetle that was known only from fossilised remains until found at Abernethy (Owen *et al.*, 1992).

22.6 The benefits of a healthy environment

This area of stunning scenery and wildlife interest is an obvious attraction to people who wish to experience this environment. From the organised coach trip to see the ospreys to the highly skilled mountaineer, off-piste skier, canoeist or hill-walker, this area is enjoyed by many thousands of visitors each year. The RSPB has an open access policy to the Abernethy Nature Reserve and many people use the well maintained passes and routes that are present on the reserve. The RSPB also tries to manage the visitor pressure by providing facilities, such as the Osprey Centre, which are in a part of the reserve robust enough to allow relatively high levels of human activity. By employing information wardens, the RSPB also provides a way for visitors to find out more about this reserve, its management and wildlife interest. The Osprey Centre attracts between 40,000 and 50,000 visitors each year and an additional 50,000 people visit parts of the reserve for the scenery, walking and wildlife. Close to Loch Garten, a series of waymarked walks is much used by visitors and local residents alike, and there is unrestricted access over an additional 100 km of tracks. The reserve is also used regularly as a venue for seminars and training courses, with delegates using local accommodation facilities. Natural history experts, both professional and amateur, also visit the reserve each year to add to our knowledge of the site, and all stay locally. From the remote mountain areas to the more accessible parts of the Cairngorm massif, a wide range of recreational, agricultural and forestry activities take place in this superlative landscape, providing economic benefits to the local area and supporting a high proportion of local employment. A study published in 1996 (Rayment, 1996) showed that the visitors to the Osprey Centre spent £1.7 million locally, supporting around 69 full time equivalent (FTE) jobs. Excluded from these figures were the other visitors who stay locally and visit the reserve, but not the Osprey Centre, during the summer season and the rest of the year. To put this into perspective, 19 out of 74 advertisements featuring holiday or guest accommodation in mainland Highland in the RSPB members' magazine, *Birds*, mentioned Loch Garten/Abernethy.

Direct on-site employment accounts for 11 FTE jobs. Other sources of employment are 1.25 FTE jobs for contractors engaged on the site; 5.1 FTE jobs supported by the local spending of staff and contractors and buying supplies; 1.4 FTE jobs with the local timber and venison dealers supported by the site's production. In total, therefore, Abernethy now supports 87 FTE jobs in the Badenoch and Strathspey economy, as well as sustaining crofting tenants, a shooting let and a sawmill. These benefits contrast with the 1.5 FTE directly employed workers at the time when Abernethy was managed as a traditional sporting estate.

References

Amphlett, A., Edgar, A., Robinson, P., Taylor, Z. & Watson, R. (2001). RSPB Abernethy Forest Reserve, Regeneration Survey 2000/2001. Unpublished report. Royal Society for the Protection of Birds, Edinburgh.

Anonymous (1995). *Biodiversity: The UK Steering Group Report - Volume 2: Action Plans.* HMSO, London.

Anonymous (1999). *UK Biodiversity Group Tranche 2 Action Plans: Volume II - Terrestrial and Freshwater Habitats.* English Nature, Peterborough.

Averis, A., Averis, B., Birks, J., Horsfield, D., Thompson, D. & Yeo, M. (2004). *An Illustrated Guide to British Upland Vegetation.* Joint Nature Conservation Committee, Peterborough.

Baddeley, J.A., Thompson, D.B.A. & Lee, J.A. (1994). Regional and historical variation in the nitrogen content of *Racomitrium lanuginosum* in Britain in relation to atmospheric nitrogen deposition. *Environmental Pollution,* **84**, 189-196.

Bain, C. (1987). *Native Pinewoods of Scotland - A Review 1957-1987.* Royal Society for the Protection of Birds, Sandy.

Baines, D. & Summers, R.W. (1997). Assessment of bird collisions with deer fences in Scottish forests. *Journal of Applied Ecology,* **34**, 941-948.

Beaumont, D., Dugan, D., Evans, G. & Taylor, S. (1994). Management of Deer for the Regeneration of Caledonian Pine Forest at the RSPB's Abernethy Forest Reserve. *RSPB Conservation Review* 8. Royal Society for the Protection of Birds, Sandy.

Darling, F.F. (1947). *Natural History in the Highlands and Islands.* Collins, London.

Dennis, R. (1995). Return of the wild. *Ecos, British Association of Nature Conservationists,* **Vol. 16 No. 2**.

Disney, R.H.L. (1988). Four species of scuttle fly (Diptera: Phoridae) from Scotland new to Britain, including three new to science. *Glasgow Naturalist,* **21**, 433-439.

Edwards, M. (2001). Survey of three Biodiversity Action Plan Bee Species (*Colletes floralis, Osmia inermis, O. uncinata*) in Scotland, 2001. RSPB, Edinburgh and SNH, Perth.

Forestry Authority (1994). *Caledonian Pinewood Inventory.* Public Information Division, Forestry Commission, Edinburgh.

Forestry Commission (2000). *Cairngorms Forest and Woodland Framework Volume I.* Forestry Commission, Edinburgh.

Forestry Commission (2003). *Introduction to the UK Woodland Assurance Standard.* Public Information Division, Forestry Commission, Edinburgh.

Gregory, R.D., Wilkinson, N.I., Noble, D.G., Robinson, J.A., Brown, A.F., Hughes, J., Procter, D., Gibbons, D.W. & Galbraith, C. (2002). The population status of birds in the United Kingdom, Channel Islands and Isle of Man: an analysis of conservation concern 2002-2007. *British Birds,* **95**, 410-448.

Hancock, M., Egan, S., Summers, R., Cowie, N., Amphlett, A., Rao, S. & Hamilton, A. (2005), The effect of experimental prescribed fire on the establishment of Scots pine *Pinus sylvestris* seedlings on heather *Calluna vulgaris* moorland. *Forest Ecology and Management,* **212**, 199-213

Holloway, C.W. (1967). *The effect of red deer and other animals on naturally regenerated Scots pine.* Unpublished PhD thesis, University of Aberdeen, Aberdeen.

Housden, S., Thomas, G., Bibby, C. & Porter, R. (1991). Towards a Habitat Conservation Strategy for Bird Habitats in Britain. *RSPB Conservation Review,* **2**, 50-53.

Karjalainen, L. & Kuuluvainen, T. (2002). Amount and diversity of coarse woody debris within a boreal forest landscape dominated by *Pinus sylvestris* in Vienansalo wilderness, eastern Fennoscandia. *Silva Fennica,* **36**, 147–167.

Kirby, P. (1992). *Habitat Management for Invertebrates: A Practical Handbook.* Joint Nature Conservation Committee, Peterborough and Royal Society for the Protection of Birds, Sandy.

Mackey, E.C., Shaw, P., Holbrook, J., Shewry, M.C., Saunders, G., Hall, J. & Ellis, N.E. (2001). *Natural Heritage Trends: Scotland 2001.* Scottish Natural Heritage, Perth

McComb, W. & Lindenmayer, D. (1999). Dying, dead and down trees. In *Maintaining Biodiversity in Forest Ecosystems*, ed. by M.L. Hunter. Cambridge University Press, Cambridge. pp. 335-372.

McVean, D.N. & Ratcliffe, D.A. (1962). Plant communities of the Scottish Highlands: a study of Scottish mountain, moorland and forest vegetation. Nature Conservancy Monographs No. 1. HMSO, London.

NCC (1986). *A method for the monitoring of native tree regeneration in relation to the presence of red deer.* Unpublished report. Nature Conservancy Council, Peterborough.

O'Sullivan, P.E. (1977). Vegetation history and the native pinewoods. In *The Native Pinewoods of Scotland*, ed. by R.G.H. Bunce & J.N.R. Jeffers. Institute of Terrestrial Ecology, Cambridge. pp. 60-69.

Owen, J.A., Lyszkowski, R.M., Proctor, R. & Taylor, S. (1992). *Agabus wasastjernae,* new record, (Sahlberg) (Col.: Dytiscidae) new to Scotland. *Entomologists Record and Journal of Variation.* **104**, 225-230.

Peterken, G. (1996). *Natural Woodland.* Cambridge University Press, Cambridge.

Rayment, M. (1996). *Abernethy Forest Nature Reserve - Its Impact on the Local Economy: A Case Study.* Royal Society for the Protection of Birds, Sandy.

Rouvinen, S., Kuuluvainen, T. & Karjalainen, L. (2002a). Coarse woody debris in old *Pinus sylvestris* forests along a geographic and human impact gradient in boreal Fennoscandia. *Canadian Journal of Forest Research.* **32**, 2184–2200.

Rouvinen, S., Kuuluvainen, T. & Siitonen, J. (2002b). Tree mortality in a *Pinus sylvestris* dominated boreal forest landscape in Vienansalo wilderness, eastern Fennoscandia. *Silva Fennica,* **36**, 127–145.

RSPB (2001). *Abernethy Forest Nature Reserve Management Plan.* Royal Society for the Protection of Birds, Edinburgh.

Scottish Natural Heritage (2001). *Natural Heritage Futures: 11 - Cairngorms Massif.* Scottish Natural Heritage, Perth.

Shirt, D.B. (1987). *British Red Data Books: 2. Insects.* Nature Conservancy Council, Peterborough.

Siitonen, J., Martikainen, P., Punttila, P. & Raugh, J. (2000). Coarse woody debris and stand characteristics in managed and unmanaged old-growth boreal mesic forests in southern Finland. *Forest Ecology and Management,* **128**, 211–225.

Smith, S. & Gilbert, J. (2001). *National Inventory of Woodland and Trees.* Forestry Commission, Edinburgh.

Soderstrom, L. (1988). Sequence of bryophytes and lichens in relation to substrate variables of decaying coniferous wood in Northern Sweden. *Nordic Journal of Botany,* **8**, 89–97.

Steven, H.M. & Carlisle, A. (1959). *The Native Pinewoods of Scotland.* Oliver & Boyd, Edinburgh.

Sykes, J.M. & Horrill, A.D. (1981). Recovery of vegetation in a Caledonian pinewood after fire. *Transactions of the Botanical Society of Edinburgh,* **43**, 317-325.

Woolgrove, C.E. & Woodin, S.J. (1996a). Current and historical relationships between the tissue nitrogen content of a snowbed bryophyte and nitrogenous air pollution. *Environmental Pollution,* **91**, 283-288.

Woolgrove, C.E. & Woodin, S.J. (1996b). Effects of pollutants in snowmelt on *Kiaeria starkei*, a characteristic species of late snowbed bryophyte dominated vegetation. *New Phytology,* **133**, 519-529.

23 Ownership in the mountains of Northern Scandinavia and its influences on management

Audun Sandberg

Summary

1. Mountain land under different types of ownership provides both harvestable and less tangible 'goods'. Most of the latter are common goods; for these, the combination of characteristics of both private and public goods creates various management problems.

2. From the 18th century in Northern Scandinavia, predators were seen as harmful, and therefore many were eradicated. The resulting 'simplified ecology' was appropriate for animal husbandry. In the late 20th century, new public goods emerged: biodiversity, ecological resilience and opportunities for recreation.

3. Mountain landscapes and animal populations are changing rapidly due to strong social forces, especially the modernisation of agriculture, animal husbandry and forestry.

4. The model of management needs to shift from one based on single sectoral interests to one based on property rights. This would lead to more integrated and sustainable management of these cultural landscapes.

23.1 Introduction

In the ongoing discussion about the fate of our mountain ecosystems, the question of property rights is often downplayed. Addressing quite simple questions, like "who owns this?" and "in what way is this owned?", often could have opened up new avenues of inquiry and generated alternative policy options. In Northern Scandinavia, the mountains are partly administered by various sectoral authorities, partly managed by government agencies and state companies, and partly utilised by farmers, herders and hikers. This resulting governance of mountain resources has traditionally been based on sector rationality, organised interests, bureaucratic systems, expertise systems and the preferences of political parties. Thus, this chapter argues, there is a need to give property rights a more prominent role in the analysis of how harvestable and recreational resources are managed for the common good.

Property rights can rightly be seen as the prime link between the biophysical world – often called the 'environment' – and the social world – often called 'society'. Every relationship between one mountain user and another in relation to a mountain resource

Sandberg, A. (2005). Ownership in the mountains of Northern Scandinavia and its influences on management. In *Mountains of Northern Europe: Conservation, Management, People and Nature*, ed. by D.B.A. Thompson, M.F. Price & C.A. Galbraith. TSO Scotland, Edinburgh. pp. 251-258.

involves some aspects of rights and duties that can be defined as some kind of property right relationship. Whether this relationship is handed down to us as *res nullius, res publica, res communes* or *dominium* is often the result of the twists and turns of a country's special property history (see Cramb, 1997). Whether a property rights arrangement should consist of all the five possible property rights (access, harvest, management, exclusion and alienation rights), or only some of these, can be seen as an ideological question, e.g. of freedom to roam versus the freedom to own property. But it can also be seen as a more practical question of which mixture of rights and duties produces the most healthy mountain ecosystem, nowadays measured in the form of biodiversity indicators and resilience in the face of external shocks. A wider interpretation of mountain policies must therefore also include the design of rights and duties in relation to mountain resources.

23.2 The nature of mountain goods

The 'goods' of the mountains include tangible, edible products like fish, game, berries and forage for grazing animals. They also include more abstract goods like the experience of the splendour and tranquillity of mountain landscapes. In recent years, a number of new goods have been identified that we expect mountains and forests to provide, for example a high biological diversity that contributes to healthy ecosystems that, in turn, serve us with a continuous flow of ecosystem services and more resilient nature able to absorb external influences. It is important to be aware that most of these goods, whether they are coming from private land, commons or public lands, have the distinct character of common goods or common pool resources (CPR). This is primarily the case for the material harvestable goods, but with increased leisure use – and accompanying congestion around popular sites ('honey pot effects') – this will, to an increasing extent, also be the case for so-called 'experience goods'.

To aid the understanding of the nature of mountain goods, it can be helpful to introduce some analytical tools. In the literature, the two crucial characteristics of common goods are that: a) it is costly to create institutions that exclude others from enjoying them, and b) one unit of a resource that is harvested by one user is no longer available to other users. This means that most mountain goods – like all common goods – have some characteristics that resemble public goods (the cost of exclusion) and some characteristics that resemble private goods (subtractability) (Ostrom, 1994). It is the combination of these two characteristics that most often creates problems for the management of mountain and forest resources, and creates overgrazing in reindeer pastures, produces over-fishing in popular trout rivers and is responsible for over-hunting in popular grouse terrains. Such problems are sometimes called 'Tragedies of the Commons', and resources that are in danger of such a tragic fate are analytically labelled 'Commons resources' (*Allmenningsressurser*) – irrespective of the status of property rights in the actual territory (Jentoft, 1998). These should not, however, be confused with the more precise legal category 'Common Property Resources' (*Res Communes*) for which, given the proper institutional arrangements, both the subtraction problem and the exclusion problem can be handled in a legitimised way.

23.3 From simplified to complex ecologies

Both expert and popular views on mountain ecology have changed during the modern age. The early modern thinking of the 17th and 18th centuries had ideas of constant progress and a rationalisation of man's relation to nature. But the mercantilist states of Scandinavia also had ideas of saving on foreign exchange by stepping up domestic food production and improving the use of domestic resources for the rapidly growing populations of these emerging nations. In northern countries, internal colonisation was one of the measures used for this purpose. Frontier human settlements were established in many parts of these countries. As in frontier settlements everywhere, the first settlers suffered serious hardships from close contact with the 'wild' ecosystems, in the form of diseases, weeds, pests, ticks, vermin or predators (Edvardsen, 2000).

In order to secure a surplus of food from agriculture and animal husbandry over and above the needs of the farm itself, the religious and scientific knowledge of the time was employed to motivate the settlers (in the language of those days this was termed 'enforcement of nature'): only by actively domesticating and refining nature through modern agricultural methods and animal husbandry could a surplus be achieved to feed the workers needed in the growing manufacturing sector. Consequently, already in 1730 and 1733, we find the first specific predator laws in Norway (frd. March 2. 1730, frd. May 8. 1733 – Kingdom of Denmark/Norway, 1795). A state-launched bounty system for eradication of predators started with wolves (*Canis lupus*) in 1730, followed by bears (*Ursus arctos*) in 1733 and a comprehensive 'Law of Eradication of Predators and of Protection of other Game' in 1845. The 1845 law explicitly included bears, wolves, lynx (*Felix lynx*), wolverine (*Gulo gulo*), golden eagles (*Aquila chrysaetos*), eagle owl (*Bubo bubo*) and hawk (*Accipiter* spp.). Twenty-nine different species were listed as harmful species to be eradicated (Kvaalen, 1997). Together these laws and regulations give a clear account of the social construction of nature at the peak of modernising optimism. Active human management of nature was thus institutionalised at an early stage, allowing the surplus production of useful creatures by the removal of harmful or useless creatures. Development was synonymous with a simplified ecology and this was perceived as a public good – an enhanced and predator-free environment.

The bounty incentive worked well in relation to its objective and, combined with a continued demand for furs, expanding human settlements and advances in shotgun and rifle technology, the outcome was a virtual extinction of predators in the course of 200 years. This era embraced more than six generations; thus we find that the early modern ideology of 'enhancement of nature' has become the 'traditional' sentiment in most contemporary rural communities. Even among the Saami reindeer herders, whose predator-evading nomadic pastoralism most probably goes back to the 15th century, the creation of 'predator-free ranging areas' is now presented as a traditional Saami value. This simplified ecology encouraged a dramatic modernisation of agriculture and animal husbandry in the Northern European mountains. It produced a vacant niche which created an opportunity for a rationalisation and expansion of animal husbandry that had not been envisaged before. In this vermin-free environment there took place an extensification of reindeer and sheep ranching that would increase the production levels dramatically – in spite of a rapidly decreasing rural population towards the end of the 20th century.

The early modern ideas of 'simplified ecologies' have now been replaced by modern ideas of biodiversity and ecological resilience. In order to understand the depth of the emotions in the type of conflicts called 'the Environmental Backlash', it is necessary to emphasise the contrast between a simplified nature yielding a harvestable surplus and a complex nature yielding resilience and robustness. Many mountain communities perceive the objective of biodiversity (as a representation of an enhanced nature) as a threat to their customary way of life (Sandberg, 1998). The idea of ecological resilience is basically the same as the supremacy of ecological complexity. When such an objective is introduced, either as policies for 'protection of endangered species' or as 'municipal biodiversity policy', it is in direct challenge to the traditional perception of ecologically simplified nature as the resource base for most of the animal husbandry and big game hunting in Northern Scandinavia. But in a number of landscapes in this northern region, for example the Salten area, the habitats are closely compressed within a short geographical distance from coastal islets to alpine glaciers (Elgersma & Asheim, 1998). Here, the objective of enhanced nature with more complexity is over-run by the ecological events themselves.

23.4 Social and ecological dynamics

Apart from the expansion of animal husbandry into vacant mountain niches, and its extensification in these, a number of social forces are rapidly changing the appearance of mountain landscapes. The most powerful of these drivers originate in the long-running processes of the modernisation and industrialisation of agriculture, animal husbandry and forestry. Fields are increasingly concentrated on flat areas and former moorland, and the harvest of firewood is dramatically reduced. Mixed grazing with livestock in hills and valley slopes has ceased, and the extensification of sheep (*Ovis aries*) herding and reindeer (*Rangifer tarandus*) nomadism has resulted in lower levels of herding and human influence of these activities in the open alpine zone. The introduction of tradeable milk quotas has led to a concentration of dairy-goat stations and to widespread amalgamation of dairy farms and consequent rural depopulation, especially in mountain valleys. At the same time, the rural community-based combined work/leisure use of mountains has changed character, and a more urban-influenced and individual leisure use is now predominant (Bærenholdt, 2002). This is less predictable and lacks the monitoring oversight of the more traditional use of the same mountain area for berry picking, hunting and fishing.

These changes have worked alongside state conservation regulations for habitat protection of pristine nature and endangered species, producing a number of strong ecosystem changes that, once started, tend to have their own dynamic. Thus, on the under-used hill and valley slopes, there is a dramatic regrowth of woodland on former pastures and former wood clearings leading to loss of the former cultural landscape. This young regrowth has resulted in an increase in wild forest herbivores like red deer (*Cervus elaphus*), roe deer (*Capreolus capreolus*) and moose (*Alces alces*), which have also benefited from the largely predator-free environments, and from new research-based shooting strategies. But eventually a growth in numbers of herbivores leads to an expansion of their wild predators, as has happened in the case of the deer to lynx and the moose to wolf relationships. Over time, the succession of vegetation in the now more homogenous forest landscapes probably will mean less favourable conditions for wild herbivores, in the face of an increased predator population and strengthened hunting interests (Sandberg, 1999).

In turn, social consequences arise from these ecological effects. The increased presence of predators in all mountain landscapes creates gradually worsening conditions for extensive sheep and reindeer husbandry, and thus adds to the downward spiral of rural depopulation. Recovery of woodland cover also results in less attractive landscapes on hills and slopes, where the mosaic of the cultural landscape earlier attracted human hikers. The modern hiker avoids dense shrubs and prefers the open alpine zone. While the present hunting bonanza is only a temporary phenomenon, the longer term social effects of the present ecological processes are grave. In spite of all of the fine objectives of ecological biodiversity and resilience, the result of this late modern ecology, created by modern citizens and nature together, might not be the intended increased complexity, but instead lead to ruin for all; herder, moose, lynx and hunter alike.

23.5 Challenges to interest-based management

The fundamental changes that these ecological and social processes bring about present challenges to the present model of management in Northern Scandinavia. This is largely a model of interest-based management of mountain and forest resources. The different agricultural industries and the important leisure and environmental organisations have, over the decades, established mutual and stable co-management and consultation arrangements with their corresponding state sector administrations. Conventionally these have been termed the forest segment, the reindeer segment, the hunting segment, etc. The sectoral segments were thought to be stable governing institutions but, like other industrial age management models, the model now shows signs of fatigue in the form of fragmentation, lack of co-ordination and lack of legitimacy from groups other than those specifically involved. It is also unable to handle an increasing number of 'cross-over problems' arising between the segments, like the effect of alpine sheep grazing on food for grouse (*Lagopus lagopus*).

The prime deficiency of this model is that it cannot manage the totality of the landscape resources, being focused on single-interest resources such as forests, pasture, game, fish, and tourist trails. Furthermore, the policies underlying the management of these crucial parts of the mountain environment are not applied directly, but indirectly through centrally-negotiated agreements between the relevant state sector and the responsible interest organisation. Most of the modernisation and rationalisation of agricultural use and animal husbandry in the uplands has its roots in the incentive systems provided through these agreements, but without the parties to the agreements being responsible for their wider ecological effects. Thus the reforestation processes mentioned above, whether by default or by human neglect, are not actions for which any responsibility is accepted; they are seen to be the fault of neither the landowner of the ground, nor the relevant authorities.

A financial support and advice system for northern agriculture and reindeer pasturing that incorporated, to a greater extent, the need for maintenance of cultural landscapes (multi-functionality) would solve some of the problems arising from the present fragmentation. However, opposition to such ideas is strong both among individual farmers and in the agricultural sector generally. Farmers' values are tied to the production of food and the management models of the industrial age, and a change to these attitudes will be difficult to achieve. In the meantime, the changing rationales for environmental policy, i.e. the virtue of a more complex understanding of nature, cause the contemporary farmer to be firmly caught in a trap of shifting public policies – termed by some a 'modernity trap'.

23.6 Claims for more rights-based management of mountain resources

With the deficiencies of the interest-based management model, there have been claims that new management systems – based on property rights – would give more integrated and more sustainable management. The rationale is that property rights in their various forms contain both rights and duties towards the resource and that the framework for property rights can be designed afresh, or be restructured to provide more accurately targeted solutions to a particular ecological problem (Rynning, 1928; Goodin, 1996). This does not imply going back to the feudal management system of common lands from the period 1500-1850 (De Moor *et al.*, 2002); there are around the world a number of long-enduring 'Common Property Regimes' that continue to manage ecosystem resources like irrigation systems, forests and pastures in efficient, sustainable and legitimate ways (Ostrom, 1994).

However, it is very rare that one can design property rights from scratch, and courts are full of cases where groups quarrel over old property rights. Yet, at certain stages in history, a window of opportunity can appear. The long constitutional process towards an agreement about Sami rights to land and water in Northern Norway is one case. Another case, which is less known, is the long-standing claim of the Northerners (*Håløyg*) to get back their 'Commons' which they claim the Danish/Norwegian King sold illegally in 1751 (Taranger, 1892; Sandberg, 2000). These Commons were later bought back by the state, and it is claimed that in the process the 'Commoners' rights' were lost. The resulting status for these lands as State Commons without commons rights is used as a rationale for exempting Northern Norway from the general Norwegian Commons legislation. Therefore, a state-owned forest company, *Statskog*, today exercises the management rights over these mountain areas alone.

With both of these claims unresolved in the North, the property rights situation there has been characterised as exceptional and temporary (Hagen *et al.*, 1999). Therefore, if the Sami land claims are resolved within the near future, it might be possible to renegotiate all property rights arrangements between communities, municipalities and the State in the north. It is probable that some form of general common rights arrangement (a 'commonification' as opposed to privatisation or nationalization) will be best suited to accommodate the Sami, community and municipal interests. If, in the future, Commons legislation (in Norway called the 'Mountain Law') is made applicable also for the North, a Mountain Council would be responsible for all use of mountain resources within the municipality (Falkanger, 1998). Thus it would be better equipped to solve inter-sectoral conflicts and also have an easier task working with private landowners than the state forest company, and therefore have a greater potential for co-ordinating hunting and fishing management with them. Thus a more integrated responsibility towards the mountain ecosystems should be possible in order to overcome the present fragmentation in the management of mountain areas.

However, if the future application of this legislation specifies a mountain council for each municipality in the north, a new kind of fragmentation might appear. While municipalities have remained small, the modern mountain user in the north is not confined to municipal entities. The functional mountain region and the extensive scale of management needed is considerably larger than a municipality, and any legitimate management system has to accommodate this.

23.7 Resource ownership and adaptive ecosystem management

The real challenge might not lie with specifying the correct administrative boundaries for future Mountain Councils. It most likely lies with the ambitions of devolution of the political responsibility for sustainable development to management levels lower than the state (Sandberg, 2002). Past conflicts over coastal protection and predator protection have shown clearly that the extensive chain of governance, from the Parliament to the individual citizen, was too long to have legitimacy in local details. Thus the Scandinavian states are looking for mechanisms by which the local or regional elected bodies can take over some of the governance of the environment, with an aim of securing local support for the late modern objectives of biodiversity and ecological resilience (Holling & Sanderson, 1996). The argument often used towards reluctant local policy makers is now less that this is part of the nation's international obligations and more that it is their responsibility to secure a continued flow of ecosystem services to their nation's inhabitants (UNEP/CBD, 1998; Direktoratet for Naturforvaltning, 1999; Edwards, S., this volume). Such ecosystem services should then include familiar common goods such as drinking water, biogenetic resources and local habitats for leisure, teaching and experience. By specifying responsibilities for the flow of goods from a particular ecosystem, it follows that regional and local government must also take greater responsibility for mountains and coasts as ecosystems. As mentioned above, there is now an opportunity to achieve some of this in Northern Norway by enacting legislation that actually gives management rights to Sami and locally- or regionally-based institutions. The difficult part of any devolution of property rights is whether also to delegate the right to exclude non-locals who do not contribute to the maintenance of an ecosystem. The exact specifications of the nature of the bundle of property rights are therefore important in any devolution, but in principle should not cause any problems since there is already a tradition for a certain price-differentiation between 'locals' and 'strangers' in terms of hunting and fishing licenses; since the basic idea of a 'Commons' does not contain any constitutional right to alienate the resource.

That leaves us with the question of who owns the ecosystem resources – or whether it is possible to own an ecosystem. In principle, an ecosystem is the most 'common' and indivisible of all natural resources, but at the same time it requires collective decisions to be maintained and to determine which ecosystem properties and services are important for humans. As in the case of unintended reforestation, nature cannot be left entirely to itself. The outcome of the present processes of devolution is most probably that owners will have to work closely with local political institutions to decide the most desirable state for the regional ecology: a question that belongs to the fascinating field of 'political ecology', which is neither just politics, nor just ecology.

References

Bærenholdt, J.O. (2002). *Coping Strategies and Regional Policies – Social Capital in the Nordic Peripheries.* Nordregio Report R2002:4. Nordregio, Stockholm.

Cramb, A. (1997). *Who Owns Scotland Now? The Use and Abuse of Private Land.* Mainstream Publishing, Edinburgh.

De Moor, M., Warde, P. & Shaw-Taylor, L. (Eds) (2002). *The Management of Common Land in North West Europe, c.1500-1850. Comparative Rural History of the North Sea Area,* CORN 8. BREPOLS Publishers, Turnhout.

Direktoratet for Naturforvaltning (1999). *Strategisk dokument for 2000-2003, med visjon for 2008, Ver. 5.1.* Direktoratet for Naturforvaltning, Trondheim.

Edvardsen, H. (2000). *Bruk og ikke-bruk av utmarka, konsekvenser for vegetasjon, i Utmark i Salten – endringer og utfordringer. Sluttrapport fra prosjektet: Friluftsliv i utmarka, "Fritidsbruk av utmarksressurser: Salten som case-område".* NF-rapport nr. 32/2000. Nordlandsforskning, Bodø.

Elgersma, A. & Asheim, V. (1998). *Landskapsregioner i Norge – landskapsbeskrivelser.* NIJOS-rapport 2/98. Norsk institutt for jord- og skogkartlegging, Ås.

Falkanger, T. (1998). Fjellovens anvendelse i Nordland og Troms, Juridisk betenkning til Norges Bondelag av 26 mai 1998.

Goodin, R.E. (Ed) (1996). *The Theory of Institutional Design.* Cambridge University Press, New York.

Hagen, S.E., Arnesen, T, Ørbeck, M. & Lien, J.A. (1999). *Evaluering av Statsskog SF.* ØF-rapport nr. 15/1999, Østlandsforskning, Lillehammer.

Holling, C.S. & Sanderson, S. (1996). Dynamics of (dis)harmony in ecological and social systems. In *Rights to Nature,* ed. by S. Hanna, C. Folke & K.-G. Mäler. Island Press, Washington, DC. pp. 57-85.

Jentoft, S. (1998). Allmenningens komedie, Medforvaltning i fiskeri og reindrift. Ad Notam, Gyldendal Forlag/MAB, Oslo.

Kingdom of Denmark/Norway (1795). Kongelige Forordninger og Aabne breve, samt andre trykte Anordninger som fra aar 1670 ere udkomne. Niels Kristensens Trykkeri, Kiøbenhavn.

Kvaalen, I. (1997). Rovdyr i risikosamfunnet? Om konstruksjon og administrasjon av Naturen. In *Miljøsosiologi - Samfunn, miljø og natur,* ed. by Ann Nilsen. Pax Forlag, Oslo. pp. 29-54.

Ostrom, E. (1994). *Neither market nor state: governance of common pool resources in the twenty-first century,* IFPRI Lecture Series No. 2. International Food Policy Research Institute, Washington DC.

Rynning, L. (1928), Allemannsrett og Særrett, Belyst ved forskjellige arter av rådighet over fremmed grunn, som ikke gaar ind under de alminnelige bruksrettigheter. Vedlegg til "Tidsskrift for Rettsvitenskap" 1928.

Sandberg, A. (1998). Environmental backlash and the irreversibility of modernisation. Presented at 'Crossing Boundaries', the seventh annual conference of the International Association for the Study of Common Property, Vancouver, British Columbia, Canada, 10th-14th June 1998.

Sandberg, A. (1999). Conditions for community-based governance of biodiversity, NF-report No 11/99. Nordlandsforskning, Bodø.

Sandberg, A. (2000). Rettighetsbasert, interessebasert eller politisk basert utmarksforvaltning - Forvaltning av nordnorske allmenninger i de 3dje årtusen, i Utmark i Salten – endringer og utfordringer, Sluttrapport fra prosjektet: Friluftsliv i utmarka, "Fritidsbruk av utmarksressurser: Salten som case-område". NF-rapport nr. 32/2000. Nordlandsforskning, Bodø.

Sandberg, A. (2001). *Institutional Challenges for Common Property Resources in the Nordic Countries.* Nordregio Report R2001:7. Nordregio, Stockholm.

Sandberg, A. (2002). Statlig miljøpolitikks regionale og distriktspolitiske effekter – intenderte og uintenderte virkninger, Notat for "Effektutvalget", Bodø 2002.

Taranger, A. (1892). Fremstilling av de Haalogalandske Almenningers retslige stilling, Juridisk betenkning av 23 April 1892, Nordland Amt.

UNEP/CBD (1998). *Report of the Workshop on the Ecosystem Approach, Lilongwe, Malawi.* Submission by the Governments of the Netherlands and Malawi. UNEP/CBD/COP/4/Inf.9.

24 The economic and socio-cultural dimensions of the mountains for the Swedish Saami reindeer herders

Nicolas Gunslay

Summary

1. The Saami reindeer (*Rangifer tarandus*) herders and herding system in Sweden are described. The environmental conditions of the mountains and the welfare of the reindeer are strongly associated, and depend on a combination of environmental and anthropogenic factors.
2. The welfare of the herding communities relies on the mountains due to their socio-economic and cultural dimensions.
3. Scientific studies and land use management policies are outlined which should benefit the future for Saami reindeer herders.

24.1 Introduction

The Saami, known as the only indigenous people of Europe, live in Northern Norway, Sweden and Finland and the Kola Peninsula of Russia. There are about 100,000 Saami irregularly distributed among these countries (Helander, 1994). In Sweden, they live mainly in the Norbotten and Vasterbotten counties, where there are estimated to be between 16,000 and 20,000 Saami, of which approximately 17% are herders. Reindeer herding areas, defined by laws and regulated by the Ministry of Agriculture, are divided in to districts called Sameby. Herding is an exclusive Saami right: non - Saami Swedes cannot be herders (Ruong, 1967). Besides this restriction, each herder belongs to a Sameby, and only members of the Sameby can own reindeer. The year of a reindeer herder is divided into eight seasons during which the semi-domesticated reindeer move freely from autumn, winter, spring and summer pastures, respectively, from forest to mountain grazing areas. The herders spend most of their time in the main village, but during the summer they move to their summer cabins in the mountains where the marking of calves takes place.

24.2 The importance of the mountains, environmental conditions

The summer grazing areas are mainly located in the sub-arctic and arctic parts of the mountains along the Swedish and Norwegian border (e.g. CAFF, 2001). The summer season is of high importance for the welfare of the reindeer; it is the time when they build up for winter. The better the summer conditions are, the lower the mortality will be during winter. In addition to abundant forage, reindeer require peaceful grazing conditions to feed

Gunslay N. (2005). The economic and socio-cultural dimensions of the mountains for the Swedish Saami reindeer herders *In Mountains of Northern Europe: Conservation, Management, People and Nature,* ed. by D.B.A. Thompson, M.F. Price & C.A Galbraith. TSO Scotland, Edinburgh. pp. 259-262.

themselves properly. Such needs imply that any potential source of stress, which will lead the herd away from feeding areas, should be minimised as far as possible for the best grazing conditions. The main factors which may disturb the peace of the herd are environmental and anthropogenic . The former include mosquitoes (Culicidae) and predators, and the latter include land use activities such as tourism, and regulations determining the size of pasture areas and their borders.

The following example illustrates the need for peaceful grazing conditions, and appropriate environmental and human-related conditions. From the vegetation aspect, a feeding area could be very good; however, the presence of predators or mosquitoes, as well as even a few tourists hiking close to the herd, would be enough to cause the herd to move, and therefore to affect their feeding needs.

Climatic conditions are also essential (see CAFF, 2001): high temperatures stress the reindeer, whereas wind contributes to reducing the effect of mosquitoes, and snow patches offer good shelter when conditions are relatively hot. Summer pasture conditions and the welfare of the herd are thus strongly related, and such a close association depends on a combination of environmental and anthropogenic factors, whose changes or variability affect the grazing conditions.

24.3 Socio-economic and social cultural dimensions of mountain areas

The summer conditions determine the welfare of reindeer and impact indirectly on the herders' welfare, which is economically partly reliant on meat production. In this system, meat production depends on the grazing conditions and especially the summer conditions (the slaughtering season takes places in the autumn). External factors which affect the summer conditions are therefore perceived as factors which determine the economic welfare of the herding units.

Mountain areas have natural resources related to Saami livelihoods, such as hunting, fishing and the picking of berries; for the herders these are complementary activities to herding, and provide either additional income or meat and fish for consumption. Land use rights are essential factors which impact negatively or positively on the herders' livelihoods, and in turn provide the mountains with an extra socio-economic dimension (see Jersletten & Konstantin, 2002).

This socio-economic dimension has to be considered in connection to common tasks and meat production. Summer in the mountains is the time where the Sameby members gather together the herd and mark the calves. The marking of the calves requires strong cooperation among the herders and their family members, not least to collect the reindeer in the mountains, to drive them to the corral and to mark the calves. It is vital for the herders that most of the calves are marked during the summer so that their ownership will not be questioned at the slaughtering time.

Being a herder is not only using the mountains as an economic resource, but is a way of life, and summer is the time when the family members move to the summer camp and live close together for a few months. Within the current herders' society and education system, their children are often away from the family. Summer therefore appears to be an essential period to strengthen the social unity of the family, and also to transmit the values, the knowledge and the history of this way of life to the youngest generations. This serves to reinforce the cultural unity and identity of the Saami herders; the mountains are a key component of their social and cultural life. Clearly, one needs to integrate socio-economic and cultural dimensions into the scientific and decision-making processes when seeking to understand the sustainability of Saami herding. Thus, it should be possible to develop an understanding of the conditions required to secure and develop the economic, social and cultural viability of the herding units. Such a holistic approach would contribute to a vastly improved mutual understanding between all concerned with the management of reindeer.

References

CAFF (2001). *Arctic Flora and Fauna: Status and Conservation.* Edita, Helsinki.

Helander, E. (1994). The Saami people, demographics, origin, economy, culture. *Diedut,* **1**, 23-34.

Jersletten, J.L. & Konstantin, K. (2002). *Sustainable Reindeer Husbandry?* University of Tromsø, Tromsø.

Ruong, I. (1967). *The Lapps in Sweden.* Swedish Institute, Stockholm.

25 Public participation for conflict reconciliation in establishing Fulufjället National Park, Sweden

Per Wallsten

Summary

1. The traditional planning process for establishing national parks in Sweden normally focuses on area design and restrictions on use inside the park. This 'outside-in process' is not sufficient for managing the conflicts between national and local interests.

2. Focus has to shift to how a national park can result in benefits for local people outside its borders. This 'inside-out process' involves a local perspective and includes socio-economic opportunities.

3. Local people have to be given confidence and support to formulate their own requirements and visions for the benefits of a national park – and the effects of the zero alternative. Long-term resources for coaching the local process are necessary for local people to trust the project and give it status and identity. This strongly contributes to local approval for the protection of the area.

4. A well-tuned, inside-out process involves the creation of formal and informal networks between local people, which is critical to encourage positive local actors and to improve their knowledge, insights and self-reliance.

5. The to improved national park process should include the necessary resources to initiate the realization of substantial parts of the local vision. This includes taking concrete physical actions, which function as strong symbols of the credibility of the process and the wider benefits of protection.

6. Potential national park opportunities need to be clearly expressed in the process. Zoning based on the Recreation Opportunity Spectrum (ROS) concept is a useful tool for planning and communicating park resources as environmental settings, visitor experiences and appropriate activities.

25.1 Introduction

Fulufjället ('Mount Fulu') is Sweden's twenty eighth and latest national park. It lies in the southern part of the Swedish high mountains, on the border with Norway. An area of 385 km² makes it Sweden's fifth largest national park. Fulufjället is a sandstone mountain with a unique morphology. An undulating bare plateau rises (Figure 25.1) from steep slopes with virgin forests. Rivers cut deep valleys and canyons in its sides. Rare low-alpine heaths and traces from ice age glaciers are also of particular value.

Wallsten, P. (2005). Public participation for conflict reconciliation in establishing Fulufjället National Park, Sweden. In *Mountains of Northern Europe: Conservation, Management, People and Nature*, ed. by D.B.A. Thompson, M.F. Price & C.A. Galbraith. TSO Scotland, Edinburgh. pp. 263-274.

Figure 25.1 Subglacial chute channels on Mt. Fulufÿället. Clear signs of previous glacial melting can be seen on the steep slopes of the mountain, and gives high geomorphological values to the National Park. (Photo: Rolf Lundquist).

Figure 25.2 Pine forest in Gölÿådalen valley on the east side of Mt. Fulufÿället. Pine forests dominate on the more level areas on both dry and wet land. (Photo: Rolf Lundquist).

Figure 25.3 Virgin forests in a sheltered location. the vegetation of deep ravines is dominated by Spruce forest with lichen-clad trees and a wealth of deadwood. (Photo: Roland Johansson).

Due to the absence of reindeer (*Rangifer tarandus*) herding, which exists in all other parts of the high mountains, Fulufjället has special botanical qualities: unique, thick lichen heaths ungrazed by reindeer make this a reference area for research on reindeer overgrazing and trampling impacts.

Many plant and animal species have their southernmost distribution in Fulufjället. A rich fauna includes golden eagle (*Aquila chrysaetos*), rough-legged buzzard (*Buteo lagopus*), common scoter (*Melanitta nigra*), and large populations of brown bear (*Ursus arctos*) and lynx (*Felix lynx*). More rarely, wolverine (*Gulo gulo*) and wolf (*Canis lupus*) occur in the area. The park provides good opportunities for visitors to experience quietness, solitude and unspoilt nature (Figures 25.2 and 25.3). Other attractions are Sweden's highest waterfall (93 m high), plus 140 km of marked summer and winter trails and several overnight huts.

Two other proposed mountain national park projects (Kirunafjällen and Jämtlandsfjällen) have been terminated in Sweden during the last 15 years due to opposition from the municipalities and local inhabitants. However, the result of the planning process in Fulufjället is now considered to be one of the best examples in Sweden of how to manage, and reconcile successfully, conflicts between local and national interests in the field of nature conservation.

25.2 Step one: Traditional process with rising conflicts

The planning process started in 1990, after the Swedish Environmental Protection Agency (SEPA), the authority responsible for establishing national parks, presented the new Swedish National Park Plan. Fulufjället was one of 20 proposed new parks, selected on the basis of important biodiversity and geological values, uniqueness and the representation of distinctive nature regions.

The Fulufjället area is sparsely populated. Around 350 inhabitants live in a few small villages around the mountain. The rate of unemployment is high: for several years many young people have had to move away to find a job. The population is thus steadily decreasing and the average age is rising. The traditional local uses of Fulufjället are hunting, fishing, snowmobiling and wild berry picking. The ability to continue to use Fulufjället in such traditional ways is considered an important part of the quality of life, especially while other conditions are hard. However, in order for Fulufjället to become a national park, such uses had to be managed and restricted so as to reduce the recreational impact and enhance natural values and visitor experiences.

The traditional planning process included providing information to the local inhabitants in large group meetings, and establishing a working group of representatives from different stakeholder groups. The conflicts between the national park plans and the interests of the local people gradually became more obvious as the regulations were formulated and clarified. Regulation of snowmobiling, fishing and the hunting of moose (*Alces alces*) and grouse (*Lagopus* spp.) was considered to have great impacts on the traditional lifestyle and could reduce the quality of life for those living around the mountain.

It was difficult to provide the different stakeholders in the area with correct information, and rumours of 'everything will be forbidden' were widespread, resulting in massive petitions. The focus was initially strongly on the restrictions that would occur inside the national park. Its local benefits were seen to be weak and unclear by the locals and the municipality, despite SEPA's efforts to inform people about it. Mistrust of the state's assurances was obvious. The situation was typical of what can be described as a traditional 'outside-in process' (Figure 25.4a).

Strong emotions were involved, including those about who should have rights to use the land – those who are 'sitting in the capital city and making decisions without understanding the reality', or the people who live in the area. This was despite the fact that most of the area was state-owned. The old conflicts between national and local interests, as well as between centre and periphery, were evident. It also appeared that resistance to the national park project contained elements of general displeasure with the local economic and living conditions, which now had an obvious, external target to be focused upon.

General meetings to exchange information were not a success because the scale of the accumulated negative opinions was too strong for balanced discussions to prevail. The local representatives in the working group resigned. The municipality politicians formulated unreasonable demands in line with the dominant (that is, most strongly expressed) local views. Another mountain national park project failure was coming closer.

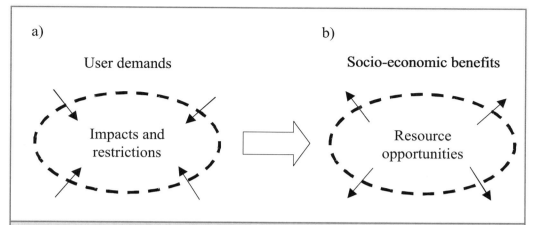

Figure 25.4 The traditional National Park planning process compared with the improved, complementary process, which led to the establishment of a new National Park in Fulufjället. a) The traditional 'outside-in' process is characterized by: 1) reducing impacts by regulating user demands inside the National Park boundary, and 2) focusing on problems. b) The improved 'inside-out' process is characterized by: 1) assessing how the area's resource opportunities can generate socio-economic benefits outside the National Park, and 2) focusing on possibilities.

25.3 Step two: Improved process with change of perspective

Recognising the situation, SEPA and the County Administrative Board of Dalarna saw the urgent need for an alternative strategy for handling the conflicts and established a new project to run in parallel with the park planning process. The 'Fulufjället Surrounding Project' was initiated in 1997. Its aim was to make the local people aware of the opportunities that the establishment of a national park could bring to them. The County Administration and SEPA had to provide the process and tools for this, but it was clear that the insights and understandings had to come from the people themselves.

The Surrounding Project implied a shift of focus, from the national to the local interest. A project leader was employed, working alone in the field, and spending much of her time talking to people living in the small villages near the mountain. It was obvious how important and appreciated it was that she visited local people, made a connection and, in conversation, learned of the reality and conditions of their life. Thus a perspective from below and inside, instead of from above and outside, was developed.

This first part ended in an unprejudiced inventory of current conditions, including demographic data (Arnesson-Westerdahl, 1998). About 40 people were visited, mainly in their homes; 27 interviews were of a more in-depth nature; and 20% of the people permanently living near the mountain in Älvdalen municipality (where all of the proposed national park is situated) participated. In total – including the other municipality, Malung, close to Fulufjället – around 10% of the permanent residents in the area were in contact with the project at this stage. All but two of the residents who worked in the area, as entrepreneurs or otherwise, were contacted. The in-depth interviews took 2-4 hours: people were allowed to talk freely about themselves and the conditions of life, including demands and suggestions for improvements. People were frank, positive and open-minded. The information they provided, their views, demands, etc., were generally similar, giving a rather accurate picture of the situation despite the limited number of interviews.

During the process, it gradually became obvious that the 'no national park alternative' with continued traditional use also had disadvantages, as the existing negative socio-economic trends would most likely go on if nothing happened. Many of the social problems found in small villages and sparsely populated areas also came to the surface, so the project leader found herself to have a somewhat 'therapeutic' function – a demanding task that made the work hard and challenging. Many telephone calls and late conversations about people's personal relations occurred. It also became obvious that the resistance to the national park had its strongest base in some older men with strong local influence.

The second part of the project assumed that a new national park in Fulufjället would be declared. From that starting point, the Surrounding Project gave local stakeholders support and the confidence to formulate their ideas of how such a national park could be beneficial for them economically, socially and in other ways, i.e. a national park as a resource for opportunities instead of as a basis for restrictions. After many discussions in different groups, inputs from other areas etc., the result was a local vision shared with many of the people living in the area (Arnesson-Westerdahl, 1999). The vision included ideas ranging from developing marketing and competence to improved roads, guided nature activities and exhibitions.

More or less informal 'Local Development Groups', with membership open to all, existed in the villages around the mountain. Most of the families living in the area were in one way or another associated with the seven groups. All of the groups, plus a group of small-scale entrepreneurs south of the mountain, participated and supported the vision. This did not necessarily mean that the groups had to commit that they were positive to the national park plans: the point was that *if* a park were to be established, this vision could/should be the reality of how it would be done. The process was open and appreciated.

The focus was now on how to use the benefits of a national park *outside* its borders, rather than the former discussion about the design and management of the park from the *inside*. SEPA had concentrated more on the local vision of, for example, a new visitor centre with local employees, new tourism facilities outside the park and better infrastructure, rather than struggling with the removal of certain trails for snowmobiles and where and when not to hunt. The 'loss' was clear – a national park with some restrictions – but the 'gain' was created by what the national park would be: a new future, with possibilities for the people to live and work in the area. As project leaders were available for discussions and coaching at all times, confidence in the state and the process replaced the mistrust.

The improved planning process, described as 'inside-out', had now created local support and thus the requirements for realizing the national park (Figure 25.4b).

25.4 Step three: Realization

An important effect of the Surrounding Project was the creation of networks between local stakeholders, making people come together and talk in new contexts. This included networks: between women (who formerly were rather invisible in the process as more 'macho' pursuits as hunting and snowmobiling were the main issues); between people with ideas for a new future; and between small-scale entrepreneurs, etc. Prerequisites for the networks were a series of actions included in the project process: the establishment of working groups on themes such as accommodation, trails, information and fishing; study tours for inspiration, with 20 people travelling to Abruzzo National Park in Italy; and education programmes and seminars, e.g. guiding in Fulufjället, foreign language in tourism entrepreneurship, marketing and local history. The activities were appreciated, had many participants and were a very visible and obvious result of the project and the benefits of the improved national park process (Pettersson, 2003).

The process to establish a formal organisation for local entrepreneurs was important. This involved meeting and networking with municipal bodies, contacts with foreign tour operators, seminars for certification, quality control, etc. The process, coached by the project leaders – now two people – resulted in ten companies in the area establishing a co-operative economic association, 'The Fulufjället Ring'. The state now had a formal partner with which it could hold further discussions on how to use the national park as a potential resource for regional and local development.

The identity gained by participating in networks encouraged people to question the common view of resistance to 'change'. They felt they were not alone in seeing the possibilities of a new national park in Fulufjället. This became obvious in the local newspaper, when one of the many negative letters to the editor was confronted by a letter from a group of other locals in support of the park's benefits. They officially expressed their opinion that a national park is beneficial for the people living in the area, and that the 'no park alternative' would result in fewer opportunities and lead to continuing socio-economic decline in the area. This was a turning point for the whole process.

This growing official local acceptance was instrumental in persuading the municipality, at SEPA's request, to give political approval to the national park concept with statutory zoning and regulations.

Contributing to the progress was the zoning structure of the national park in the management plan (Naturvårdsverket, 2002). It was based on the Recreation Opportunity Spectrum (ROS) concept (Clark & Stankey, 1979; Hendee & Dawson, 2002), and included establishing a pristine core zone, with the removal of former hunting, fishing and snowmobiling. This was balanced with more heavily used recreational activity zones, where some local needs could be fulfilled (see map in Figure 25.5). The ROS zoning idea was introduced early as a basis for the process and was very important in communicating with the different stakeholders. It was transparent, and both theoretically and practically comprehensible by potential users of the park. The ROS defined environmental settings, appropriate activities and experience values of different zones (see Figure 25.6). Thus the advantages of how the resources in the area would be used became obvious to people, and also that both national and local interests would be provided for in the park – even if spatially distributed and not in the same places.

The zone concept and park regulations also fulfil the ecological objectives of biodiversity preservation. Important habitats for carnivores and raptors are protected from disturbance through the regulation of rock climbing and the absence of visitor infrastructure. Further prohibition of reindeer herding in the park, with a consequent prohibition of all-terrain motorcycles when collecting reindeer, will preserve the unique heath vegetation.

Once support for the park was evident, SEPA intensified the formal national park process, with strong local support, and started to improve the visitor infrastructure with investments and concrete actions in and around the park. This was not just the beautiful words of actions by others in the future: the strategic effect was important. Bulldozers, helicopter transport and construction are strong symbols of change and the prospect of new jobs! SEPA had sound funding for this work, partly due to European Union (EU) funding,

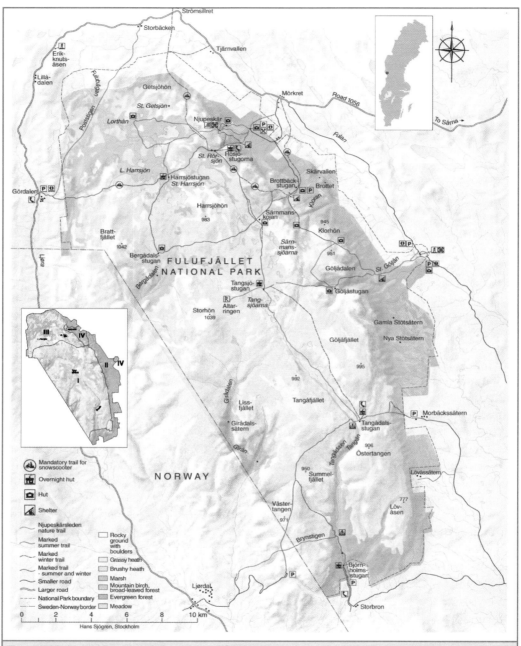

Figure 25.5 Map of Fulufjället National Park. ROS Zones, as seen on the small inset map: I) Undisturbed zone (60% of the park area), core area with no hunting, fishing, liming, snowmobile trails or aircraft landing. II) Low activity zone (15%), includes moose-hunting. III) High activity zone (25%), includes fishing, liming (to reduce acidification) and snow-mobile trails. IV) Structure zone (<1%), includes roads and larger visitor establishments.

and made the largest investment to date in Sweden in improving the visitor quality in a new national park (c. 3.5 million Euro), including a new visitor centre, new and restored huts and trails, and interpretive information, etc.

	Zone I Undisturbed	Zone II Low activity	Zone III High activity	Zone IV Structures
Human influence	Low --			High
Extent of visitor services	Low --			High
Signs of other visitors	Few --			Many
Visitor encounters	Few --			Many
Potential for experience				
• solitude	High ---			Moderate/low
• silence	High ---			Moderate/low
• unspoiled nature	High ---			Moderate/low

Figure 25.6 Example of factors defining values in different ROS zones.

Briefly, the three phases of the Surrounding Project cost c. 0.5 million Euro (€) – with half of the sum coming from the EU Structural Funds. This is about 5% of the total for establishing the national park. SEPA considers the price for the improved process well worth its outcome. Without it, SEPA would most likely not have been able to create a national park in Fulufjället, with consequent negative long-term national implications for national parks in the mountains, and their ability to be accepted locally and play a role in regional development.

Many of the priority needs identified in the local vision are now realized, such as: the visitor centre, which provides employment for a local manager; jobs for local leaseholders; a restored restaurant; a fishing camp inside the park; repair of the worn-out road around the mountain; and, after many decades of former fruitless complaints, the establishment of mobile phone communications, and education programmes. Marketing from the area is developing fast in and outside Sweden, resulting in new interest from tourists and commercial tour operators. Stakeholders around the mountain have joined together to form the above-mentioned economic association in order to strengthen the co-operative working. The possibilities created by the national park for the economic and socio-economic development of the Fulufjället area are obvious to all. Tourism facilities outside the park are starting to improve and some new small-scale tourism companies have been founded.

To meet the demands from local users for hunting and snowmobiling, special projects and conditions were developed. The hunting of small game will be excluded from the park but, during a ten-year transitional period, will be permitted for local people under special conditions in certain zones. The local hunters have support for developing an education programme for moose hunting in parts of the national park, for the first time in a Swedish national park. Most of the hunters are present or former locals. Also, SEPA has provided the local snowmobile association with funding for building a new snowmobile trail outside the park, as a compensation for the restrictions inside its core zone. These requirements have given further credibility to the process and a feeling among people that local claims are being respected, if not yet completely fulfilled.

The inclusion of the perspectives of local people consequently also strongly influenced SEPA's official book on the national park (Lundquist, 2002, 2005). The book, well appreciated locally, is thus both a part and a result of the 'inside-out' process.

His Majesty the King of Sweden finally opened Fulufjället National Park on 17th September 2002. It was a successful public event, with great symbolic importance and high media coverage. Former local opponents figuratively reached out their hands to jointly make something good out of the new situation. The day felt not like the end of a 12-year process, but more like a start of something new – or, to be more explicit, as the 'end of the beginning' for the Fulufjället area in terms of ecological as well as socio-economic sustainable development.

The challenge now is for the local community to take advantage of the opportunities created by the new national park. SEPA is convinced that the improved planning process has contributed to conditions to encourage this.

25.5 The future

The national park is now complete, and the Surrounding Project will be transferred gradually to the municipality and the local economic association. New EU funding is being sought for project development, strategy work and actions.

Local stakeholders will participate in the 'management council' for the national park that will be established, to implement good communication, inspiration and information. Organized tourism inside the national park is controlled by the park regulations, but the aim is to be generous to sustainable tourism, which is in line with the conservation objectives: park managers will formulate strategies for this.

SEPA carried out a major study of the visitors to the Fulufjället area before the national park was realized (Hörnsten & Fredman, 2002). Quantitative and qualitative visitor data were inventoried as a basis for management planning and as a baseline study. This was repeated last year and is the first 'before and after' study of visitor perceptions and needs around the establishment of a national park in Sweden (Fredman et al., 2005). SEPA will be able to follow and quantify what happens with the visitor numbers, origin, expectations, experiences, and preferences etc., for follow-up and as a management tool. As tourists' preferences and actual use will be clarified, the study will be very useful for local tourism enterprises. The European Tourism Research Institute, ETOUR, is undertaking this work, which is a good example of using research to guide protected area management and planning.

Due to the innovative processes used, Fulufjället is now established as one of Europe's first three certified Protected Area Network (PAN) Parks. PAN Parks (www.panparks.org), an initiative from the Worldwide Fund for Nature, is a marketing and communication tool intended to support nature areas in Europe. A basic objective of the initiative is to encourage sustainable tourism development, providing socio-economic development at a local level, which in turn gives reasons for nature protection and enhancement. Benefits for the Fulufjället stakeholders will be practical support, knowledge sharing, partnerships with small business owners and marketing at a European level: all PAN Park actions that are in line with the Surrounding Project. Europe's first 'PAN Parks Village' with tourism cabins was built in 2004 close to the park, another important symbol of the new job possibilities and other benefits. A joint project involving the county, PAN Parks and the local entrepreneurs is the beginning of the establishment of a tourism strategy in the surrounding area.

The work with Fulufjället national park will also have wider benefits. SEPA considers that the Fulufjället process can serve as a model for establishing future national parks in Sweden. In its description of a new policy for nature conservation, the Swedish government emphasizes the Fulufjället process as a good example of how nature conservation can contribute to regional development, thus strengthening the policy implementation.

References

Arnesson-Westerdahl, A. (1998). Fulufjällets omland. Länsstyrelsens i Dalarna rapport 1998:11 (The Surrounding Project part 1). County Administration Board, Dalarna, Sweden. (In Swedish).

Arnesson-Westerdahl, A. (Ed) (1999). Fulufjällsringen. En vision och framtidsstrategi. Länsstyrelsens i Dalarna rapport 1999:14 (The Surrounding Project part 2, Vision and Strategy for the Future). County Administration Board, Dalarna, Sweden. (In Swedish).

Clark, R. & Stankey, G.H. (1979). The recreation opportunity spectrum: a framework for planning, management and research. General Technical Report PNW-GTR-98. US Department of Agriculture Forest Service, Pacific Northwest Research Station, Portland, Oregon.

Fredman, P., Hörnsten Friberg, L. & Emmelin, L. (2005). Friluftsliv och turism i Fulufjället. Före - efter nationalparksbildningen. Naturvårdsverket, rapport 5467. Dokumentation av de svenska nationalparkerna, nr 18. Swedish Environmental Protection Agency, Stockholm.

Hendee, J.C. & Dawson, C.P. (2002). *Wilderness Management: Stewardship and Protection of Resources and Values.* Fulcrum Publishing, Golden, Colorado.

Hörnsten, L. & Fredman, P. (2002). Besök och besökare i Fulufjället 2001. En studie av turismen före nationalparksbildning. (Visits and visitors in Fulufjället 2001). Etour rapport U 2002:6 och Naturvårdsverkets rapport 5285-3. European Tourism Research Institute, Östersund. (In Swedish with English summary).

Lundquist, R. (2002). Fulufjället – nationalpark i Dalafjällen (Fulufjället – national park in the mountains of Dalarna). Naturvårdsverket, Stockholm. (In Swedish).

Lundquist, R. (2005). Fulufjället – national park in the mountains of Dalarna. Naturvårdsverket, Stockholm.

Naturvårdsverket (2002). Skötselplan för Fulufjället nationalpark (Management plan for Fulufjället national park). Rapport 5246. Swedish Environmental Protection Agency, Stockholm. (In Swedish - an English summary is unpublished).

Pettersson, B. (2003). Fulufjällets omland etapp III, slutrapport. Länsstyrelsen i Dalarna, Miljövårdsenheten, rapport 2002:20 (The Surrounding Project part 3, Final Report). County Administration Board, Dalarna, Sweden. (In Swedish).

26 Establishing the Cairngorms National Park: lessons learned and challenges ahead

Murray Ferguson & John A. Forster

Summary

1. The Cairngorms are Britain's premier mountain range in terms of the extensive area of high ground, the distinctiveness of the landscape and the concentration of biodiversity. The mountains themselves have shaped the cultural heritage of the area.
2. Several strategic initiatives to co-ordinate and integrate the management of this special area have been put in place over the last decade, culminating in the designation of Scotland's newest National Park in 2003.
3. The process used to establish the Park involved lengthy dialogue amongst a wide range of interested parties, over a number of years. The experience of engaging with people in local communities is described and the principal lessons learned are summarised.
4. The Park Authority faces several challenges in terms of the way communities are involved in the management of the area. A strict approach to the definition of sustainable development is recommended and perspectives are given on what this will mean for sustainable communities and sustainable housing.

26.1 Introduction

In global terms, Scotland has come relatively late to the subject of National Parks, with the relevant legislation being passed only in 2000. But there are advantages in being such a late arrival, particularly because we have had the opportunity to learn from the extensive collective experience of managing protected areas in both Scotland and internationally. At the time when this conference was held, there was only one National Park in Scotland, in Loch Lomond and the Trossachs, but Scottish Ministers were preparing to table their final proposals for the second, in the Cairngorms.

This chapter focuses on the engagement of local communities in the preparations leading to the designation of the Cairngorms National Park, firstly by examining the processes that have been used to establish the Park and secondly by considering some of the challenges that will lie ahead.

Ferguson, M. & Forster, J.A. (2005). Establishing the Cairngorms National Park: lessons learned and challenges ahead. In *Mountains of Northern Europe: Conservation, Management, People and Nature*, ed. by D.B.A. Thompson, M.F. Price & C.A. Galbraith. TSO Scotland, Edinburgh. pp. 275-290.

26.2 The Cairngorms – why they are special

Located in the eastern Scottish Highlands (see Figure 26.1), the Cairngorms are Britain's premier mountain range in several respects. First, in a British context they are distinctive because of their scale – both extent and altitude – having the largest continuous area of ground above 1,000 m in the United Kingdom and around 50 'Munros', peaks over 914 m (3,000 feet) (Scottish Natural Heritage, 2002). The mountains and surrounding straths, as defined by the area of the Cairngorms Partnership (the management unit in place at the time of the conference for the integrated management of the mountains), extends to over 6,500 km².

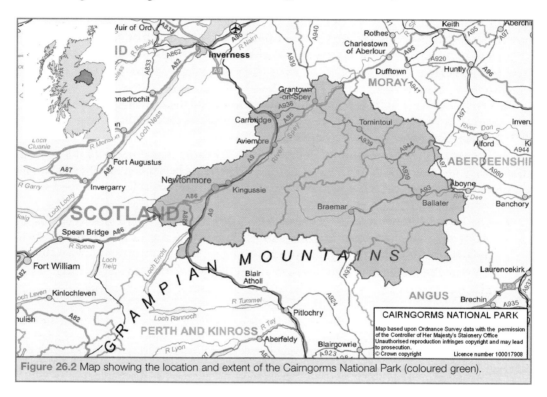

Figure 26.2 Map showing the location and extent of the Cairngorms National Park (coloured green).

Second, the landscape is distinctive. The core of the Cairngorms is granite and the landscape is dominated by rolling plateaux, studded with extensive boulder fields and cut by deep, often steep-sided, glacial troughs (Turnbull Jeffrey Partnership, 1996). The diversity of earth heritage features makes the Cairngorms an exceptional resource for the study of long-term landscape evolution. Indeed these features form the basis for the Government's nomination of the area as a candidate World Heritage Site (Department of Culture, Media and Sport, 1999).

Third, the mountains are a biodiversity hotspot, supporting a range of species and habitats that are either rare or unusual in Scotland. The characteristic special features include the wind-blasted habitats. Their arctic-like credentials are demonstrated by snow bunting (*Plectrophenax nivalis*), our most northern land bird, found here further south than anywhere else in world. Lower on the slopes the largest remnants of Scotland's native pinewoods are to be found. No less than one quarter of all of the UK's priority Action Plan species are to be found here (Cosgrove, 2002). For many of these species, the Cairngorms hold a significant proportion of the total UK population and, for a few, the entire population.

But the mountains are special not just for their natural qualities. The area has a rich cultural heritage, often seen in the shape of the land or its vegetation cover. Importantly, the area is home to around 17,000 people (Cairngorms Partnership, 1996a). The economy is based on a combination of farming, forestry, and sport shooting but with a high reliance on service sectors with heavy dependence on tourism and public sector employment.

The combination of exceptional landscape quality, outstanding biodiversity and relative remoteness has long been a source of attraction for those seeking outdoor recreation. The area is now popular for many informal outdoor pursuits including hill-walking, ski mountaineering and both summer and winter climbing, and is home to the national outdoor training centre at Glenmore Lodge and many other outdoor activity centres (Taylor & MacGregor, 1999). The area also contains three of Scotland's five downhill skiing facilities and provides many opportunities for diverse kinds of recreation at lower altitude. A comprehensive recent account of the special qualities of the Cairngorms is set out in Gimingham (2002).

26.3 Special management needs

Those managing the Cairngorms mountains to protect these special qualities have had to deal with a number of difficult issues over the years (Gimingham, 2002) but it is only possible to give three brief examples here.

- Grazing by red deer (*Cervus elaphus*) at high densities has prevented tree regeneration.
- There has been pressure for expansion of downhill skiing facilities into sensitive upland sites of high nature conservation value.
- The development of bulldozed estate tracks and other man-made artefacts has in places compromised the sense of wildness that is valued by many people.

Other issues affect local people directly. For example, in common with most upland areas in the UK, there has been a major decline in agricultural employment over the last 40 years. There is also a lack of affordable housing in some communities (Cairngorms Partnership, 2002) while tourism, a very important economic component of the area, is associated with many low-paid, short-term jobs (Copus *et al.*, 1999). Over a long period of time, concerns have been expressed about lack of involvement of local people in decisions that affect the area (Cairngorms Partnership, 1996b).

Against this background it is important to bear in mind three underlying factors.

- The bulk of the land in the Cairngorms area is owned privately. Only a relatively small area is in public ownership or is owned by non-governmental organisations. The establishment of the National Park does not change the pattern of land ownership directly.
- The majority of the people who live in the area depend on the natural resources only indirectly. Relatively few people actively manage the land or can influence its management.
- There have been problems over the years with co-ordination of the organisations who are variously responsible for the area. For example, administration of the mountains is divided amongst five local authorities.

26.4 Management initiatives

Several initiatives have attempted to address these management needs over the years and significant progress has been made in the last decade, particularly through the work of the Cairngorms Partnership which was established in 1994. The Partnership, comprising public agencies, local authorities, communities and landowners, was established around a limited company, funded by the Scottish Office (now Executive), with a small team of staff co-ordinating a voluntary programme of action. A Management Strategy was published in 1997 (Cairngorms Partnership, 1997).

The passing of the National Parks (Scotland) Act 2000 in September 2000 heralded a new era for the management of Scotland's most special places. The Act does not create National Parks itself, but it gives the Scottish Parliament the power to do so if it wishes. Scottish National Parks have four aims – conservation, sustainable use, recreation, and sustainable economic and social development of communities as set out in Box 26.1.

Box 26.1 The aims of Scottish National Parks

- To conserve and enhance the natural and cultural heritage of the area;
- To promote sustainable use of the natural resources of the area;
- To promote understanding and enjoyment (including enjoyment in the form of recreation) of the special qualities of the area by the public; and
- To promote sustainable social and economic development of the area's communities.

The emphasis of the legislation is on achieving the integration of these four aims through a National Park Plan which will be prepared by a new, publicly funded organisation for each park, the National Park Authority. The Authority is charged with ensuring that the four National Park aims are collectively achieved in a considered way. If there is a conflict between the aims, then the Authority must give greater weight to conservation and enforcement of the natural and cultural heritage. This integration of conservation with socio-economic factors chimes strongly with ideas about so-called 'new conservation' and the paradigm shift that has taken place over the last two decades concerning the role of people in the management of protected areas (Brown, 2002).

26.5 The processes used in establishing the National Park

When Ministers made a formal proposal for the Cairngorms National Park, they appointed Scottish Natural Heritage (SNH) as official advisor. In approaching this work, SNH emphasised the importance of involving local people in the process and drew upon the good practice set out in a report which they had commissioned earlier (Govan *et al.*, 1998). A national consultation exercise was planned (Scottish Natural Heritage, 2000) with a number of different elements as shown in Box 26.2.

The consultation sought responses around five broad themes, which included the principle of the proposed Park, the location of the boundary and the powers of the Park Authority. Special efforts were made to stimulate local participation. To plan for this, SNH met representatives of Community Councils and Associations in the Cairngorms to review the success of previous consultations.

Box 26.2 Main elements of the SNH consultation exercise

- Consultation document – to provide background information on Scottish National Parks and describe the main issues on which views were sought.
- Summary consultation leaflet – to provide a summary of information and a simple response form.
- Information packs and leaflets – including a leaflet to address questions that had been posed by young people in the Cairngorms area during the 1998 consultation, and information in Gaelic and for the visually impaired.
- Displays and posters – to use at public meetings and open events.
- SNH web site – to provide widespread open access to information.
- Think-net discussion web site – to promote online discussion with a follow-up conference.
- Public open meetings in six Scottish cities – to engage those far from the Park in the debate.
- A telephone help-desk and meetings on request with any interested party – to encourage a sense of openness.
- Advertisements in the local and national press and a media strategy that resulted in at least 105 articles and items on television and radio to raise awareness and to promote participation.

Overall the view was that while previous approaches, using independent facilitators to organise meetings in villages halls, had been moderately successful, it would be possible to make significant improvements. There was a strong desire for more local control of the methods used and a feeling amongst community representatives that if they were given adequate resources and training they could ascertain community views more effectively than had been achieved before. Community representatives also emphasised the importance of working through existing democratic structures, especially the 26 Community Councils and Associations in the area (Forsyth & Downie, 2000).

SNH accepted these views and offered to pay the expenses of local facilitators who would be managed directly by the Councils. SNH also provided training in facilitation methods and provided additional information about the issues associated with the proposed Park. For those communities that did not want to take on this role, SNH offered to pay for an independent facilitator and organised public meetings along more traditional lines.

Once the consultation started, the most obvious effect of the increased community control was the use of a much wider range of consultation methods specifically designed to meet local needs. These methods included coffee mornings in private houses, meetings for specially targeted groups (e.g. shepherds and gamekeepers), telephone surveys of every household and meetings with community groups that were known to be influential locally, such as Parent Teacher Associations and Village Hall Committees.

Reports were written in a standard format and, before they were sent on to SNH, each report had to be endorsed by the relevant Community Council. Once they had been received, all of the reports (and all of the other responses received during the consultation) were made available for public inspection in several SNH offices.

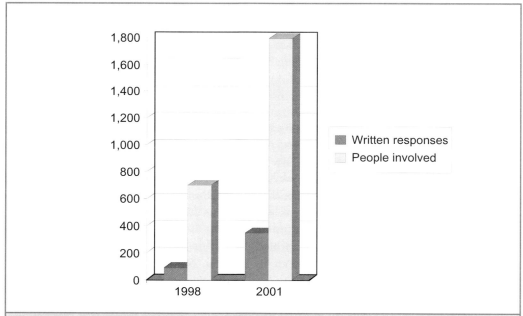

Figure 26.2 Comparison between the local level of participation in the consultation exercises relating to the National Park in the Cairngorms. Source: SNH (1999, 2001) and Rettie (2001).

Overall, local participation increased significantly compared with previous similar consultations, as shown in Figure 26.2. But though the volume of responses was encouraging, the most significant difference was in the quality of participation. The community-led facilitation provided repeated opportunities for discussion of the complex issues over a sustained period (20 weeks). This led to an enhanced level of thought and discussion which was reflected in the reports received by SNH. In general, SNH formed the view that those communities which took control of the consultation were able to undertake a more thorough consultation process compared with those that did not.

Based on these findings, SNH submitted a report to Ministers in August 2001 (Scottish Natural Heritage, 2001) that analysed the arguments put forward and made a set of recommendations. As an indication of the commitment to local communities, the SNH Chairman, John Markland, met community representatives in the Cairngorms on the same day as the report was submitted to Ministers so that they could hear about the recommendations first hand and learn why they had been made.

Several assessments were made of the SNH consultation exercise, notably an independent evaluation by a social anthropologist from St Andrew's University who had experience of National Parks in Canada. Rettie (2001) described the process as 'unique' and said there was unanimous agreement amongst the residents that she had interviewed that people had had a fair opportunity to register their opinions. She concluded

"... success was achieved through innovative processes that increased participation and contributed towards the accumulation of social capital in the communities."

However, there was general agreement among all the parties involved that it would be possible to improve on the process next time around. Community representatives and SNH agreed a number of recommendations for the future, including that:

- there should be more lead-in time for preparation;
- SNH publications, especially the main consultation document, should be written more clearly and simply;
- the training of facilitators should be longer and more targeted; and
- there should be greater efforts to involve young people.

In May 2002, Ministers, having considered SNH's advice, issued further proposals in the form of a Draft Designation Order (DDO) for the Park. Some of the proposals caused considerable concern to communities and there was the perception that information gathered so painstakingly during the previous consultation had been disregarded. For example, communities had expressed the view that the Park boundary should not divide communities but in the DDO it was drawn on detailed maps through the middle of several villages. A further concern was that little or no explanation had been given for the reasoning behind the Ministers' decisions, for example to exclude certain areas from the Park (Cairngorms Community Councils Group, 2002a).

26.6 Lessons learned

Many lessons can be drawn from this experience and we have selected only a few under four broad themes.

26.6.1 Success was possible only as part of a larger process

The consultation exercise in 2000/01 was successful only because it was part of a longer-term process. SNH and others had already reviewed good practice in working with communities as part of earlier work on National Parks (Govan *et al.*, 1998). When the consultation was being planned there was a strong network of Community Councils and Associations already in place as part of the Cairngorms Partnership. Also, the community representatives had experienced, and learned from, similar consultation exercises about the Cairngorms held previously (Cairngorms Community Councils Group, 2002b). SNH therefore had a ready-made, experienced network with which to work.

The benefits of the local knowledge and social networks which communities were able to exploit in bringing people into the debate was very significant. A government agency such as SNH simply could not have replicated the process if acting alone, no matter the quantity of resources devoted to the exercise.

The experience suggests that a community's experience of a previous consultation affects their attitude to subsequent ones. And if those conducting a consultation do not listen to the messages received, or provide feedback to those who have given their views on why decisions have been reached, the process of developing community participation next time around will be that much harder.

Stoll-Kleemann & O'Riordan (2002) have advocated that if effective community engagement is to be achieved, committed management attention is needed over a sustained period of time. The legislative framework provided by the National Park should help to provide the staff and financial resources that will be necessary for this long-term approach.

26.6.2 Delegation of control over the consultation was not straightforward

The process of delegating control over the methods used in the consultation to community representatives sometimes led to difficult issues for SNH. For example, there were minor concerns that:

- some communities used misleading information to attract people to meetings;
- there was sometimes bias in the questionnaires that were used to seek residents' views; and
- some consultative processes did not enable everyone in the community to express their views.

At the same time, some individuals were critical of the methods used in their own, or adjacent, communities and urged SNH to intervene.

Faced with this situation, SNH staff decided to take a relatively relaxed approach and resisted the temptation to intervene in the process too strongly. Given the nature of the final outcomes, this approach seemed to pay off. However, more time for planning and better training of facilitators could have helped ensure that some of these issues did not arise. SNH considers that it will be important for community representatives to appreciate the nature of these concerns in planning for future consultations.

26.6.3 Consultation and community capacity building took place at the same time

The approach to the community consultation allowed two important processes to take place simultaneously: the collection of the views of local people on the proposals, and the building of community capacity. Community facilitators received some training but, more importantly, they gained practical experience in their own communities of organising meetings, facilitating discussions and writing reports. This is the 'learning by doing', or on-the-job capacity building, that IUCN - The World Conservation Union recognises as being so important (e.g. Borrini-Feyerabend, 1997)

In comparison to previous consultations, democratic structures were strengthened. For example, some community representatives said they had gained new insights into their own communities and how they worked. New networks and social contacts were made at community level, and one community even reported that discussion was so intense it was difficult to get farmers to leave the meetings at the end of the evening.

Capacity-building took place within SNH too: those staff who were involved learned new skills, gained confidence and made new contacts.

26.6.4 Community views are diverse and difficult to summarise

The views received on the proposals from any one community were diverse, not only on the substance of the consultation but also on the consultation process used. This diversity of views was not always appreciated, even by those who sat through meetings where a diversity of views was expressed. Similarly, even where communities did come to a consensus view on a particular issue, it sometimes differed markedly from views received from other communities.

At times, the facilitators who were responsible for writing the reports for SNH struggled to cope with this complexity. The ability to summarise the variety of views heard and still convey the richness of the messages received, the arguments behind the view and who held them is a key capacity that requires to be developed further. Future training and development of facilitators needs to take this into account.

For its part, SNH endeavoured to respect the complexity of the situation and correct any impression of there being a consensus where none existed. And a key part of the consultation was to ensure that all the responses and reports of meetings were available to Ministers and to anyone who wished to see them, so that any bias or misrepresentation that had been introduced through the summarising process could be exposed.

At a wider level Agrawal & Gibson (1999) have described the dangers for community-based conservation of assuming the existence of a 'mythic community', based on small, integrated, locally evolved norms, and the risks of ignoring the differences within and between communities. They advocate instead a stronger emphasis on the divergent interests of multiple actors within communities and the various institutions that affect them. Such an approach may well be fruitful in the new National Park.

26.7 Future challenges for the Park

This part of the chapter is about the future of the Park, viewed from the perspective of its communities, and particularly about the aspirations of communities and the way in which they may be achieved. The attitudes and ambitions of all stakeholders have significance, but it will be the processes established by the Park Authority to relate to its communities that are of particular significance here.

The aims of Scottish National Parks are presented in Box 26.1 while the work undertaken by Community Councils, including their aspirations for the Park, has been well described by the Cairngorms Community Councils Group (2002b). All four National Park aims are important, but from a community perspective the fourth aim is of particular interest. This aim has been said to give the Authority a licence to promote economic and social development and consequently some commentators have expressed concern that this may lead to conflict with the aims for protection of landscape and wildlife (e.g. Watson, 2001).

However, it can be argued that there are two reasons why the promotion of this fourth aim by the Park Authority should not be pursued in such a way as to cause unacceptable pressure on the special qualities of the area. In the first place, there is the overriding constraint imposed by Section 9 of the Act, that

"… the general purpose of a National Park authority is to ensure that the National Park aims are collectively achieved in relation to the National Park in a co-ordinated way."

Provision is also made in the legislation for 'greater weight' to be given to the conservation aim by the Park Authority if there appears to be conflict between the aims.

In addition, a closer reading of the fourth aim shows that it does not refer to development at any price – it is development that is to be 'sustainable'. At the present time the word 'sustainable' is applied with a wide range of meanings. For example, applied to a community or a business it is frequently used in the sense of surviving over a period of time. In that sense, nothing is being conveyed about the nature of that survival or the nature of the relationships that might exist between the business and its environment or society. We argue that in the context of a National Park, the meaning of the word 'sustainable' must be interpreted strictly. Sustainable development in our view should be development that, in the definition of the World Commission on Environment and Development (UNCED, 1987)

"... seeks to meet the needs and aspirations of the present without compromising the ability to meet those of the future."

In the context of sustainable communities and economies, the UNCED goes on to say that

"... policy makers guided by the concept of sustainable development will necessarily work to assure that growing economies remain firmly attached to their ecological roots and these roots are protected and nurtured so that they may support growth over the long term. Environmental protection is thus inherent in the concept of sustainable development."

If genuine commitment could be obtained to development that was in accordance with this approach, it might go some way towards ensuring that all aims of the Park could be achieved in ways that were mutually compatible. The first challenge for the Park Authority is therefore to ensure that it adopts and promotes a concept of sustainable social and economic development that is firmly rooted in a profound understanding of the concept of sustainability. Some of the general principles that might be applied by the Park Authority to achieve this are set out in Stoll-Kleemann & O'Riordan (2002).

The next part of the chapter examines, through two examples, the kind of policies that might result if these principles described above were adopted.

26.8 Sustainable communities

The Community Councils Group in 2002 produced a statement setting out some characteristics of a sustainable communities (Box 26.3). These characteristics are similar in many respects to those which have been set out as necessary to secure sustainable rural livelihoods by the UK Department for International Development (Carney, 1998).

We would like to single out three aspects for discussion. Firstly, the concept that ways of earning a living should be compatible with the maintenance of biodiversity and natural resources. This principle is included because of the communities' recognition that throughout the area of the Park, people's long-term livelihoods and well-being depend on the maintenance of natural resources and biodiversity. Communities perceive that there is little benefit to be obtained by sacrificing these assets of the Park for short-term advantage.

The second principle to be stressed is the need for communities to have access to the land and buildings that they need in order to live and work. The great majority of the land in the Cairngorms National Park is owned and managed by a relatively small percentage of the population. This can create difficulties in terms of the supply of land around some settlements and, where the existing owners are unwilling to sell land or release buildings, people's aspirations may be thwarted. Unless these constraints are addressed, communities will find it difficult to develop in a sustainable manner.

Box 26.3 Characteristics of a sustainable community

In a sustainable community there will be:

- a range of ages and occupations;
- livelihoods compatible with maintenance of biodiversity and natural resources;
- livelihoods economically sustainable;
- a range of networks and support for people (high levels of social capital); and
- opportunities for communities to be involved in decisions that affect them.

In a sustainable community people will be able to obtain access to:

- education and training;
- health services;
- affordable housing;
- work with reasonable pay and conditions; and
- resources (finance, land, etc.) to undertake livelihoods.

Finally, it is important to draw attention to the need for investment in the provision of education and training that will enable local people to acquire the skills which enable them to take advantage of the new opportunities that the Park will bring. This investment will be the means by which the capacity of local communities can be enhanced so that they will be able to start new sustainable businesses, get high quality jobs, adjust their existing businesses to become sustainable and conform to expectations in the Park.

26.9 Sustainable housing

The second example demonstrates how the adoption of sustainable development policies could enable communities' aspirations to be met and made compatible with the maintenance and enhancement of biodiversity and natural resources. Many different areas of activity could be chosen, but the provision of affordable housing is particularly relevant and illuminating.

It is clear that those interested in the conservation of biodiversity are concerned about the potential impact of new housing development, as it is frequently associated with an increase in the demand for water, an increase in the volume of effluent, increased traffic, habitat destruction and increased production of waste (Smith *et al.*, 1998). All of these lead to a degradation in the quality of the Park's natural resources and biodiversity. But so far as communities are concerned, the provision of new, affordable housing ranks very high in their list of priorities. Housing therefore seems, at least superficially, to be an aspect of development that will lead to inevitable conflict between communities and the conservation of biodiversity.

What can sustainable development contribute to the resolution of this conflict? There are numerous design approaches that set out the means by which sustainable housing can be developed (e.g. Smith *et al.*, 1998; Barton, 2000). Good designs for sustainable housing should take account of both the needs of a community and the need to minimise the impact on the resources of the Park. A list of some of the features that should be included is set out in Box 26.4.

If these ideas were fully implemented it should be possible to build more houses with lower running costs and less pressure on natural resources. If combined with siting in appropriate locations and sensitivity to landscape considerations, this approach would truly be a 'win-win' situation.

The Cairngorms Affordable Housing Strategy (Cairngorms Partnership, 2002) has made a good start by including within it a commitment that new housing should be based on sustainable designs. The strategy recognises the difficulty of implementing such a proposal and therefore includes a recommendation that developers should have access to a sustainable design service. It is to be hoped that the new National Park Authority and local authorities, which will have responsibility for devising and implementing housing policies, adopt this approach.

Box 26.4 Some features of sustainable housing

Scheme design

- Accessible from services/public transport
- Employment nearby
- Gardens for home-grown food and composting
- Conversion or re-use, rather than new build
- Accessible open space

Building design and construction

- Low water use, and use of water recycling
- Sensitive management of run-off
- Low impact sewage systems
- Energy efficient
- Use life cycle analysis to promote durability and reduce maintenance
- Minimal amount of excavation
- Minimise new materials use – re-cycle material where possible
- Minimise construction waste
- Local labour and materials

26.10 Conclusions

This chapter has ranged widely but at its heart has been a concern about communities and their role in the establishment and management of protected areas. We have highlighted some of the lessons that can be taken from the work undertaken to date and stressed the need for strong bonds between sustainable community development and the area's natural resources.

Finally, we believe that the establishment of the Park provides the opportunity for a welcome focus on communities. Indeed, so central do we see communities to the success of the Park that we believe that a prosperous, empowered, confident, engaged community provides an important part in securing the long-term maintenance of the special features of the Cairngorms. Indeed we would propose that the stated purpose for the International Year of the Mountains could be turned around so that it reads:

"To promote the wellbeing of mountain and lowland communities and thereby ensure the conservation and sustainable development of mountain regions."

Along with many others we await with interest to see how the new Park Authority engages with these issues and so helps to secure a sound future for the Cairngorms and the people who live there.

26.11 Postscript

The Cairngorms National Park was formally opened on 1 September 2003, and covers an area of 3,800 km^2, making it by far the largest National Park in the UK. The Park was officially opened by Liz Hannah, John Muir's great great granddaughter (Muir was a Scot who founded the concept of National Parks when he moved to the United States) and by Allan Wilson, Deputy Minister for Environment and Rural Affairs at the Scottish Executive.

The Park Authority has a Board of 25 members, most of whom live in or close to the Park. An innovative approach has been taken to the involvement of local people in that five Board members are directly elected by people living in the Park. The remainder are appointed by Scottish Ministers, 10 on the nomination of local authorities. The early work of the Park Authority has focussed on recruitment of staff, establishment of new procedures, building relations with partner organisations and commencement of work on the new Local and National Park Plans.

To assist with this and other important work, the Park Authority has formed a number of new Advisory Forums which facilitate the development of policies and strategies whilst engaging with a wide range of stakeholders. For example, working with one such Forum on visitor services and tourism, the Authority has prepared a Sustainable Tourism Strategy and Action Plan. On the basis of this stratagy the Park has been assessed by Europarc and awarded the European Charter for Sustainable Tourism in Protected Areas. Meanwhile, the Community Councils and Associations in the area have formed an Association of Cairngorms Community Councils, funded by the Park Authority but separate from it, to provide a forum for involvement of communities in the Park.

The Park is of course still young and it will take some time for the Authority to establish itself. Only then can a proper evaluation be made as to whether this new approach to the intergrated management of one of Scotland's finest environments is succeeding.

Further details about the Park and the Park Authority can be found at www.cairngorms.co.uk.

References

Agrawal, A. & Gibson, C. (1999). Enchantment and disenchantment: the role of community and natural resource conservation. *World Development*, **27**, 629-649.

Barton, H. (2000). Conflicting perceptions of neighbourhood. In *Sustainable Communities*, ed. by H. Barton. Earthscan, London. pp. 3-18.

Borrini-Feyerabend, G. (Ed.) (1997). *Beyond Fences: Seeking Social Sustainability in Conservation*. IUCN, Gland.

Brown, K. (2002). Innovations for conservation and development. *The Geographical Journal*, **168**, 6-17.

Cairngorms Community Councils Group (2002a). Evidence submitted to the Rural Development Committee of the Scottish Parliament, (paper RD/02/24/6), Kingussie, 11 October 2002. Available from the Scottish Parliament's web site at www.scottish.parliament.uk/business/committees/historic/x-rural/papers-02/rap02-24.pdf.

Cairngorms Community Councils Group (2002b). Local Communities and the Cairngorms National Park: The People and the Place. Cairngorms Community Councils Group, Grantown-on-Spey.

Cairngorms Partnership (1996a). Cairngorms Assets. Cairngorms Partnership, Grantown-on-Spey.

Cairngorms Partnership (1996b). A Vision for the Future: a Cairngorms Partnership Working Paper. Cairngorms Partnership, Grantown-on-Spey.

Cairngorms Partnership (1997). Managing the Cairngorms: the Cairngorms Partnership Management Strategy. Cairngorms Partnership, Grantown-on-Spey.

Cairngorms Partnership (2002). An Affordable Housing Strategy for the Cairngorms. Cairngorms Partnership, Grantown-on-Spey.

Carney, D. (1998). *Sustainable Rural Livelihoods; What Contribution Can We Make?* Department for International Development, London.

Copus, A.K., Gourlay, D., Petrie, S., Cook, P., Palmer, H. & Waterhouse, T (1999). Land use and economic activity in possible National Park areas in Scotland. Scottish Natural Heritage Review No. 115.

Cosgrove, P. (2002). *The Cairngorms Local Biodiversity Action Plan*. Cairngorms Partnership, Grantown-on-Spey.

Department of Culture, Media and Sport (1999). *World Heritage Sites: the Tentative List of the United Kingdom of Great Britain and Northern Ireland*. Department of Culture, Media and Sport, London.

Forsyth, B. & Downie, A. (2000). Cairngorms National Park: Community Councils Support. Unpublished report. Cairngorms Partnership, Grantown-on-Spey.

Gimingham, C. (Ed.) (2002). *The Ecology, Land Use and Conservation of the Cairngorms*. Packard, Chichester.

Govan, H., Inglis, I., Pretty, J., Harrison, M. & Wightman, A. (1998). Best practice in community participation in National Parks. Scottish Natural Heritage Review No. 107.

Rettie, K. (2001). The Report on the Proposal for a National Park in the Cairngorms: an Independent Assessment of the Consultation on the Proposed National Park for the Cairngorms. Scottish Natural Heritage, Perth.

Scottish Natural Heritage (1999). National Parks for Scotland: Scottish Natural Heritage's Advice to Government. Scottish Natural Heritage, Perth.

Scottish Natural Heritage (2000). A Proposal for a Cairngorms National Park. Scottish Natural Heritage, Perth.

Scottish Natural Heritage (2001). The Report on the Proposal for a National Park in the Cairngorms. Scottish Natural Heritage, Perth.

Scottish Natural Heritage (2002). Natural Heritage Futures: The Cairngorms Massif. Scottish Natural Heritage, Perth.

Smith, M.A.F., Whitelegg, J. & Williams, N. (1998). *Greening the Built Environment*. Earthscan, London.

Stoll-Kleemann, S. & O'Riordan, T. (2002). Enhancing biodiversity and humanity. In *Biodiversity, Sustainability and Human Communities: Protecting beyond the Protected*, ed. by S. Stoll-Kleemann & T. O'Riordan. Cambridge University Press, Cambridge. pp. 295-310.

Taylor, J. & MacGregor, C. (1999). Cairngorms Mountain Recreation Survey 1997-1998. Scottish Natural Heritage Research Survey & Monitoring Report No. 162.

Turnbull Jeffrey Partnership (1996). Cairngorms Landscape Assessment. Scottish Natural Heritage Review 75.

Watson, A. (2001). The Cairngorms National Park: a political fix. *Leopard Magazine*, **281**, 31.

World Commission on Environment and Development (1987). *Our Common Future.* Oxford University Press, Oxford.

27 Two models of National Parks: ethical and aesthetic issues in management policy for mountain regions

Alan P. Dougherty

Summary

1. Although a sound grounding in science (the term is used in its broadest sense) is essential to policy formulation in respect of upland land use, this chapter will explore the role of philosophy as the necessary bridge between science and policy.
2. Note is taken of Hume's dictum that it is a fallacy to go from an 'is' to an 'ought' and thus the basis of decision making must be ethical.
3. A comparison is made between the North American 'wilderness' model of National Parks with the Anglo-Welsh and, now, Scottish model of National Parks as 'cultural landscapes'. Neither model is taken to be unproblematic and will be critiqued to expose some of the ethical and aesthetic issues raised by these contrasting visions of National Parks in mountainous areas.
4. In the wider North European context, the idea of differing park models appropriate to areas of varying character is raised. The notion of appropriate and authentic engagement with wilder places is introduced.

27.1 The role of environmental philosophy as the bridge between science and policy

Science (expert knowledge, including the contribution of the arts) can inform us about the ecological history of the landscape, explain the workings of its present day ecosystems, and suggest the various likely outcomes of different land management in the future. What science cannot do alone is determine choices for the future. These are choices about how we ought to proceed, and raise ethical and other considerations.

Attempting to address such issues is one of the roles of applied philosophy in the sphere of public policy making. The techniques of applied philosophy (Pratt *et al.*, 2000; Brady, 2003) can also be used to

- carry out conceptual analysis of the concepts involved in the debate;
- tease out the prior values that underpin various arguments;
- establish if contrasting viewpoints are commensurable or mutually exclusive; and
- identify and explore the aesthetic basis as an important value dimension.

Dougherty, A.P. (2005). Two models of National Parks: ethical and aesthetic issues in management policy for mountain regions. In *Mountains of Northern Europe: Conservation, Management, People and Nature*, ed. by D.B.A. Thompson, M.F. Price & C.A. Galbraith. TSO Scotland, Edinburgh. pp. 291-294.

27.2 Hume's is/ought dictum

The 18th century Scottish philosopher Hume (Hume, 2003, p. 258) asserted that there is no cogent reasoning that proceeds simply from an 'is' to an 'ought'; that the appeal to things as they are, as explained by science or suggested by tradition, cannot alone be the basis for decisions about how we ought to proceed. Therefore, during environmental policy making, ethical considerations must be applied additionally to our scientific knowledge. For example, science might indicate that, if continued, certain actions would change the ecosystem in a particular way, but it is an ethical consideration to decide that policy should be altered for the benefit of future generations. It would be a further ethical choice to extend consideration to non-human species or even non-sentient components of the environment, such as rocks and water. Deep ecology (Naess, 1989) is perhaps the most widely known exposition of the extension of ethics beyond our own species but various philosophies regarding human behaviour in relation to nature are being developed (Pratt *et al.*, 2000).

27.3 A comparison of the North American 'wilderness' and Anglo-Welsh and Scottish 'cultural landscape' models of National Parks in mountain regions

27.3.1 Key features

In North America, park land is often held in common (governmental) ownership, whilst in the UK both private and charitable organisation ownership is typical, together with other land held by governmental agencies and departments, for example, the Forestry Commission.

Patterns of ownership (or, more precisely, the rights and obligations vested in such) can, and have, influenced upland management. A full discussion is not possible in the context of this chapter but it might be noted that one finds little in the major philosophical writings supportive of the large-scale private ownership of land.

Although there is some debate as to the extent to which indigenous populations have adapted their environment, mountain National Parks in North America can be typified as comprising wilder, less humanised, landscapes. Within these parks, the current notion of carrying capacity is applied sometimes and, consequently, access is often limited.

In contrast, United Kingdom are located in predominantly humanised landscapes. Some forms of self-selective access 'restriction', for example 'the long walk in principle' and the non-provision of car-parking or footbridges, have been applied in the UK but access is generally less restricted, excepting onto some private land and wildlife conservation sites. The Countryside and Rights of Way Act 2000 should (subject to possible site restrictions because of conservation or land management concerns) improve access to privately owned uplands in England and Wales. In Scotland, the Land Reform (Scotland) Act 2003 has legally enshrined responsible access to land and water.

27.3.2 Contrasting ethical foci

Within the North American model the wilderness debate is a primary issue. A scientific understanding of climax ecosystems facilitates the ethical notion of integrity, and constrains the range of possible 'oughts'. In the light of scientific accounts, Leopold (1949, p. 224) asserted that

"A thing is right when it tends to preserve the integrity, stability and beauty of the biotic community. It is wrong when it tends to do otherwise."

This can be adopted as a basis for management that is grounded in the integrity of ecosystems.

In contrast, the Anglo-Welsh and Scottish model does not broach the wilderness debate as a central issue and the question of natural integrity tends not to be addressed. Cultural considerations predominate over ecological ones, and little emphasis tends to be given to the environmental factors that constrain the development of cultural landscapes. Fundamental questions as to the appropriateness of humanising actions in particular ecosystems tend to be avoided.

27.3.3 Contrasting aesthetic foci

In both models, aesthetic and ethical discussions become intertwined. The North American model is characterised by an aesthetic in which integrity with the whole environment is a key concept. Although ecosystems are characteristically dynamic, they have the potential to endure. An aesthetic (or, indeed, ethical) account based upon integrity with nature also has the potential to endure.

Aesthetic quality in the context of the Anglo-Welsh and Scottish model is judged with respect to contextual integrity, understood in the light of historical context, but only with reference to a selective version of the cultural landscape, for example, the working agricultural landscape (of a particular period) but not a contemporaneous one such as active mining or quarrying. An aesthetic account grounded upon integrity with the human community is susceptible to varying cultural trends, for example, changes in agricultural practice or taste in landscape, and thus prone to be relatively ephemeral.

27.4 Differing park models for varying wildness

Mountain-based National Parks include landscapes of varying character and wildness. It might be timely to consider less rigid management models, which whilst guided by nature, move beyond the potential sterility of approaches that define wilderness as an absolute but unrealistic concept. This is unrealistic in the sense that it is likely that the whole of nature on Earth is subject to, at least, unintentional human influences, so an absolute definition of wilderness becomes problematic. Thus it is unreasonable to treat the uplands as if isolated from the rest of environment and human activity; but pandering to the ephemeral basis of cultural landscape models can also be mistaken for their tendency to subjugate the part played by environmental constraints on culture.

Instead, we ought to consider appropriate and authentic engagement with specific environments. Guided by nature, such a model should engender behaviour appropriate to the well being of the ecology and wildness of an area.

Authentic engagement is a defining characteristic of such behaviour. Guignon (2004) constructs a strong case against the individualised and introspective take on authenticity that is common to much philosophical treatment of the concept. In contrast, he posits authenticity of behaviour with respect to other humans as a relational concept. Such an account of authenticity can be extended to human relationship with the environment. Meeting nature more on its own terms, engagement is regarded as authentic when not mediated overly by technology (Dougherty, in review). It is also guided by nature and, thus, despite the dynamics of ecosystems but because of their potential to endure, should be less prone to being ephemeral.

Whilst allowing an ethical and aesthetic respect for nature to be its central principle, this model not only has the potential to respond to varying degrees of past and present human interaction with the uplands but also to guide current and future engagement with mountain areas.

References

Brady, E. (2003). *Aesthetics of the Natural Environment.* Edinburgh University Press, Edinburgh.

Dougherty, A.P. (in review). Aesthetic and ethical issues concerning sport in wild places. In *Sporting Danger: Philosophies of Adventure Sports*, ed. by S. Eassom. Routledge, London.

Guignon, C. (2004). *On Being Authentic.* Routledge, London.

Hume, D. (2003). *A Treatise of Human Nature. Book III, Part I, Section I.* Dent, London.

Leopold, A. (1949). *A Sand County Almanac and Sketches Here and There.* Oxford University Press, Oxford.

Naess, A. (Translated by C. Rothenburg) (1989). *Ecology, Community and Lifestyle: Outline of an Ecosophy.* Cambridge University Press, Cambridge.

Pratt, V. with Howarth, J. & Brady, E. (2000). *Environment and Philosophy.* Routledge, London.

28 Community land ownership in Scotland: progress towards sustainable development of rural communities?

Aylwin Pillai

Summary

1. Sustainable development is the underlying principle behind the community right to buy in Part 2 of the Land Reform (Scotland) Act, which received Royal Assent on 25th February 2003.
2. Community land ownership, achieved through the community right to buy, is intended to promote sustainable development by empowering communities to manage the land according to their own economic, social and environmental objectives. However, the success of community ownership will depend on the safeguards put in place to ensure that a balance is achieved between these goals.
3. Existing experience suggests that, while community land ownership does offer communities considerable opportunities, it is difficult to balance the economic development of communities and the protection of the environment in remote areas of rural Scotland.

28.1 The relationship between community land ownership and sustainable development

The principle of sustainable development is said to underpin the community right to buy in Part 2 of the Land Reform (Scotland) Act 2003 (Scottish Executive, 2001). When the Land Reform Policy Group was set up to consider land reform, it posited that the overriding objective should be to remove the land-related barriers to sustainable development. It identified barriers including the concentrated pattern of land ownership and the monopoly power of the Highland estates. It concluded that a community right to buy would empower communities, diversify land ownership and use, and facilitate progress towards the sustainable development of rural communities (Land Reform Policy Group, 1998, 1999).

There is an important relationship between community land ownership and sustainable development because ownership gives communities greater control over land use and management. Ownership empowers communities to take decisions according to their own economic, social and environmental objectives. However, it is questionable whether empowerment alone is sufficient to advance sustainable development. Safeguards must be put in place to ensure that communities will not merely satisfy short-term economic goals at the expense of environmental or social objectives.

Pillai, A. (2005). Community land ownership in Scotland: progress towards sustainable development of rural communities? In *Mountains of Northern Europe: Conservation, Management, People and Nature*, ed. by D.B.A. Thompson, M.F. Price & C.A. Galbraith. TSO Scotland, Edinburgh. pp. 295-298.

28.2 The Land Reform (Scotland) Act Part 2: The Community Right to Buy

The community right to buy gives community bodies, incorporated as companies limited by guarantee, a right to buy land. The community company must register an interest in the land before it can exercise the right when the landowner decides to sell. Safeguards at each stage of the process should ensure that the community's plans for the land are compatible with sustainable development. However, the Act has serious weaknesses in relation to sustainable development.

The Act provides that Ministers shall not consent to the exercise of the right to buy unless they are satisfied that what the community proposes to do with the land is compatible with furthering the achievement of sustainable development. However, the Act contains no definition of sustainable development. An earlier draft of the legislation defined sustainable development as "development calculated to provide increasing social and economic advantage to the community and protect the environment" (Land Reform (Scotland) Bill, 2001). This provides no insight into how economic, social and environmental objectives are to be balanced or whether trade offs can be made where objectives conflict. There has, as yet, been no indication of assessment criteria from the Scottish Executive so that at present it appears that the effectiveness of this safeguard will depend upon the subjective assessment of Ministers.

By using the vehicle of the company limited by guarantee, a key safeguard on the powers and limitations of communities can be placed in the company constitution. Initially the Land Reform (Scotland) Bill required sustainable development to be included in the memorandum and articles of association as the *main* purpose of the company. However, the Act as passed has dropped this requirement. It is sufficient that Ministers are satisfied that the main purpose of the body is consistent with furthering the achievement of sustainable development.

In any case, without a clear definition of sustainable development, such a requirement is unlikely to restrict unsustainable activities of communities. Even if a community does make a commitment to sustainable development in its constitution, it may interpret this principle from a purely economic perspective. For example, the objects of one community company comprise the promotion of "sustainable development including agriculture, silviculture, arts and crafts and other economic activities" (Isle of Eigg Heritage Trust company constitution, undated).

28.3 Existing experience of community land ownership

The establishment of community land ownership initiatives is a growing phenomenon throughout rural Scotland. Recent examples include the purchases of the Isle of Gigha, and the Amhuinnsuidhe and North Harris Estate. The Community Right to Buy will provide opportunities for further community purchases.

The existing experience of well-established communities provides good evidence of the long-term effects of community ownership for sustainable development. Many existing schemes have been established in ways that the legislation will continue to encourage, e.g. by using the company limited by guarantee as the ownership vehicle; gaining funding for the purchase or post-purchase projects from funding bodies already in place to support the objectives of the legislation, such as the Scottish Land Fund and the Highlands and Islands Enterprise Community Land Fund; and by using the advice and technical support of Highlands and Islands Community Land Unit which has been available since its inception in 1997. Two brief examples illustrate this.

28.3.1 The Isle of Eigg Heritage Trust

"The objectives of the Isle of Eigg Heritage Trust are to provide security of tenure, manage and develop the island in a sustainable way, create employment, training opportunities, make opportunities available for individual development and try to develop Eigg in a way that doesn't destroy the natural heritage … Sustainable development is the ultimate aim … It has any number of shapes and forms but it is all about creating opportunities for people" (Fyffe, 2002; M. Fyffe, pers. comm; Pillai, 2004).

The community of Eigg has made significant progress in its economic, social and environmental objectives. The community's key achievements include granting of security of tenure to tenants, establishment of a trading subsidiary company, establishment of Eigg Tearooms and Eigg Construction (which has renovated three houses), and the creation of 14 full- and part-time jobs. The Trust has initiated schemes for waste management, woodland regeneration and hydropower. There is increased awareness that the environment is one of the community's main assets and should be protected.

28.3.2 The Assynt Crofters Foundation

In contrast, John Mackenzie has outlined the difficulties of ownership for achieving sustainable crofting communities. These difficulties apply equally to community land ownership. The limitations of land as an asset in terms of economic viability can be acute in remote areas of rural Scotland. Efforts to achieve progress are hampered by the lack of integration of conservation and economic development policies. Communities require high levels of financial and technical assistance (Mackenzie, 1998).

28.4 Conclusion

The Community Right to Buy aims to promote sustainable rural development by empowering communities to manage land according to their own objectives. Yet the success of the community right to buy as a tool for achieving this policy goal will depend on effective safeguards to ensure a balance between economic, social and environmental objectives. Disappointingly, the Act makes a weaker commitment to sustainable development than earlier drafts. It is debatable whether the Act can be effective without a clear definition of sustainable development by which Ministers can assess an application to exercise the right and by which communities can define their objectives.

The legislation lacks a long-term vision for future generations since the activities of the communities post-purchase are left unregulated, with no monitoring or overseeing body. Existing examples of community ownership in Scotland highlight the possible achievements but also illustrate that there are difficulties. In particular it is doubtful whether the high level of technical and financial assistance required by such schemes can be sustained in the long-term for every community ownership initiative.

References

Fyffe, M. (2002). Isle of Eigg Heritage Trust Progress Report. Unpublished report. Isle of Eigg Heritage Trust, Isle of Eigg.

Isle of Eigg Heritage Trust (undated). Company Constitution: Memorandum and Articles of Association. (Registered Company no. 170339.)

Land Reform (Scotland) Act 2003.

Land Reform (Scotland) Bill (2001). (Session 1 SP Bill 44B).

Land Reform Policy Group (1998). *Identifying the Problems.* The Stationery Office, Edinburgh.

Land Reform Policy Group (1999). *Recommendations for Action.* The Stationery Office, Edinburgh.

Mackenzie, J. (1998). Business Planning: The Assynt Experience. In *Social Land Ownership: Eight Case Studies from the Highlands and Islands of Scotland, (Vol.1)*, ed. by G. Boyd & D. Reid, The Not-for-Profit Landowners Project Group, Inverness. pp. 22-29.

Pillai, A. (2004). *The Community Right to Buy: Progress Towards Sustainable Development?* PhD Thesis. University of Aberdeen, Aberdeen.

Scottish Executive (2001). *Land Reform (Scotland) Bill Policy Memorandum.* The Stationery Office, Edinburgh.

29 Tourists, nature and indigenous peoples - conservation and management in the Swedish Mountains

Robert Pettersson & Tuomas Vuorio

Summary

1. Many tourists have visited Saami attractions when travelling in Northern Sweden, and a large number of them is interested in doing so in the future.
2. There is a gap between Saami tourism supply and demand.
3. Visitors in the Södra Jämtlandsfjällen area were classified by a purism scale comprising neutralists (68%), purists (17%) and urbanists (15%).
4. Provision of information is seen as the best management action to avoid ground damage, while visitor fees and limitations in the right of common access are less favoured solutions.

29.1 Introduction

For the last 100 years, culture- and nature-based tourism has been part of the Swedish rural economy (Heberlein *et al.*, 2002). This is particularly the case in the peripheral and sparsely populated areas in the Swedish mountains, which stretch for over 1,000 km along the border with Norway.

The northern mountains are the homeland of the indigenous Saami peoples, who call their land Sápmi. Lately the Saami of Northern Sweden have begun to engage in tourism, partly due to the structural changes in reindeer herding (Pettersson, 2004). The Saami share their land with tourists attracted by Saami culture, activities and outstanding nature.

While tourism may be attractive for some interests as a means of economic development for remote communities, others are concerned about the effects on fragile habitats and culture. The expansion of tourism has lately raised questions about land use, policy and sustainability. Various solutions have been put forward over the years to address the effects of more visitors. However, little focus has been paid to the visitors' point of view. Data on outdoor recreation and tourism are needed in many phases of the planning process: environmental impact assessment, spatial planning in relation to the utilization and management of the area and the implementation of the plan.

Drawing from two case studies carried out in Northern Sweden (Tourism in Sápmi: Müller & Pettersson, 2001; Pettersson, 2002; and Visitors in Södra Jämtlandsfjällen: Vuorio *et al.*, 2000), this chapter describes visitors' opinions on culture and nature-based tourism.

Pettersson, R. & Vuorio. T. (2005). Tourists, nature and indigenous peoples - conservation and management in the Swedish Mountains. In *Mountains of Northern Europe: Conservation, Management, People and Nature,* ed. by D.B.A. Thompson, M.F. Price & C.A. Galbraith. TSO Scotland, Edinburgh. pp. 299-302.

29.2 Tourism in Sápmi

Indigenous tourism is an expanding sector in the growing tourism industry. The Saami people living in Sápmi in Northern Europe have started to engage in tourism, particularly in view of the rationalised and modernised methods of reindeer herding. Saami tourism offers job opportunities and enables the spreading of information. On the other hand, Saami tourism may jeopardise the indigenous culture and harm the sensitive environment in which the Saami live.

In the research project; 'Tourism in Sápmi', the Saami tourism entrepreneurs, their business services and location were studied in relation to tourists' attitudes towards Saami tourism. The predicted discrepancy between the attractions on offer and the tourists' expectations have been tested in two studies focusing on Saami tourism and its supply and demand.

29.2.1 Saami tourism supply

For some people visiting Sápmi the information gained from tourist brochures and other sources is sufficient to satisfy their curiosity. Others choose to visit Saami culture, and these people will have the opportunity to obtain first-hand experiences. However, if the motivation to travel to the destination is to be realised, the tourist attraction has to be easily accessible for the visitor.

A mapping of the Saami tourist companies in Northern Sweden shows a number of relatively young tourist attractions connected to Saami culture (Müller & Pettersson, 2001). These attractions are often close to municipal and tourist centres where it is possible to utilise the existing infrastructure of road network, accommodation, food and other supplies. Saami tourist attractions are thus relatively

Figure 29.1 A Saami in a richly coloured dress, and a reindeer. (Photo: B. Lind)

easily accessible, but this easy access may demand some preparation from the visitor, not least because of the fact that many of the Saami attractions are only open for booked groups.

Altogether there are today about 40 Saami tourist entrepreneurs in the Swedish part of Sápmi, from which about 20 run their tourist enterprises parallel to their reindeer herding (see Figure 29.1). Besides the entrepreneurs, it is also possible to experience Saami culture at museums, outdoor attractions, events and through craft sales.

29.2.2 Saami tourism demand

By using the Stated Preference (SP) method, tourists' attitudes and preferences were measured for three main attributes: i) the activities offered (supply); ii) the prices of these activities; and finally iii) the access to the activities (Pettersson, 2002). A questionnaire survey was carried out in July 2000 among existing and potential visitors to Sápmi travelling in the north of Sweden.

The study shows that there seems to be a considerable potential market for these kinds of activities, and that there is a discrepancy, in some respects, between the activities offered and those demanded. For instance, the most wanted service (a mountain walk with a Saami guide) is least on offer.

29.3 Visitors in Södra Jämtlandsfjällen

This study was carried out as a part of the deeper comprehensive plan for the area Södra Jämtlandsfjällen (Vuorio *et al.*, 2000; Denstadli *et al.*, 2001). Several methods were used to estimate present patterns of use such as tent counts from the air, self-registration and questionnaires. In the mountains, there has been only limited knowledge about the people that are using these areas for tourism and outdoor recreation (Denstadli *et al.*, 2001; Heberlein *et al.*, 2002; Vuorio, 2003). The need is to provide the planning system with relevant information, that is information that can be used for predicting different reactions to different management actions, and to help resolve conflicts (Vuorio, 2003).

29.3.1 Classifying the visitors

It is quite clear that outdoor recreational and tourists have different interests and needs concerning 'nature without human influence'. Their perception of crowding, the effects of contact with other people and of the concept of untouched nature differ greatly. For management and planning it is very interesting to know what kinds of qualities people are seeking and appreciate. Tourism that is intended to build on the wilderness experience should make efforts to gather information about customers' attitudes and expectations.

By asking a set of questions about different preferences for unspoiled nature or places with wilderness characteristics, it is possible to get a good picture of individuals' preferences on a purism scale (Kaltenborn & Emmelin, 1993; Fredman & Emmelin, 2001). The object is to understand visitors' general ideals, not just their expectations of a specific area. Characteristic questions are, for example, about visitors' attitudes towards marked trails, huts, other visitors, different restrictions, etc. The purism scale is a one dimensional addition of answers on all of these questions. The neutralists (68%) is the group lying inside the standard deviation, the purists (17%) and the urbanists (15%) are the groups outside this range.

29.3.2 Visitors' attitudes towards management

In general visitors are happy with the existing tourism establishments and the quality of service. More information on these areas is needed for visitors as the best way to minimise wear and tear on the area. Actions for nature conservation, building of footbridges, limiting of entry to the sensitive areas, and the control of the number of visitors during the sensitive periods were also seen as positive actions. Rigid restrictions to managing environmental disturbance were seen as less desirable. The most negative reactions were caused by the suggestion of introducing fees for visitors and limitations to the right of common access. Restrictions that would especially affect visitors' own activities were seen in a more negative light. The study was an important part of the comprehensive planning process carried out by local and regional authorities. The results have been used in management of Södra Jämtlandsfjällen as one guideline for changes in service, infrastructure and information.

29.4 Conclusion

Tourism, based on nature and culture, is a growing sector in the world tourism industry. A well developed tourism plan puts focus on sustainability, especially in areas with fragile nature and culture, such as the Swedish mountains. There is often a tendency in tourism planning towards a production rather than a consumption perspective. The two studies presented here show that there is a lot to gain by considering the opinions of the visitors in the area.

References

Denstadli, J.M., Lindberg, K., Vuorio, T. & Fredman, P. (2001). Residents in Södra Jämtlandsfjällen – attitudes toward wind power, national park designation and tourism development. ETOUR Working Paper 2002:3. ETOUR, Östersund.

Fredman, P. & Emmelin, L. (2001). Wilderness purism, willingness to pay and management preferences. A study of Swedish mountain tourists. *Tourism Economics*, **7**, 5-20.

Heberlein, T.H., Fredman, P. & Vuorio, T. (2002). Current tourism patterns in the Swedish Mountain Region. *Mountain Research and Development*, **22**, 142-149.

Kaltenborn, B.P. & Emmelin, L. (1993). Tourism in the High North: management challenges and recreation opportunity spectrum planning in Svalbard, Norway. *Environmental Management*, **17**, 41-50.

Müller, D. & R. Pettersson, R. (2001). Access to Saami tourism in northern Sweden. *Scandinavian Journal of Hospitality and Tourism*, **2001:1**, 5-18.

Pettersson, R. (2002). Saami tourism in northern Sweden: measuring tourists' opinions using stated preference methodology. *Tourism and Hospitality Research*, **3**, 357-369.

Pettersson, R. (2004). Saami tourism in northern Sweden – supply, demand and interaction. Scientific book series V 2004:14. Doctoral thesis. ETOUR, Östersund.

Vuorio, T. (2003). Information on recreation and tourism in spatial planning in the Swedish mountains – methods and need for knowledge. Licentiate Dissertation Series 2003:03. Blekinge Institute of Technology, Karslkrona.

Vuorio, T., Emmelin, L. & Göransson, S. (2000). Vandrare i Södra Jämtlandsfjällen – underlag för översiktlig planering. ETOUR Working paper 2000:11. ETOUR, Östersund.

PART 5:
Prospects

Adult golden eagle (*Aquila chrysaetos*) hunting over Glen Spean, Scottish Highlands, UK (Photo: Laurie Campbell).

PART 5:

Prospects

Virtually every day we see or hear about news on global climate change. For instance, on 11 August 2005 half of the front page of *The Guardian* (Sample, 2005) was devoted to the accelerated thawing of Siberia's peat bog permafrost. This ecosystem, equivalent in size to that of France and Germany combined, will release methane as it melts, contributing to the already relatively high methane levels found in the atmosphere (compared with the levels measured earlier in the last century). Also in August 2005, the journal *Nature* featured on its front cover the collapse of a significant part of the Larsen ice shelf of the Antarctic Peninsula (see Domack *et al.*, 2005); this break up is unprecedented during the Holocene. The UK Natural Environment Research Council has produced a valuable briefing paper on the scientific certainties and uncertainties regarding climate change (Anon., 2005). That paper highlights the fact that, over the last century, the average global surface temperature rose by around 0.7°C, with continents in the northern hemisphere warming the most: 1998, 2002 and 2003 were the warmest years since 1860, the earliest year for which a precise global estimate of temperature is possible.

Hallanaro & Usher, in the opening chapter in this final part of the book, show how the prospects for the uplands of Northern Europe would have been quite different when viewed from different times in the past. They reflect on the major phases of tree cutting and removal, farming activities, grazing by large herbivores, pollution and tourist development. They close their chapter with a consideration of climate change, and provide graphic examples of changes in human influences on the biodiversity of mountain areas.

Over the last five years there has been growing interest in the development of policies specific to mountain areas (e.g. European Commission, 2004). In this context, Aalbu (Chapter 31) describes recent work to define the mountain municipalities in Europe that could be the subject of policies focused on mountain areas. It is possible that further European policies will be devised to influence sustainable development in mountain areas through the European Union's Structural Funds and Common Agricultural Policy.

The final chapter in this book on National Parks concerns the establishment of the Vatnajökull National Park in Iceland. Benediktsson & Þorvarðardóttir (Chapter 32) describe the prospects for the designation of this park in a process which began in 2002. The chapter reveals the many challenges involved in reconciling nature conservation, and economic, cultural and ideological elements. Interestingly, this research has shown that, so far as local communities around the Vatnajökull are concerned, the conservation of biodiversity and geodiversity is not a major issue in terms of National Park designation. This does appear to reflect a theme emerging from many of the other chapters: that biodiversity is viewed increasingly as an asset, which attracts people into areas to generate some much needed income for local people.

Higgins (Chapter 33) examines outdoor recreation and education development in Scotland. He points to the significant revenue generated in the Scottish rural economy by outdoor recreation. He notes with concern, however, that outdoor education provision has been in decline in recent decades, although there is now a basis for some modest growth in this area.

In looking ahead at the prospects for mountain areas, it is important that we have information and data drawn from local, national and regional global sources. Heal *et al.* (Chapter 34) describe the northern hemisphere network of terrestrial field sites which is being used to develop our understanding of long-term ecological and environmental processes in mountain and Arctic areas. This network, referred to as SCANNET, provides a facility for students, researchers, managers and policy makers to access information and expertise at field sites throughout Northern Europe. The authors provide pointers to research that develops scenarios for climate change regionally and seasonally, and it will be important to see how this work contributes to policy development.

The final four chapters embrace the diverse challenges ahead in conserving and managing mountain areas. Carver (Chapter 35) focuses on the concept of wilderness restoration, and seeks to identify appropriate areas using a variety of datasets. His work underscores the technical and philosophical difficulties of trying to identify mountain areas with the aim of restoring habitats and even whole landscapes to near-natural conditions. Grabherr (Chapter 36) considers priorities for the conservation and management of the natural heritage across Europe's high mountains. Like Carver, Grabherr considers the importance of areas that are encouraged to function as dynamic, natural ecosystems. However, he also highlights the importance of cultural landscapes, noting that the cultural heritage of mountain areas is as much at risk as the natural heritage. Grabherr's research on the impacts of climate change on vegetation in the Alps reveals that species richness has increased at high altitude because more species have moved upwards in response to changes in conditions. Pepper & Moen provide some personal reflections on the conference as a whole. They point to the importance of mountain areas for people living in lowland as well as upland parts of Europe, and they highlight the need to take a more integrated approach in protecting, managing and using mountain areas.

In the concluding chapter, Galbraith & Price reflect on the key issue of adopting an integrated, multidisciplinary approach to managing mountain environments. As the diverse content of this book reveals, the great challenge in moving forward is to ensure that our understanding of environmental change is informed by research which describes the scientific, social and cultural drivers. The contributors to this book have demonstrated the diversity of approaches needed to achieve this, in order that we can care for Europe's mountain areas and bring enduring benefits for people and nature.

References

Anonymous (2005). *Climate Change. Scientific Certainties and Uncertainties.* Natural Environment Research Council, Swindon.

Domack, E., Duran, D., Leventer, A., Ishman, S., Doane, S., McCallum, S., Amblas, D., Ring, J., Gilbert, R. & Prentice, M. (2005). Stability of the LB ice shelf on the Antarctic Peninsula during the Holocene epoch. *Nature*, **436**, 681-685.

European Commission (2004). *Mountain Areas in Europe: Analysis of Mountain Areas in EU Member States, Acceding and Other European Countries.* Directorate-General for Regional Policy, European Commission, Brussels.

Sample, I. (2005). Warming hits 'tipping point'. *The Guardian*, August 11 2005, 1-2.

30 Natural heritage trends: an upland saga

Eeva-Liisa Hallanaro and Michael B. Usher

Summary

1. Throughout Northern Europe, the uplands have been affected by the cutting and removal of trees and shrubs, either to create farmland or to produce timber products, and they have been further influenced by large domestic, feral or wild herbivores. The landscapes that we see today are therefore a reflection of at least 1,000 years of human activity.

2. In Scotland, there have been declines in the extent of moorland, peatland and semi-natural upland grasslands. Some species, such as the red grouse (*Lagopus lagopus scoticus*), have apparently declined. Many Arctic and Boreal plants have reduced ranges in Britain as a whole, but not in Scotland.

3. Mountain birch (*Betula* spp.) is a keystone species in Fennoscandia, but large areas can be devastated by outbreaks of the autumnal moth. Reindeer (*Rangifer tarandus*), with three subspecies, one of which has been domesticated, are an important economic resource.

4. In both Fennoscandia and Scotland, large herbivores are affecting semi-natural vegetation communities. Overgrazing occurs in many places and attempts are being made to assess and quantify the extent of the damage. There are concerns about the genetic constitution of the wild species, with hybridisation of red deer (*Cervus elaphus*) and Sika deer (*Cervus nippon*) in Scotland and of wild and domesticated reindeer in Fennoscandia.

5. Careful planning, possibly new legislation and careful management of the nature resource are all required to control a spectrum of modern pressures including pollution by toxic chemicals, tourist developments and the use of off-road vehicles.

6. There may be a need for sufficient sites to assess the effects of climate change on both upland biodiversity and upland ecosystem processes.

Hallanaro, E.-L. & Usher, M.B. (2005). Natural heritage trends: an upland saga In *Mountains of Northern Europe: Conservation, Management, People and Nature,* ed. by D.B.A. Thompson, M.F. Price & C.A. Galbraith. TSO Scotland, Edinburgh. pp. 307-324.

30.1 Introduction

The uplands of Northern Europe, including the Arctic fells and the Arctic upland plateaux of Fennoscandia and Iceland, are among the least productive natural environments in Europe. They are also thought to be some of the least utilised, and thereby some of the most pristine, areas. They contain relatively few signs of human activity, such as houses, roads, managed fields or polluted lakes and rivers. There is, nevertheless, a long tradition of human habitation in these areas and an equally long tradition of the utilisation of their natural resources.

The Icelandic Sagas, written mainly in the 13th century but containing accounts of events that had happened mostly in the 10th and 11th centuries, recount how the settlers of this arctic-alpine island cut or burned local birch woods to provide space for farming, for support for buildings, for use as fuel, and later to make charcoal. Wood-cutting is mentioned so often that it seems clear that the farmers and their workers spent considerable time undertaking this activity (Kristinsson, 1995). The destruction of woodland was completed by allowing uncontrolled sheep grazing all over the country.

Iceland may be an extreme example, but similar developments also took place in other uplands of Northern Europe, not least in Scotland. Even the Arctic fells of Fennoscandia (Norway, Sweden, Finland and the adjacent areas of Russia) have been affected by both wood-cutting and grazing by either sheep or reindeer. Moreover, there is an even longer tradition of hunting and fishing in these areas, and as late as the early 18th century many of the Sámi (Saami) of Fennoscandia were still hunters and fishermen, with wild reindeer being their most important prey. People from further afield also enjoyed the offerings of nature; many hunting and fishing expeditions were made to the remote upland and Arctic areas which had become known for their natural riches. Today, these areas draw even more people, including tourists who want to enjoy the natural beauty of the uplands.

The aim of this chapter is to explore some of the trends that have occurred, and are occurring, in these Arctic and upland areas. Having pinpointed some of the main trends, two case studies, which explore the effects of overgrazing by large herbivores and the pressures caused by tourism and pollution, are presented.

30.2 Significant trends in the natural heritage

30.2.1 Trends in Scotland

Historically, the major change in the Scottish uplands has been the descent of the upper limit of tree growth and the almost complete disappearance of both a natural treeline and the scrub communities. For example, in the Cairngorm Mountains only remnants of a natural treeline are to be found on Creag Fhiaclach at about 650 m: this is thought to be the best example of a natural treeline in the UK (McConnell & Legg, 1995). The woolly willow (*Salix lanata*) is one of 12 willow species that occur in the Scottish uplands, but now occurs in only 12 locations. One of these populations (in Glen Clova) has over 100 individuals, four populations (if you can call them that!) have just a single individual, and the other seven populations have between two and 100 individuals. This contrasts with the abundance of the species over much of Iceland, but can such a comparison be valid? Recently it has been estimated that willow scrub, in which woolly willow would occur, might have occupied at least 2,000 ha in Scotland (Anon., 1998). Why has there been this trend, resulting in the almost complete loss of upland scrub?

The effects of grazing animals, especially sheep (*Ovis* spp.) and red deer (*Cervus elaphus*), together with the cutting of trees, are likely to be the primary causes of these historical changes (Scott, 2000). While the few remaining scattered trees in the uplands are capable of producing viable seed (Stewart *et al.*, 2000), the seedlings are grazed when they reach the height of the surrounding vegetation, thus ensuring that there is no woodland regeneration. In the last decade or two, land management in a few areas has aimed to reduce the grazing pressure and has thus allowed the natural regeneration of trees and shrubs to succeed.

As well as this major decline in woody vegetation, other changes are occurring in the Scottish uplands, as documented by Mackey *et al.* (1998, 2001). From the 1940s to the 1980s, it is estimated that the extent of heather moorland declined by about 23%, of peatland (predominantly blanket bog) by 21%, and of semi-natural upland grassland types by about 10%. These changes are predominantly due to afforestation, drainage and agricultural improvement. During the last decade of the 20th century, the declines in the extent of peatland and semi-natural grasslands appear to have been halted, but the extent of heather moorland declined by a further 5%. While these changes in upland land cover were occurring, there have been increases in the numbers of large herbivores. Between 1950 and 1990, it is estimated that the sheep population had increased by 32% and that, over approximately the same period, the red deer population had at least doubled. With enclosure of land for both afforestation and agricultural improvement schemes, there has clearly been a very considerably increased grazing pressure on the unenclosed upland habitats in Scotland.

The effects of the changing land cover on many species have been quite marked. The development of the sporting estates during the 1800s (Wightman & Higgins, 2001) was from only six or seven 'deer forests' in 1811, increasing to between 130 and 150 by the end of the century, with a further increase to 183 by 1957. This resulted in the predators of grouse (*Lagopus lagopus scoticus*) and young red deer being sought out and killed. Estate records, where they exist for the 19th century, demonstrate the large number of predators, or species suspected of being predatory, that were killed. One estate, quoted by Smout (2000) recorded killing 381 'hawks' in just one year, 1829–30. Whilst this slaughter reduced population sizes, it is possible that the only upland species to become extinct in Scotland was the red kite (*Milvus milvus*). This species did, however, just survive in the upland areas of central Wales, where four to ten pairs bred during the first four decades of the 20th century, with the numbers subsequently increasing to about 20 pairs by the late 1960s (Snow, 1971). With re-introduction programmes in England (with birds translocated from Spain) and in Scotland (with birds from Sweden), and with a continued increase in the Welsh population, the number of pairs of red kites in Great Britain can now be numbered in the hundreds, but most of these occur in the lowlands rather than in the uplands.

It is difficult to give quantitative data about changes in the population size of the red grouse in Britain (this subspecies is endemic to the British Isles). On any grouse moor the population tends to cycle, but it is known that between the 1950s and the 1990s the number shot on Scottish moors declined, particularly from the early 1970s onwards

(Mackey *et al.*, 2001). Over an approximate 20-year period from about 1970, the breeding range of 31 upland bird species in Scotland has been estimated. Nine species expanded their range by at least 10%, 15 species changed by less than 10%, and 7 species showed a range contraction of more than 10%. During this period the number of male dotterel (*Charadrius morinellus*) sitting on nests in montane Scotland is estimated to have declined by 23% (Mackey *et al.*, 2001).

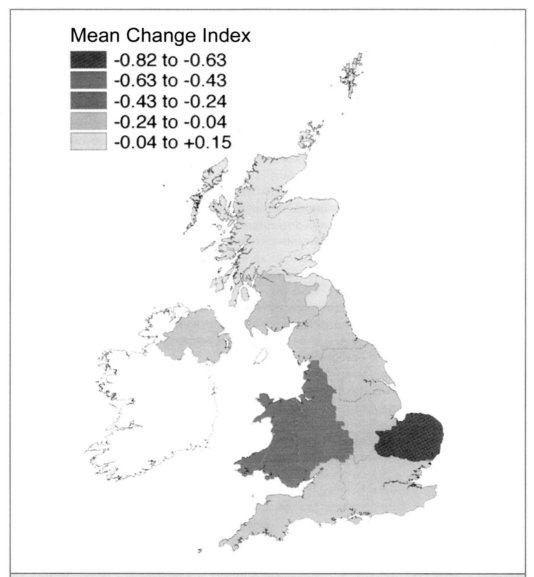

Mean Change Index
- -0.82 to -0.63
- -0.63 to -0.43
- -0.43 to -0.24
- -0.24 to -0.04
- -0.04 to +0.15

Figure 30.1 The mean change index for plant species native to the United Kingdom with northern (Boreal and Arctic) distributions. The index compares the relative change in the number of 10 km grid squares in which each species was recorded during the first recording period (1930-1969) and the second recording period (1987-1999). A negative index value indicates a relative decline in frequency. The illustration is taken from Preston *et al.* (2002), which is Crown copyright.

There are less precise data for most plant species. Between the 1940s and 1980s, the extent of bracken (*Pteridium aquilinum*) apparently expanded by 79%, though Mackey *et al.* (1998) consider that this might be an over-estimate. The atlas of vascular plants in the British Isles (Preston *et al.*, 2002) provides data on changes in distribution and abundance from approximately the 1950s to the 1990s. Plants characteristic of the Arctic and Boreal biogeographical zones were those that showed the greatest decline, with an average change index in the United Kingdom being -0.31 (Figure 30.1). There are, however, strong regional differences, with the greatest declines in species such as lingonberry (*Vaccinium vitis-idaea*) and hare's-tail cottongrass (*Eriophorum vaginatum*) in eastern England and near the border between England and Wales. Scotland, apart from the south west, has shown little change in these species.

The United Kingdom's Biodiversity Action Plan contains individual plans for seven upland habitats and for 40 upland species that occur in Scotland (Usher, 2000). All of these species have shown substantial declines since the 1970s, and work is now in progress to halt further decline and to reverse the trends. The particular point about these plans (Table 30.1) is that more than half of the species are non-vascular plants, a taxonomic grouping for which Scotland, in a European context, is particularly species rich. Although relatively little is known about the ecology of many of these mosses, liverworts and lichens, it has been predicted that many of them are vulnerable to the potential effects of climate change, and it is also known that large herbivores can adversely impact them (Fryday, 2001). Further change in the Scottish uplands seems inevitable.

Table 30.1 Within the United Kingdom's Biodiversity Action Plan there are separate plans for about 400 species, of which 226 species either currently occur, or have occurred historically, in Scotland. The table shows a taxonomic breakdown of the 40 Species Action Plans that relate to species that occur in the Scottish uplands, heaths and bogs (data compiled from Usher, 2000).

Taxon	Number of plans	Example
Birds	2	black grouse (*Tetrao tetrix*)
Snails	2	Geyer's whorl snail (*Vertigo geyeri*)
Insects	8	slender Scotch burnet moth (*Zygaena loti scotica*)
Vascular plants	7	woolly willow (*Salix lanata*)
Mosses and liverworts	11	Skye bog-moss (*Sphagnum skyense*)
Lichens	9	alpine sulphur-tresses (*Alectoria ochroleuca*)
Fungi	1	white-stalk puffball (*Tulostoma niveum*)

30.2.2 Trends in the Nordic Countries

In much of the Nordic area, the mountains and the Arctic are inseparable. Northward of about latitude 60°N, the climatic effects of high altitude and high latitude gradually become mixed, until by about 70°N the Arctic effect takes over (Körner, 1995). In between these latitudes, it is not always clear whether vegetation should be classified as alpine or Arctic, as it is usually influenced by both high altitude and high latitude, often augmented by the oceanic effects of climate. Consequently, inland in southern Norway at latitudes 60 to

62°N, the climatic treeline lies between 1,100 and 1,200 m, while along the coast of northern Norway the trees peter out at an altitude of only 200 m (Moen, 1998). In Iceland, the altitudinal limit of the native birch woodland ranges from over 550 m in the most continental areas down to about 200 m or less in the extreme oceanic areas (Kristinsson, 1995).

In Scandinavia, the treeline was at its highest soon after the most recent glaciation, about 9,000 years ago; since then the limit of Scots pine (*Pinus sylvestris*) has descended by around 500 m as the climate has become colder and more oceanic, and Norway spruce (*Picea abies*) and birch (*Betula* spp.) have become more prominent in the upland areas (Kullman, 1998). The position of the treeline has also been influenced by human activity. Felling of timber and intensive grazing have locally caused the treeline to descend, whereas in some other areas trees have been planted higher up on the mountains than they would have established themselves naturally.

The forests near the northern timberline have been exploited quite heavily in many parts of Northern Europe. Unlike in Asia or North America, large settlements and infrastructural development reached up to, and beyond, the treeline, making these forests open to exploitation. In Iceland, the destruction of native birch woodland started soon after the island was first settled in the 9th and 10th centuries and by 1400 had already resulted in a situation where wood had to be transported from more remote regions (Kristinsson, 1995). Today, only 5% of Iceland's former forested area remains. On the Kola Peninsula, felling, forest fires and industrial air pollution have damaged coniferous forests since the 1930s and significantly increased the role of birch stands which form pioneer communities on burnt and cut areas, and which are more tolerant of air pollution. Today they account for 26% of the total forest area of the Murmansk region (Neshatayev & Neshatayeva, 1993).

A wide zone dominated by mountain birch (*Betula pubescens*) normally runs along the northern treeline of Fennoscandia, whereas in most parts of Siberia and North America the boreal coniferous forests tend to merge directly into the tundra. Large stands of mountain birch are found over a range of treeline environments, from the cold, drier plateaux of northern Fennoscandia to coastal areas, where the winters are milder and snow often lies deep. Such continuous belts of mountain birch woodland are sensitive to various natural threats, particularly insect damage. Sudden local eruptions in the population of the autumnal moth *(Epirrita autumnata)* can periodically affect birch stands; in a particularly severe attack, in 1965, caterpillars defoliated about 5,000 km² of the mountain birch woodland of northern Finland with the result that trees died over an area of about 1,000 km², dramatically reducing the extent of the mountain birch zone (Lehtonen, 1981; Hanhimäki, 1989). Such damage, together with grazing animals and climatic influences, has resulted in an 'anthropo-zoogenous' timberline pattern where the uppermost tree stands are open and savanna-like (Oksanen *et al.*, 1995).

The large reindeer (*Rangifer tarandus*) herds of the Fennoscandian uplands have always been accompanied by several predatory mammals and birds, such as the wolf (*Canis lupus*), the wolverine (*Gulo gulo*), the Arctic fox (*Alopex lagopus*) and the golden eagle (*Aquila chrysaetos*). They all either hunt reindeer or make use of reindeer carcasses. Since the dawn of reindeer husbandry, the wolf and the wolverine, in particular, have been the arch enemies of reindeer herders, and eagles have been despised as they kill new-born calves, especially those in poor condition born after a particularly severe winter. The persistent persecution

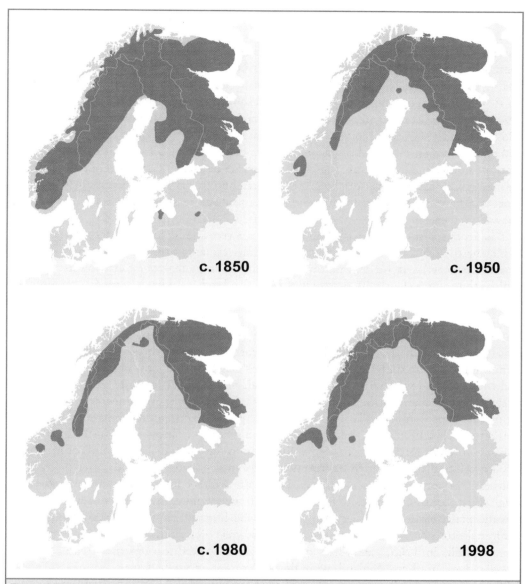

c. 1850

c. 1950

c. 1980

1998

Figure 30.2 In the 19th century wolverines lived as far south as Latvia, but since then their range has receded northwards. Despite being protected in the Nordic countries, their numbers have not increased significantly although their range has increased slightly since the 1980s. Only areas with permanent wolverine populations are shown on this series of four maps, taken from Hallanaro & Pylvänäinen (2002).

of all three of these animals led to a severe decline in their populations and an almost complete disappearance of wolf and wolverine from the upland areas of Norway, Sweden and Finland in the 20th century (Figure 30.2). The Arctic fox shared much the same fate, probably partly because it used to be hunted for its fur, but also because there were fewer carcasses available for it due to the smaller numbers of larger predators. Today, there are just a few hundred Arctic foxes left in north-western Fennoscandia, mostly in Sweden and

Norway, and another population of less than 1,000 animals in the north-easternmost corner of the Kola Peninsula (Hallanaro & Pylvänäinen, 2002).

Another upland species which has become endangered throughout northern Fennoscandia is the lesser white-fronted goose (*Anser erythropus*). It is an exception among the wild geese of Northern Europe, in that its numbers failed to recover during the latter part of the 20th century. The species became progressively rarer until, in the 1980s, it effectively became extinct in northern Sweden and Finland, where thousands of these geese used to breed (Madsen *et al.*, 1999). The main reason for this dramatic decline may lie along its migration route and on the wintering grounds rather than in the breeding grounds of the Fennoscandian uplands. Unlike other geese nesting in Fennoscandia, lesser white-fronted geese migrate south-east to Kazakhstan and onwards, most likely towards wintering areas around the Caspian Sea, mainly in Azerbaijan, but possibly as far as the Black Sea or the Mediterranean. It is likely that these geese are still shot both in their resting places along this route and in their wintering grounds (they are very similar in appearance to the white-fronted goose (*Anser albifrons*), which is seen as fair game in those regions). An even more significant negative factor may be the natural and man-made habitat changes in the birds' wintering grounds.

In order to help northern Fennoscandia's nesting lesser white-fronted geese to recover, they have been bred in captivity, with the goslings later released into the wild. Swedish biologists have further tried to ingrain new behaviour patterns into young geese by transferring clutches of eggs laid in captivity to incubating barnacle geese (*Branta leucopsis*), which are known to spend the winter in Western Europe. When the barnacle geese migrate south-westwards to Western Europe, the lesser white-fronted goslings accompany them, and are effectively programmed to follow this safer migration route for the rest of their lives. The project has enjoyed some success, as several pairs of lesser white-fronted goose are now breeding in Sweden.

30.3 The effects of grazing animals

The most important grazing animal in most of the uplands of Fennoscandia is the reindeer: first wild reindeer and, later, increasing populations of semi-wild or domesticated reindeer. Reindeer herding started along the present border of northern Sweden and Norway more than 1,000 years ago and was probably developed by combining the Sámi's reindeer hunting skills with the Norwegians' experience of sheep herding. It was not, however, until the 16th century that reindeer herding became common in the Sámi communities. Today, there are two major concerns regarding reindeer herding in Fennoscandia: excessive reindeer populations resulting in severe overgrazing, and the risk of cross breeding between various populations of wild and domesticated reindeer.

As reindeer were amongst the first animals to spread into Fennoscandia after the latest glaciation, distinct local races have evolved, each adapted to local conditions. The picture has been complicated by introductions of reindeer from behind the White Sea to the Kola Peninsula and by the movement of both wild and domesticated animals from one place to another within Fennoscandia. As a result there are at least three places where there is a risk of inter-breeding (Figure 30.3). First, in southern Norway, where the wild population of about 35,000 animals of tundra reindeer (*R. t. tarandus*) lives partly in the same area as semi-domesticated reindeer. Second, in eastern Finland where wild forest reindeer (*R. t. fennicus*)

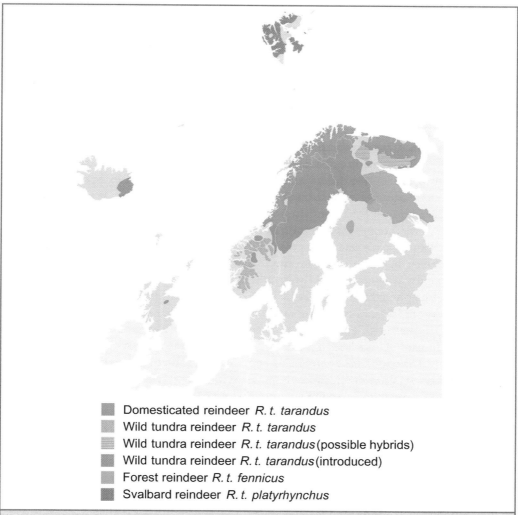

Domesticated reindeer *R. t. tarandus*
Wild tundra reindeer *R. t. tarandus*
Wild tundra reindeer *R. t. tarandus* (possible hybrids)
Wild tundra reindeer *R. t. tarandus* (introduced)
Forest reindeer *R. t. fennicus*
Svalbard reindeer *R. t. platyrhynchus*

Figure 30.3 A map showing the ranges of the three subspecies of the reindeer: the tundra reindeer, the forest reindeer and the Svalbard reindeer. The wild tundra subspecies has been extensively domesticated, and has also been introduced to both Iceland and Scotland (and to South Georgia in the Southern Hemisphere). The illustration is taken from Hallanaro & Pylvänäinen (2002).

live to the south of the area where semi-domestic reindeer are raised. Third, in southern parts of the Kola Peninsula where wild forest reindeer spreading from the south may eventually reach the area now occupied by the Kola reindeer of mixed origin (Hallanaro & Pylvänäinen, 2002).

In Scotland, there are similar concerns about hybridisation, but they are about two species of deer, the native red deer (*Cervus elaphus*) and the introduced Sika deer (*Cervus nippon*). Sika deer were introduced to Scotland from the late 1800s as a decorative species in lowland estates and parks (Anon., 1997). From at least ten places of introduction, the deer escaped and have spread so that their populations are now present over at least a third of Scotland's red deer range. Although the Sika deer is predominantly a woodland species, it will inter-breed with red deer, and such hybrids are now found widely on mainland Scotland. Because hybridisation will eventually lead to a loss of genetic integrity of the Scottish red deer, sanctuaries have been established on some of the more isolated Hebridean islands. On these islands, it is known that no Sika deer or hybrids are present, and it has been made illegal to introduce any species of genus *Cervus* to these islands. Despite these concerns about hybridisation, the problems of overgrazing by any of the species, or their hybrids, remain.

In Finland, overgrazing by reindeer takes place on both summer pastures, which lie high up on the Arctic fells, and on the more forested winter pastures. In the winter grazing areas, the lichen cover has typically thinned throughout the regions where reindeer husbandry is practised, and in the worst-affected areas the lichen has been almost completely nibbled away or trampled, exposing the bare ground. The numbers of reindeer increased rapidly in Norway, Sweden and Finland from the 1970s to the 1990s, when some herders started to provide extra winter fodder for their animals both to compensate for the lack of natural food and in an attempt to increase the profitability of keeping reindeer. Today, however, attempts are being made to limit the numbers of reindeer in all three countries so that reindeer husbandry becomes more sustainable.

Recent legislation in Scotland introduced the concept that action could be taken on land where deer grazing was causing a severe impact on the natural heritage. This leaves unanswered the question about how 'severe' impacts can be distinguished from any other sort of lesser impact. Guidelines have been drawn up that can be used in such assessments (Table 30.2). Although these remain to some extent subjective, they do highlight the detrimental impacts that deer can have on many different vegetation communities, and presumably on the invertebrate animals supported by those communities. Grazing pressure by large mammalian herbivores is a continuum from none at one extreme to something at the other extreme where virtually all of the palatable plants are eaten down to ground level. How this continuum can be divided into categories such as 'none', 'slight', 'moderate', 'heavy' and 'severe' will always remain, to some extent at least, subjective.

As well as the kind of approach outlined in Table 30.2, other approaches are possible. Morellet *et al.* (2001) developed a browsing index, which they then applied to populations of the roe deer (*Capreolus capreolus*). Modelling approaches have been used so as to understand the interactions between a number of herbivore species and vegetation responses to their grazing (e.g. Blatt *et al.*, 2001). However, there is abundant evidence that large herbivores have their major effect on woody vegetation. After years of intense grazing pressure, largely by red deer, on the island of Rum in the Hebrides, Virtanen *et al.* (2002) recorded only two saplings of rowan (*Sorbus aucuparia*) in exclosures or areas where deer were culled. Hester *et al.* (2000), however, hypothesised that the creation of gaps by red deer would enhance the establishment of birch seedlings. Their experiments did not support this hypothesis on moorland dominated by heather or purple moor grass (*Molinia caerulea*), but they suggested that it might hold true on steeply sloping ground.

30.4 Contemporary pressures on the upland

Despite the various pressures and negative trends, the uplands of northern Europe have remained some of the region's best preserved habitats. This is because of a smaller human habitation pressure and fewer land use demands than on other landscapes. Even chemical

Table 30.2 An example of guidance that has been developed for assessing the degree of damage being caused to the natural heritage of Scotland by large, mammalian herbivores. Source: Anon., 2002.

Indicators of an adverse impact	Additional indicators of a severe impact
Woodland	
1. Mature trees with a distinct browse line	-
2. Mostly older age classes of trees present, with no signs of regeneration over the last 20 years	Structure of woodland very heavily skewed towards oldest age classes, and no signs of regeneration for many decades
3. Bark stripping occasional	Bark stripping frequent
4. Low growing trees and bushes pruned to give a 'topiary' effect	-
5. Shrub layer either absent or moribund	-
6. Tree seedlings and saplings present, but not or hardly projecting above the ground layer. The only evidence of regeneration growing is in areas inaccessible to large, herbivorous mammals	Tree seedlings browsed soon after germination, and hence virtually no evidence of tree or shrub regeneration
7. More palatable shrubs and herbs confined to areas inaccessible to large, herbivorous mammals	-
8. Common mosses often more abundant than vascular plants	Moss layer patchy or absent due to trampling or desiccation
Dry heaths	
1. 'Topiary', 'drumstick' or 'carpet' growth forms of heather (*Calluna vulgaris*) evident	These growth forms of heather frequent and almost everywhere
2. Amount of heather stem breakage as a result of trampling noticeable (if supplementary feeding occurring, more than 50 m distant)	Amount of heather stem breakage widespread throughout the area
3. More than 2/3 (67%) of long shoots of heather and blaeberry (*Vaccinium myrtillus*) browsed (assessed in March or April, after the winter)	Similar end of winter sampling indicates that up to 4/5 (80%) of long shoots are browsed
4. Broad bands (more than 10 m) of heavily browsed dwarf shrubs around grassland	Everywhere there are obviously heavily browsed dwarf shrubs
5. Trampled, bare ground evident	Trampled, bare ground widespread
Wet heath and blanket bog	
1. More than 2/3 (67%) of long shoots of heather and blaeberry browsed (assessed in March or April, after the winter)	Similar end of winter sampling indicates that up to 85% of long shoots are browsed
2. Browsed shoots conspicuous on bog myrtle (*Myrica gale*). Signs of browsing on bearberry (*Arctostaphylos uva-ursi*), crowberry (*Empetrum nigrum*), cross-leaved heath (*Erica tetralix*) and lingonberry (*Vaccinium vitis-idaea*)	Extensive browsing of bog myrtle, and obvious signs of browsing on bearberry, crowberry, cross-leaved heath and lingonberry
3. Only limited flowering of cottongrasses (*Eriophorum* spp.) and cloudberry (*Rubus chamaemorus*) during April to July	At same time of year, virtually no flowering of cottongrasses or cloudberry
4. Bare peat surface evident	Bare peat surface widespread

pollution has usually been less severe, with some notable exceptions such as the surroundings of the large metallurgical plants of the Kola Peninsula. There, however, is always the threat of further pollution from radionucleides (the major incident at Chernobyl in 1986 affected upland areas in Norway, Scotland, England and Wales).

One major concern has been the dispersion of toxic chemicals, particularly heavy metals and persistent organic pollutants (POPs), into the Arctic and their accumulation in Arctic food chains. There is evidence that pollutants emitted in warmer parts of the world, perhaps even in equatorial regions, gradually make their way towards the polar regions, condensing and re-evaporating on the way. Having spread into Arctic regions, they begin to move up the food chains, accumulating in species at the top of these chains, such as birds of prey. Various symptoms have been induced, but the most serious manifestations are reproductive disorders. Research into birds of prey nesting in Arctic regions has shown that the smallest concentrations of toxins are found in gyrfalcons (*Falco rusticolus*) and other non-migratory birds that mainly feed on animals which themselves also live in the Arctic the whole year round, whereas higher levels are found in migratory birds of prey (March *et al.*, 1998). This indicates that Northern Europe's Arctic terrestrial environments are relatively free of these pollutants. It is the slowness of natural processes and the longevity of some Arctic species, as well as their ability to build up large quantities of fat in their bodies, that make Arctic ecosystems particularly vulnerable to toxic chemicals.

Over recent decades, the concentrations of toxic chemicals in birds of prey have generally declined, and the birds have begun to recover from the reproductive disorders which previously afflicted them. The problem has not, however, disappeared altogether. In the Alta region of northern Norway, for example, locally nesting merlins (*Falco columbarius*) have recently been observed to suffer from high concentrations of DDT, with their eggshells an average of about 10% thinner than normal healthy eggs (Nygård *et al.*, 1994).

Thanks to their perceived pristine character, there is an increasing interest in using the uplands for tourism and recreational activities. Hotels, ski centres and both publicly and privately owned log cabins and holiday homes have become more common in the fells of Norway, Sweden and Finland, whereas in the uplands of Iceland tourism and the use of

Table 30.3 Snowmobiles began to become more common in the Nordic countries in the 1970s. They are mainly used in the northern and upland areas where snow stays on the ground for several months each year. Despite the snow cover, snowmobiles can still damage the vegetation as they compress the snow and wear it too thin to protect the plant communities and soil beneath. The table gives information about the occurrence of snowmobiles in five Nordic areas. Source: Hallanaro & Pylvänäinen (2002).

Country	Total number	Vehicles per 1,000 km²	Vehicles per 1,000 inhabitants
Sweden	194,000	431	22
Finland	89,179	264	17
Norway	40,914	126	10
Iceland	2,906	28	11
Svalbard	1,281	20	366

4-wheel drive vehicles have become more popular in recent decades. In Scotland, the number of people walking the 284 'Munros' – peaks over 3,000 ft (914 m) in height – increases year by year, as does the number of people who have climbed all of 'The Munros' (Bennet, 1999; Mackay, this volume). Such developments have contributed to more roads, snowmobile tracks, hiking trails and upland paths, all of which break up previously continuous areas of undisturbed habitat. Walkers and off-road vehicles can cause harmful erosion away from established paths and tracks.

Off-road vehicles (Table 30.3) are used for many purposes: by reindeer-herders and shepherds, by hunters, berry-pickers and tourists, simply for getting from one place to another whenever the roads are in bad condition, or for short cuts. In Northern Europe, the use of off-road vehicles in summer is especially common in Iceland, where inland roads are often impassable, while snowmobiles are most common in northern Scandinavia. So far there are still relatively few off-road vehicles in either Scotland or Russia, but in Russia any vehicles that do exist are often very heavy, such as caterpillar tractors, causing severe damage to the vegetation cover.

Legislation plays an important role in limiting off-road traffic. In Norway, all driving off-road is illegal throughout the year, whereas in Sweden off-road driving is allowed on snow, and snowmobiles can be used everywhere, except in national parks and other protected areas. In Finland, the landowner's permission is normally required for snowmobiling. In Finland, Sweden and Norway, special snowmobile routes have been set up to concentrate this form of off-road traffic and prevent more widespread damage and disturbance.

30.5 Discussion

There truly has been an upland saga in Northern Europe. Following the last Ice Age, the climate set the scene, with long and generally cold winters and short, cool summers. This is modified by the Atlantic Ocean, so that along the western seaboard the winters are less cold, but the summers are generally cooler than in more continental areas. Given these factors, biological productivity on land tends to be low and decomposition of organic material, and hence the recycling of nutrients, tends to be slow.

People, who arrived in these northern and upland areas at various times since the last glaciation, have had a profound influence. The background on which the saga has been written reflects the low productivity of the land, the higher productivity of the waters, and the efforts of the people to make a living in an inhospitable environment. The earliest negative influence was the cutting of trees and shrubs. This was followed by the introduction or domestication of large herbivores which could convert the vegetation into useful products such as hides, meat and milk. More recently, there have been problems with recreation in these upland areas which, paradoxically, are often perceived by the visitors to be pristine. All of these influences have been local, affecting small areas near to where the activities have occurred. But there are other factors that have a more global influence, or which have arisen from outside the local area where they have impacts. Pollution is perhaps the largest of these, but associated with it is the future change in the climate of this planet.

The effects of overgrazing have been discussed in section 30.3. They are clearly rather subtle. At first, the changes in the vegetation community being grazed are small, and there is plenty of scope for arguments about whether the vegetation is being overgrazed or not, or even whether the grazing is increasing biodiversity! This is well illustrated in Table 30.2, which shows that it is difficult to give clear rules about when grazing is having an adverse impact on vegetation, and when that grazing pressure is so severe that the changes are no longer reversible when the pressure is released. Much of Iceland acts as a reminder of what can happen. Previously vegetated upland areas have become cold, sandy deserts, now almost devoid of vegetation. Arnalds *et al.* (2001) make a 'gloomy' prognosis when they conclude "the spread of sandy deserts in Iceland is a continuous environmental threat. The ecosystems which are lost are fully vegetated and rich ecosystems which have high biological value and are important for water cycling. The sandy deserts lack water-holding capacity ..." (Arnalds *et al.*, 2001, p. 370). The clearance of trees and shrubs, and then the pressures of overgrazing, can obviously have a very unpleasant end point, both for local human populations and for biodiversity in general.

Some of the pressures due to recreation and tourism have been reviewed in section 30.4. These effects can be direct or indirect. The direct effects are easy to see — the creation of tourist infrastructure, erosion along tracks and paths, and damage to soils and vegetation by a variety of off-road vehicles. An excellent example of some direct effects comes from the Everest region of Nepal. Bhuju & Rana (2000) counted the number of pieces of litter (which included plastic wrappers, biscuits, metal cans and chewing gum) along trekking routes. The average was one piece of litter every 27.6 human paces (which they equate to every 14 m). It is perhaps surprising that this quantity of litter can be found in an area designated as a national park and as a World Heritage Site, dropped presumably by people who were sufficiently environmentally aware to have travelled to this remote and beautiful area. What is not recorded in the Nepalese study is the indirect effect of the litter on scavenging bird species, attracting them along the trails, and their subsequent impact on other bird species. It is, however, known that litter in the Highlands of Scotland can attract scavenging birds (notably crows and gulls) into the uplands, and then that these species affect locally nesting bird species, such as the dotterel (see section 30.2.1).

Another aspect of pollution in the uplands, not discussed in section 30.4, is eutrophication, in which there is an increasing deposition of nitrogen, derived often from the lowlands. This has the potential to change the balance of productivity and the nature of the vegetation (see Britton *et al.*, this volume). The nitrogen content of modern *Racomitrium lanuginosum* moss is about three times that of museum specimens collected a century ago (Baddeley *et al.*, 1994). Experiments in Norway on the growth of heather have indicated that, at higher nitrogen levels, shoot growth starts nearly two weeks earlier in the year, making the plants more susceptible to frost damage (Saebø *et al.*, 2001). Also, during the winter, nearly half of the plants at the higher nitrogen level died, whereas all of the plants at the lowest nitrogen level survived. Kirkham (2001), working in northern England, considered that the nitrogen:phosphorus ratio was important, and in situations where phosphorus was limiting, due to increased nitrogen availability, the purple moor grass would be at a competitive advantage to heather. Experimental work of this nature clearly indicates that heather, a typical plant of the uplands, could be under threat from eutrophication, as it is in the lowlands of Europe (Smidt, 1995).

The effects of pollution and the changes in the world's climate are undoubtedly going to have a continuing effect on the uplands. It is difficult to predict how climate change may affect upland land use, because of the mix of environmental, social and economic aspects, but some general statements about the drivers of change (Usher, 1992) can be made. An increase in temperature might allow both arable crops and intensively managed grasses to be grown at higher altitudes. An increase in precipitation and an increase in cloudiness might hinder the ripening of crops and make grassland use more difficult. An increase in average windspeed, with the associated increase in storminess, might have dramatic effects on tree stability, and hence on a forest industry. Hence, new systems of both agriculture and silviculture will be needed both to embrace the advantages of climate change and to circumvent the disadvantages. However, one prediction that can be made with a degree of certainty is that there will be increased pressure on the semi-natural ecosystems of the uplands.

Climate change will certainly affect many species that are characterised by Arctic or alpine distribution patterns, pushing them to higher altitudes and possibly pushing the more southerly populations to extinction. An interesting study by Phoenix *et al.* (2001), in northern Sweden, indicated that there could also be an effect on the leisure activities of people. One of their findings was that there would be a reduced production of blaeberries (*Vaccinium myrtillus*); the picking of such forest fruits is a traditional leisure activity for many people from the towns and cities of the Nordic nations. It is possible that the upland ecosystems may be amongst the most sensitive to environmental change and Debinski *et al.* (2000) suggested that montane meadows may be useful indicators of this change. It behoves us to consider carefully whether in Northern Europe we have sufficient monitoring to detect the effects of a changing climate on our upland ecosystems. In Scotland, the terrestrial Environmental Change Network includes only three upland sites, one truly upland in the Cairngorm mountains, and two at lower altitudes at Glensaugh and Sourhope. Is this small upland network, supplemented by the Moor House–Upper Teesdale site in England and the Y Wyddfa (Snowdon) site in Wales, really sufficient to determine the effects of climate change on the uplands of the United Kingdom?

There are, however, also some positive signs. People have created the problems, and people must now attempt to find solutions. Probably the two key measures are planning and management. The first step is understanding what the adverse trends are and why they are occurring. This review indicates that we now have a considerable knowledge of both the 'what' and the 'why', and we now need to focus more on the 'how'; how do we plan so that the future impacts are reduced and how do we manage so that the resource that remains can be maintained or, preferably, enhanced? These are questions that need to be addressed, and there is some evidence, as in the biodiversity planning summarised in Table 30.1, that a start has been made, but there is still much that needs to be done.

The upland areas of the Nordic countries now have a considerable protection status compared to more productive habitats, such as lowland forests and wetlands. Most of the large protected areas in these countries are located in either mountainous or far northern areas, often close to or above the treeline. In Finland, for example, protected areas of various kinds cover about a third of the country's northernmost province (Lapland), whereas the figure for the entire country is around 10%. By far the largest protected area in the whole of Europe is the National Park of North and East Greenland, which has nearly 8% of the total extent of protected areas on the planet. The second largest is the Northeast Svalbard Nature Reserve. A start has been made in Scotland, with the establishment in 2002 of its first national park, Loch Lomond and The Trossachs, and the second, in the Cairngorm Mountains, in 2003.

As we bring our modern day 'saga' to a close, we can look to the words of a modern day 'sage', Professor P.S. Ramakrishnan.   He said "I realized that the traditional ecological knowledge (ecological and social) on which the livelihood concerns of these societies [north-eastern hill regions of India] rests is multidimensional. This traditional holistic wisdom is based on the intrinsic realization that man and nature form part of an indivisible whole, and therefore should live in partnership with each other" (Ramakrishnan *et al.*, 2000). He goes on to expand these concepts, relating them to the functioning of natural systems, but clearly this is in a situation where there is still a strong connection between people and the land on which they live. Although this is a huge topic, in Northern Europe how much of our traditional ecological knowledge, acquired over the millennia since people first colonised these northern and upland environments, has been lost? Is there still a chance for further exploration of the processes that make these upland ecosystems function efficiently, in both space and time, both to conserve their total biodiversity and to support the economy of their human inhabitants?

References

Anonymous (1997). *A Policy for Sika Deer in Scotland.* Deer Commission for Scotland, Inverness.

Anonymous (1998). *UK Biodiversity Group Tranche 2 Action Plans: Volume 1 – Vertebrates and Vascular Plants.* English Nature, Peterborough (see pp. 221-223).

Anonymous (2003). *Deer (Scotland) Act and Damage by Wild Deer to Woodland and the Natural Heritage.* Deer Commission for Scotland, Scottish Natural Heritage & Forestry Commission, Inverness.

Arnalds, O., Gisladottir, F.O. & Sigurjonsson, H. (2001). Sandy deserts of Iceland: an overview. *Journal of Arid Environments,* **47**, 359-371.

Baddeley, J.A., Thompson, D.B.A. & Lee, J.A. (1994). Regional and historical variation in the nitrogen content of *Racomitrium lanuginosum* in Britain in relation to atmospheric nitrogen deposition. *Environmental Pollution,* **84**, 189-196.

Bennet, D. (Ed) (1999). *The Munros, 3rd Edition.* Scottish Mountaineering Trust, Edinburgh.

Bhuju, D.R. & Rana, P. (2000). An appraisal of human impact on vegetation in high altitudes (Khumbu Region) of Nepal. *Nepal Journal of Science and Technology,* **2**, 101-106.

Blatt, S.E., Janmaat, J.A. & Harmsen, R. (2001). Modelling succession to include a herbivore effect. *Ecological Modelling,* **139**, 123-136.

Debinski, D.M., Jakubauskas, M.E. & Kindscher, K. (2000).   Montane meadows as indicators of environmental change. *Environmental Monitoring and Assessment,* **64**, 213-225.

Fryday, A.M. (2001). Effects of grazing animals on upland/montane lichen vegetation in Great Britain. *Botanical Journal of Scotland,* **53**, 1-19.

Hallanaro, E.-L. & Pylvänäinen, M. (2002). *Nature in Northern Europe – Biodiversity in a Changing Environment.* Nordic Council of Ministers, Copenhagen.

Hanhimäki, S. (1989). Induced resistance in mountain birch: defence against leaf-chewing insect guild and herbivore competition. *Oecologia,* **81**, 242–248.

Hester, A.J., Stewart, F.E., Racey, P.A. & Swaine, M.D. (2000). Can gap creation by red deer enhance the establishment of birch (*Betula pubescens*)? Experimental results within *Calluna*- and *Molinia*-dominated vegetation at Creag Meagaidh. *Scottish Forestry,* **54**, 143-151.

Kirkham, F.W. (2001). Nitrogen uptake and nutrient limitation in six hill moorland species in relation to atmospheric nitrogen deposition in England and Wales. *Journal of Ecology,* **89**, 1041-1053.

Körner, C. (1995). Alpine plant diversity: a global survey and functional interpretations. In *Arctic and Alpine Biodiversity: Patterns, Causes and Ecosystem Consequences,* ed. by F.S. Chapin & C. Körner. Springer-Verlag, Berlin. pp. 48-62.

Kristinsson, H. (1995). Post-settlement history of Icelandic forests. *Búvísindi, Icelandic Agricultural Sciences*, **9**, 31–35.

Kullman, L. (1998). Aktuella vegetationsförändringar i södra delen av fjällkedjan. In *Hållbar Utveckling och Biologisk Mångfald i Fjällregionen, Rapport från 1997 års Fjällforskningskonferens*, ed. by O. Olsson, M. Rolén & E. Torp. FRN rapport serie 98:2. Forskningsrådsnämnden, Stockholm. pp. 160-169.

Lehtonen, J. (1981). Kasvillisuuden muutokset tunturimittarin aiheuttaman tuhon jälkeen. *Luonnon Tutkija*, **85**, 23–126.

Mackey, E.C., Shewry, M.C. & Tudor, G.J. (1998). *Land Cover Change: Scotland from the 1940s to the 1980s*. The Stationery Office, Edinburgh.

Mackey, E.C., Shaw, P., Holbrook, J., Shewry, M.C., Saunders, G., Hall, J. & Ellis, N.E. (2001). *Natural Heritage Trends: Scotland 2001*. Scottish Natural Heritage, Perth.

Madsen, J., Cracknell, G. & Fox, T. (1999). *Goose Populations of the Western Palearctic: a Review of Status and Distribution*. Wetlands International, Wageningen and National Environmental Research Institute, Rönde.

March, B.G.E., de Wit, C.A. & Muir, D.C.G. (1998). Persistent organic pollutants. In *AMAP Assessment Report: Arctic Pollution Issues*. Arctic Monitoring and Assessment Programme, Oslo. pp. 183-372.

McConnell, J. & Legg, C. (1995). Are the upland heaths in the Cairngorms pining for climate change? In *Heaths and Moorland: Cultural Landscapes*, ed. by D.B.A. Thompson, A.J. Hester & M.B. Usher. HMSO, Edinburgh. pp. 154-161.

Moen, A. (1998). *Nasjonalatlas for Norge: Vegetasjon*. Statens kartverk, Hønefoss.

Morellet, N., Champely, S., Gaillard, J.-M., Ballon, P. & Boscardin, Y. (2001). The browsing index: new tool uses browsing pressure to monitor deer populations. *Wildlife Society Bulletin*, **29**, 1243-1252.

Neshatayev, V. & Neshatayeva, V. (1993). Birch forests of the Lapland State Reserve. In *Aerial Pollution in Kola Peninsula: Proceedings of the International Workshop*, ed. by M.V. Kozlov, E. Haukioja & V.T. Yarmishko. Apatity, St. Petersburg. pp. 328–338.

Nygård, T., Jordhøy, P. & Skaare, J.U. (1994). *Environmental pollutants in merlin in Norway*. Norwegian Institute for Nature Research Forskningsrapport No. 56.

Oksanen, L., Moen, J. & Helle, T. (1995). Timberline patterns in northernmost Fennoscandia. *Acta Botanica Fennica*, **153**, 93–105.

Phoenix, G.K., Gwynn-Jones, D., Callaghan, T.V., Sleep, D. & Lee, J.A. (2001). Effects of global change on a sub-Arctic heath: effects of enhanced UV-B radiation and increased summer precipitation. *Journal of Ecology*, **89**, 256-267.

Preston, C.D., Telfer, M.G., Arnold, H.R., Carey, P.D., Cooper, J.M., Dines, T.D., Hill, M.O., Pearman, D.A., Roy, D.B. & Smart, S.M. (2002). *The Changing Flora of the UK*. Department for Environment, Food and Rural Affairs, London.

Ramakrishnan, P.S., Chandrashekara, U.M., Elouard, C., Guilmoto, C.Z., Maikhuri, R.K., Rao, K.S., Sankar, S. & Saxena, K.G. (Ed) (2000). *Mountain Biodiversity, Land Use Dynamics, and Traditional Ecological Knowledge*. Oxford and IBH Publishing, New Delhi.

Saebø, A., Haland, A., Skre, O. & Mortensen, L.M. (2001). Influence of nitrogen and winter climate stresses on *Calluna vulgaris* (L.) Hull. *Annals of Botany*, **88**, 823-828.

Scott, M. (2000). *Montane Scrub: Preasarnach na Beinne*. Scottish Natural Heritage, Perth.

Smidt, J.T. de (1995). The imminent destruction of northwest European heaths due to atmospheric nitrogen deposition. In *Heaths and Moorland: Cultural Landscapes*, ed. by D.B.A. Thompson, A.J. Hester & M.B. Usher. HMSO, Edinburgh. pp. 206-217.

Smout, T.C. (2000). *Nature Contested: Environmental History in Scotland and Northern England since 1600*. Edinburgh University Press, Edinburgh.

Snow, D.W. (Ed) (1971). *The Status of Birds in Britain and Ireland.* Blackwell, Oxford.

Stewart, F.E., Hester, A.J., Swaine, M.D. & Racey, P.A. (2000). The effects of altitude on birch seed germinability at Creag Meagaidh, Scottish Highlands. *Scottish Forestry*, **54**, 17-23.

Usher, M.B. (1992). Land use change and the environment: cause or effect? In *Land Use Change: the Causes and Consequences*, ed. by M.C. Whitby. HMSO, London. pp. 28-36.

Usher, M.B. (Ed) (2000). *Action for Scotland's Biodiversity.* Scottish Executive, Edinburgh.

Virtanen, R., Edwards, G.R. & Crawley, M.J. (2002). Red deer management and vegetation on the Isle of Rum. *Journal of Applied Ecology*, **39**, 572-583.

Wightman, A. & Higgins, P. (2001). Sporting estates and outdoor recreation in the Highlands and Islands of Scotland. In *Enjoyment and Understanding of the Natural Heritage*, ed. by M.B. Usher. The Stationery Office, Edinburgh. pp. 171-176.

31 Delimitation of mountain areas for the purpose of developing EU policies

Hallgeir Aalbu

Summary

1. Calls for the EU to develop policies for mountain areas have increased in recent years, as such areas are perceived to have specific geographical handicaps with regard to economic development.
2. The European Commission funded a study to investigate the development challenges faced by mountain areas, and to compare these regions with other regions in the current and future EU. The first step of the study was the to delineate the mountain areas to be included.
3. The methodology used is based on altitude, slope and climate, and identifies 33,800 out of 125,700 European municipalities in which more than 50% of the territory is considered mountainous.

31.1 Regional policies for mountain regions?

Over the last few years, increasing attention has focussed on the need for specific policies to support the socio-economic development of mountainous regions. Three European countries (France, Italy and Switzerland) have an integrated cross-sectoral mountain policy, while a number of additional countries have elements of mountain policies particularly within their national agricultural or environmental policies (EC, 2004). Organisations such as Euromontana and the European Association of Elected Representatives from Mountain Areas (AEM) have even argued for the introduction of designated European Union (EU) interventions in support of economic development in such mountainous regions.

Mountainous regions are, in general, highly attractive and are perceived as having good development opportunities. So why should they be the subject of attention and particular scrutiny with regard to regional policies? The Second Report on Economic and Social Cohesion (EC, 2001) identified three types of areas with specific geographic features: mountain areas, coastal and maritime areas, and islands. It notes that:

"Mountainous areas represent geographical barriers... While some mountainous areas are economically viable and integrated into the rest of the EU economy, most have problems, as witnessed by the fact that more than 95% of them (in terms of land area) are eligible for assistance under Objectives 1 or 2 of the Structural Funds" (EC, 2001, p. 35).

Aalbu H. (2005). Delimitation of mountain areas for the purpose of developing EU policies (2005). In *Mountains of Northern Europe: Conservation, Management, People and Nature,* ed. by D.B.A. Thompson, M.F. Price & C.A. Galbraith. TSO Scotland, Edinburgh. pp. 325-334.

Objective 1 regions have less than 75% of EU average GDP per capita (in Purchasing Power Parities). Objective 2 regions are urban or rural regions with specific challenges concerning industrial restructuring.

What factors define mountainous regions? Two broad issues are important. The first is, of course, the impact of topography and climate. Economic activity is restricted where the terrain is particularly rough and slopes are steep. High altitudes mean lower temperatures and shorter growing seasons. People live in the valleys, and travelling from one valley to the next may, in some places, be difficult. In general, accessibility may be low even where distances as the crow flies are not great.

The second factor is that many mountain regions form parts of national peripheries. They are sparsely populated, with long distances and poor access to national and European markets. Mountains often constitute national and regional borders: the border region is where the road ends. Mountain regions are often located on the margins of national economic systems.

Both factors are valid for mountain regions in Northern Europe. Nordic and British mountainous areas have rough terrain and face difficult climatic conditions. Many are indeed peripheral in a European context, and most are peripheral nationally. Few people live in these upland areas, and those who do are often confronted by long distances separating them from Europe's larger markets, as compared to most other regions in Europe. For many years, these areas have received attention through various forms of regional development policy.

However, the situation is not similar in all European mountain areas. In parts of the Alps, for example, as well as in certain other ranges, the main challenge for spatial planning is the pressure on vulnerable ecosystems due to tourism, as well as increased transport flows and growing populations (e.g. Fredman & Heberlein; Mackay; Thompson *et al.*, this volume).

31.2 What is a mountain region?

The definition of a mountainous region is rather more difficult to provide than one would at first think. In a socio-economic context, the question is: when do rough terrain and harsh climate constitute a handicap to economic development?

A global map of mountain areas was published by the United Nations Environment Programme - World Conservation Monitoring Centre (UNEP-WCMC) in 2000 (Kapos *et al.*, 2000). In the construction of this map, different delimitation criteria were tested and agreement was reached on a map that looked reasonable for the whole world. The European part of this map is shown in Figure 31.1. This definition builds upon a combination of criteria for slope and altitude.

A further interlinked factor is that of climate, which is a product of altitude and geographic location. Climate generally becomes more adverse with increasing altitude, while also becoming relatively more extreme – at the same altitude – as we move further north. For policy purposes, is it therefore necessary to include criteria that can easily mirror such climatic differences, as illustrated by the following definition used by the European Commission (EC, 1999).

1. Mountain areas shall be those characterised by a considerable limitation of the possibilities for using the land and an appreciable increase in the cost of working it, due:
    - to the existence, because of the altitude, of very difficult climatic conditions the effect of which is substantially to shorten the growing season,
    - at a lower altitude, to the presence over the greater part of the area in question of slopes too steep for the use of machinery or requiring the use of very expensive special equipment, or
    - to a combination of these two factors, where the handicap resulting from each taken separately is less acute but the combination of the two gives rise to an equivalent handicap.

SAC
Rural Policy
Group

Figure 31.1 Mountains of Europe. Map prepared by S. Blyth, United Nations Environment Programme - World Conservation Monitoring Centre, Cambridge, based on Kapos *et al.* (2000).

2. Areas north of the 62nd Parallel and certain adjacent areas shall be treated in the same way as mountain areas.

This broad definition is designed with agriculture policy in mind. The lack of objective criteria is, however, striking – the handicap of mountainous regions is defined relative to other areas, and all areas north of a certain geographical line are included.

National definitions do exist, but are different from country to country. The main criterion for defining mountainous regions is that of elevation, while for many countries this is further supplemented by consideration of slope. While in the UK a mountainous area must be more than 240 m above sea level, the threshold is 900 m in Austria and 1,000 m in Spain (EC, 2004).

This illustrates that mountainousness in a socio-economic context is something that is rather relative. Thus far there is no common European definition of mountainous areas that is suitable for regional policy purposes.

31.3 Can we identify areas that may be eligible for European mountain policies?

In the EU, regional development policies are the combined responsibility of the Member States and the European Commission. If mountain policies are to be part of this, it may be necessary to establish a common definition of handicapped regions due to topographic and climatic reasons.

Geographically-differentiated policies for economic development are usually implemented at the regional level and are valid for specific administrative units such as counties or municipalities. Such regional policies exist in all European countries, including those without mountains. These policies are based on an analysis of the situation and challenges in each region of a country. The targeting of EU Structural Fund policies is based on a comparison of regions, with Gross Domestic Product (GDP) per capita as the most important indicator of economic development.

Regional policies are implemented for administrative regions. So, for specific socio-economic mountain policies at the European level, a common definition of administrative mountain regions would become necessary. To combine the natural geographic map with the map of administrative regions is not easy, however. This is illustrated by Copus & Price (2002) in a pilot study commissioned by Euromontana together with Highlands and Islands Enterprise. Their approach was to overlay the map of administrative regions at NUTS 3 level with the WCMC map referred to above, and to calculate the degree of mountainous area within each administrative region. 'NUTS' is the French acronym for 'Nomenclature of Territorial Statistical Units'. NUTS 0 is the national level (countries), NUTS 3 is the regional level (often counties) and NUTS 5 is the local level (municipalities or communes). Figure 31.2 reveals the results. Administrative regions are not homogeneous with regard to landscape, nor are they economically homogeneous. Therefore, the NUTS 3 level is inappropriate for the delimitation of mountainous regions (Copus & Price, 2002). This is not surprising, since administrative regions at this level most often include both mountain and foothill areas, cities and their hinterlands, and both the core and its periphery.

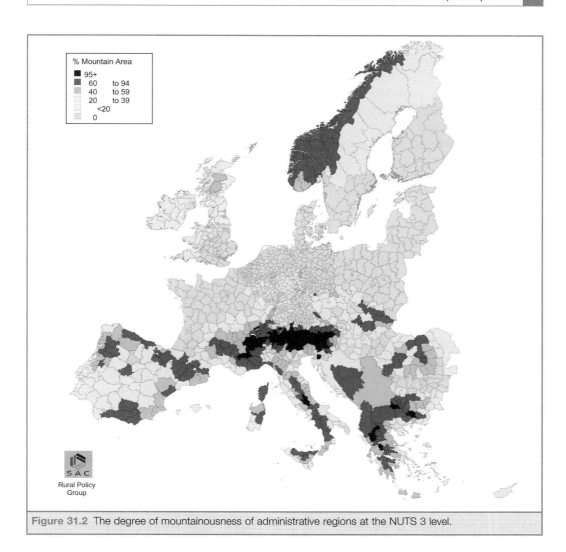

Figure 31.2 The degree of mountainousness of administrative regions at the NUTS 3 level.

31.4 A study of mountainous regions in Europe

To address such issues requires more detailed analysis, which was undertaken by a research consortium of 21 partners, led by Nordregio, which was awarded the contract to carry out a study entitled "Analysis of mountain areas in the European Union and in the applicant countries" by the European Commission's Directorate-General for Regional Policy (DG Regio). The geographical area encompasses 29 countries: the 15 EU members, the (then)10 acceding countries (which joined in may 2004), the two accession countries (Bulgaria and Romania), and Switzerland and Norway (who joined the study at their own expense). This study had three parts.

31.4.1 Delineation of mountain areas

The first challenge was to define mountainous regions in a way that can be applied across the whole of Europe. The objective of this delineation exercise was to circumscribe areas where the natural context for economic and social development is characterised by a high degree of topographic roughness. This was done on the basis of calculations of altitude, slopes and climate.

In order to allow for statistical analyses, the physical delimitation of mountainous areas was approximated to NUTS 5 entities (municipalities), i.e. the lowest possible geographical level.

31.4.2 Database construction and analysis of the social and economic situation of mountain areas

A socio-economic analysis must build upon comparative data. The principal sources of data for statistical analysis were Eurostat and its geographic information system for the Commission (GISCO). However, these databases do not have many data on the NUTS 5 level.

Some indicators can, however, be calculated from GISCO raster data. The climatic database offers typologies of contexts for human activities within mountain areas. From land use data, the study identified different types of agriculture and forests, as well as the extent of mountain towns and cities. Analyses made from such data must be precautionary, as national methodologies regarding the identification and classification of land uses remain quite heterogeneous.

Available network models for the European territory were used to calculate accessibility to markets, services and population. These models can be used at a local scale, identifying the level of service provision within reach of specific mountain localities and the local markets available for urban centres. At a wider scale, they provide an indication of the position of mountain areas within European production centres and markets.

Other important indicators were gathered directly through a network of national experts, who collected data for a limited number of indicators at the NUTS 5 level, including active population in the primary, secondary and tertiary economic sectors; unemployment rates; number of jobs; and indicators for tourism. Based on available data, the study established three typologies of European mountain areas showing a striking diversity of situations with regard to social and economic capital, access to services, and land use and land covers across Europe. These typologies will be of value in identifying regions with similar challenges.

It is important to point out that 'mountainousness' is a handicap only in some regards. Tourism is thriving in many mountain areas. Quality of life is a major determinant in the development of tertiary and technological activities. One major challenge of the study was therefore to analyse mountain areas at a discrete enough level to be able to appreciate the localised contrasts, while at the same time encompassing 29 countries whose area has seldom been covered by previous joint socio-economic analyses.

31.4.3 Policy review

The third part of the study was to describe the specific mountain policies – where they exist – and national and European policies with particular significance for mountain regions. European policies have influenced the development of mountainous areas for a long time, both through the Structural Funds and via the Common Agricultural Policy (CAP). At the national level, numerous aspects of public policy – agriculture, forestry, environmental management, rural development, tourism and business development – have specific and implicit impacts on mountainous areas.

31.5 The first results - the delimitation of mountain municipalities

The physical definition of mountain municipalities was the starting point for the analysis of the social, economic and environmental specificity of mountainous areas. The definition is based on altitude, topography and climate. A Digital Elevation Model was used to calculate slopes and local elevation ranges.

A climate contrast index was calculated for all of Europe, as an indicator of areas where the climate itself imposes severe restrictions on human activity. The threshold chosen was that of the value of the peaks of the Alps. This in effect only adds areas in the northern part of Scandinavia to the overall set of mountainous areas included in the study.

The following areas are included:
- elevation above 2,500 m;
- elevation 1,000-2,500 m + slope >5° (1,000-1,500 m) or >2° (above 1,500 m);
- elevation 300-1,500 m + local elevation range >300 m (7 km search radius);
- elevation 0-300 m and standard deviation >50 m (3 km search radius); and
- independent of elevation, areas with climate contrast index <0.25.

Municipalities are considered mountainous if these criteria apply over more than 50% of their territory.

'Mountainousness' presupposes that the topographic or climatic constraints extend over a significant area. Consequently, isolated patches of less than 5 km² are excluded, as are groups of municipalities with a total area of less than 50 km². Similarly, non-mountainous groups of municipalities forming an enclave of less than 50 km² within a mountain range were assimilated, allowing the inclusion of wide valleys.

The results are shown in Table 31.1 and Figure 31.3. A total of 33,800 out of 125,700 municipalities are defined as mountainous. For the 29 countries covered by the study, the

Table 31.1 Population and area covered by mountain municipalities, in % of national population/area for 'EU27+2' (the 29 countries included in the study). Top 10 countries + Finland, Sweden and the UK. Source: EC (2004).

Country	% of population in mountain municipalities	% of area covered by mountain municipalities
Switzerland	89.7	90.7
Slovenia	82.1	78.0
Norway	63.8	91.3
Slovakia	52.8	62.0
Greece	50.7	77.9
Austria	49.8	73.4
Bulgaria	45.6	53.3
Spain	38.9	55.7
Italy	34.1	60.1
Portugal	27.2	39.1
Finland	12.0	50.8
Sweden	6.9	50.6
UK	4.5	25.5

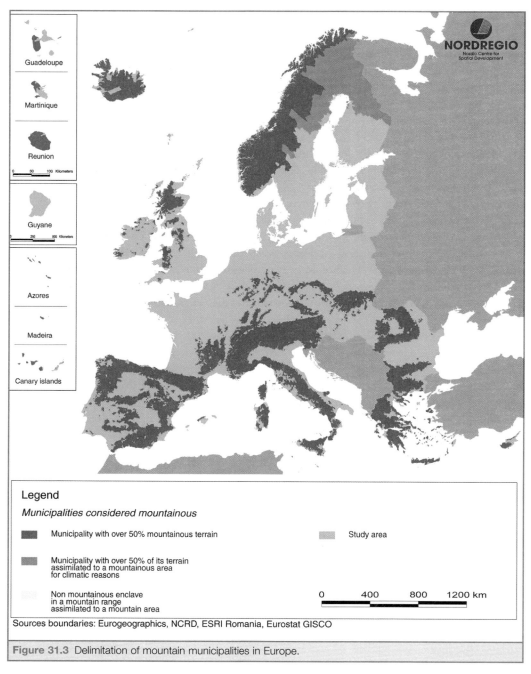

Figure 31.3 Delimitation of mountain municipalities in Europe.

share of mountain area ranges from 0 (Denmark, The Netherlands, Estonia, Latvia, Lithuania and Malta) to 78%-91% (Norway, Switzerland, Slovenia and Greece). The share of population living in mountainous municipalities is 50%-90% in Switzerland, Slovenia, Slovakia and Greece. In absolute terms, mountain area populations are largest in Italy (19 million), Spain (16 million), Germany (8 million) and France (8 million) (EC, 2004).

For most countries, this delimitation corresponds rather well to national perceptions of mountain areas. However, for countries such as Switzerland, Norway, Austria, Finland and Sweden, this definition significantly exceeds national ones. For other countries, such as Poland, Bulgaria and Romania, however, this definition produces an area smaller than the one used in the countries concerned.

References

Copus, C. & Price, M. (2002). *A Preliminary Characterisation of the Mountain Areas of Europe.* Euromontana, Brussels and Highlands and Islands Enterprise, Inverness.

EC (1999). Council Regulation (EC) No 1257/1999 on support for rural development from the European Agricultural Guidance and Guarantee Fund (EAGGF).

EC (2001). *Unity, Solidarity, Diversity for Europe, its People and its Territory – Second report on Economic and Social Cohesion.* European Commission, Luxembourg.

EC (2004). *Mountain areas in Europe: Analysis of mountain areas in EU Member States, Acceding and other European countries.* Directorate-General for Regional Policy, European Commission, Brussels.

Kapos, V., Rhind, J., Edwards, M., Price, M.F. & Ravilious, C. (2000). Developing a map of the world's mountain forests. In *Forests in Sustainable Mountain Development: A State-of-Knowledge Report for 2000*, ed. by M.F. Price & N. Butt. CAB International, Wallingford. pp. 4-9.

32 Frozen opportunities? Local communities and the establishment of Vatnajökull National Park, Iceland

Karl Benediktsson & Guðríður Þorvarðardóttir

Summary

1. In Iceland, preparations got underway in 2002 for the designation of Vatnajökull National Park. The local communities involved saw opportunities for tourism as the main rationale for the park. This was also central to the argument of the Ministry for the Environment which, in order to avoid both complications of ownership and confrontations with the interests of the power industry, decided to draw the park boundary to include initially only the Vatnajökull ice cap, together with two existing protected areas.

2. The designation of conservation areas is only partly about nature: it is just as much about cultural and ideological constructs – and realpolitik. National parks are best viewed not as purified 'natural' spaces, but as hybrid spaces at the interface of the social and the natural, their creation frequently involving contests and conflicts between various actors. To a large extent, the 'success' of a park hinges on the outcome of such processes, in which local communities are fundamentally important players.

3. An ongoing research project on local communities around Vatnajökull shows that conservation of biodiversity and/or geodiversity is not a major issue in local discourses on national park designation: it is construed largely as a marketing manoeuvre. Without the inclusion of some of the marginal mountainous areas, the opportunity from a conservation point of view is largely missed; some of these areas contain unequalled landscape diversity or expanses of wilderness, whereas others are nationally or internationally acknowledged sites of biological importance. While opportunities for local tourism entrepreneurs are enhanced by the park's creation, a more expansive designation would genuinely serve both the interests of conservation and economic development in the local communities.

32.1 Introduction

National parks have long been seen as jewels in the crown of nature conservation. Yet their creation has often been clouded in controversy, relating to differing interests and even differing world views of various individuals and social groups: nature conservationists, state bureaucrats, hydropower development enthusiasts, tourists, national politicians – and local

Benediktsson, K. & Þorvarðardóttir, G. (2005). Frozen opportunities? Local communities and the establishment of Vatnajökull National Park, Iceland. In *Mountains of Northern Europe: Conservation, Management, People and Nature*, ed. by D.B.A. Thompson, M.F. Price & C.A. Galbraith. TSO Scotland, Edinburgh. pp. 335-348.

people. In other words, national parks are contested spaces and their success or lack thereof is, to a large extent, related to the outcomes of such contests.

The designation of national parks was put squarely on the political agenda in Iceland in the late 1990s. The country's first national park was created in 1928 and two others were added in 1967 and 1973. There then followed a long interval with little activity, but as the *fin de siècle* drew closer, the idea of national parks (NPs) again gathered momentum.

Snæfellsjökull NP was designated in the year 2001, in an area which had in fact been discussed as a potential conservation area since the early 1970s (Preparatory Committee for National Park Designation, 1997), and in 1999 the national parliament made a resolution about a new national park which would include the ice cap of Vatnajökull (Alþingi, 1999, see Figure 32.1). A few months later, the Minister for the Environment announced her intent to designate this new park in 2002, to coincide with the UN's International Year of Mountains. At the time of writing (mid-2003), the Vatnajökull National Park project is still at a preparatory stage. Committee meetings and consultations are proceeding slowly. Official reports have been produced (Minister for the Environment, 2000; Preparatory Committee for the Establishment of Vatnajökull National Park, 2002) but there is a long way to go before the actual physical planning of the park. One interesting feature is how

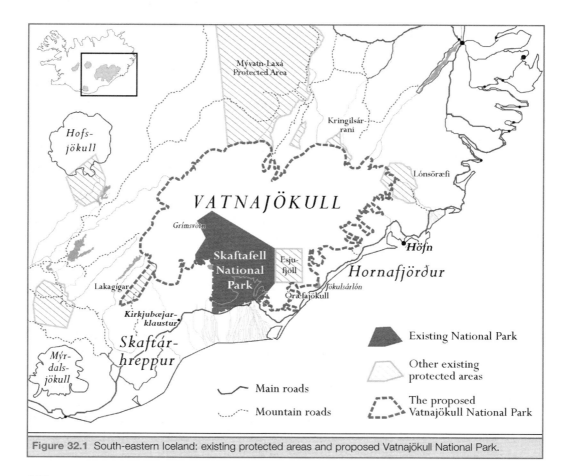

Figure 32.1 South-eastern Iceland: existing protected areas and proposed Vatnajökull National Park.

tourism is being presented by all parties as one of the main reasons for designation – for some, almost the sole reason, it would seem.

Following a discussion regarding the 'social nature' of parks in general, this chapter presents some results from a research project which was started in 2001 in order to map out and analyse the sentiments of the local communities towards the national park project, and to study how the new park would slowly take shape – as a conservation area, as a tourist destination, and as an identity project for local populations. A preliminary opinion survey was undertaken, followed up by qualitative interviews among people in the districts to the south of the proposed park.

32.2 National parks as 'social nature'

At the outset, it is necessary to make some observations about the theoretical directions guiding this work. Much academic research on conservation issues, including national parks, has been in the mould of 'resource management', which entails an objectivist standpoint. This certainly applies to the work of natural scientists, who have advanced the notions of biodiversity and geodiversity to enable the 'objective' mapping of conservation values. Similarly, social scientists who deal with conflict resolution in conservation projects most often proceed through an 'objective' analysis of competing stakeholder interests and suggest ways in which benefits can be maximized and costs minimized.

What is not acknowledged in most conventional accounts is the mutual constitution of the 'natural' and the 'social' in issues of conservation, as indeed in other issues where nature is central. In recent years, an insightful critique has been advanced, by scholars from many disciplines (cf. Latour, 1993; Cronon, 1995; Soper, 1995; Macnaghten & Urry, 1998; Castree & Braun, 2001), of the nature–society dichotomy which has dominated mainstream thinking on matters of nature. An attempt has been made to 'socialize the natural' (Castree, 2001). It follows that, instead of expecting to find fixed truths about the value of ecosystems and landscapes, we should examine closely the 'production of nature' in the discourses and social practices surrounding particular conservation projects. Seen in this way, a national park is a profoundly hybrid space: simultaneously natural, social and cultural (cf. Mels, 1999).

Park designation is not solely about the production of (social) nature, however. It is also about the ways in which individual and territorial identity is refashioned through park projects. There is wide agreement about a renewed importance of local identity in the face of globalisation. The inhabitants of a particular locality have to be more reflexive – more open to scrutinizing their own current position and possible futures – than at the heyday of industrial capitalism (Ray, 1999; Aarsæther & Bærenholdt, 2001). Often this means formulating a territorial identity which is different from that of the past, while building on that past. In an 'economy of signs' (Lash & Urry, 1994) the association of territory with particular products has become much more important (cf. Ray, 1998; Ilbery & Kneafsey, 1998; Jenkins, 2000; Kneafsey, 2000, 2001).

Tourism is one of the most obvious economic strategies requiring territorial identification. The tourist destination product is marketed and consumed as a *local* product. Its integrity depends to a large extent upon the strength of local identity. A locality that does not have a well-defined identity and is able to present itself outwards as such is less likely to compete successfully with other tourist destinations.

National parks and tourism have gone hand in hand since the first national parks were established in the late 19th century (Butler & Boyd, 2000). However, this relationship has not always been harmonious (cf. West & Brechin, 1991; Machlis & Field, 2000). For those charged with the task of planning, it is therefore essential to understand the metabolism of this uneasy symbiosis in the particular context in question.

32.3 Vatnajökull and its surroundings

Vatnajökull itself contains landscapes of extraordinary geological importance. Together with its outlet glaciers, the ice cap covers an area of some 8,300 km². Subglacial geothermal and volcanic activity make it unique among the world's major ice caps and have offered scientists excellent opportunities to study the interaction between these two important elements in the earth's geological history (cf. Björnsson, 1988).

Apart from this, the temperate Vatnajökull is considerably 'warmer' and thus wetter at its base than its polar relatives. Many outlet glaciers and numerous large glacial rivers (*jökulsá*) bear witness to this, together with spectacular periodic surges of some of the glacier tongues, and floods of catastrophic proportions (*jökulhlaup*) in certain rivers.

Several nunataks pierce the ice cap, the most important of which are included in the nature reserve of Esjufjöll. Covering an area of 27,000 ha, this reserve has been protected since 1978 due to its pristine vegetation, free from the influence of humans and their grazing animals. Adjacent to Vatnajökull are numerous other areas considered to be of high conservation value. Some are already protected (Nature Conservation Agency, 1996, cf. Figure 32.1), including Lakagígar, an area of 16,000 ha to the SW of Vatnajökull, which was designated as a natural monument in 1975. This is a 25 km long row of volcanic craters, created by an eruption which started in June 1783. Apart from creating a vast lava field of 565 km² that totally changed the landscape of a large farming region, its climatic consequences led to the death of a fifth of Iceland's population and well over half of all domestic animals. This area thus has deep historical resonances, as well as providing textbook examples of volcanic landforms. Lónsöræfi to the east of the glacier – an area of 32,000 ha protected as a nature reserve in 1977 – is characterized by varied and colourful geological formations that are the deeply eroded remains of volcanoes, and rich mountain flora, including natural birch woods. Within the reserve are also two deserted farms which were inhabited during the late 19th century. Kringilsárrani to the north is also a nature reserve (from 1975), as an important calving area for reindeer. In addition it contains some unique periglacial formations. However, its area was recently reduced in size by the Minister for the Environment (2003) to clear the ground for a very large and controversial hydropower development project (see below). Last but not least, Iceland's largest existing national park, Skaftafell, is the product of Vatnajökull, geologically speaking. The park, designated in 1967, covers an area of 160,000 ha. It contains a large part of Vatnajökull itself, together with mountains, valleys and a large section of outwash plain (*sandur*). The landscape is sculpted by glacial action and water erosion. Valley glaciers and glacial rivers are prominent features. The geothermal area Grímsvötn in Vatnajökull is inside the park. On the old farming property of Skaftafell itself, the exceptionally lush vegetation contrasts sharply with the ice and black sands (Figure 32.2). Adjacent to Skaftafell NP is the strato-volcano Öræfajökull, Iceland's highest mountain (2,119 m). It has erupted twice in historical times, in 1362 and 1727, with severe local consequences.

Figure 32.2 A view of parts of Skaftafell National Park and Öræfajökull. Old farms with hayfields to the left, surrounded by birch woodland. The strato-volcano Öræfajökull is at the back, with Iceland's highest peak (2,119 m) to the right. (Photo: Snævarr)

Scores of other mountainous areas and additional tracts of undisputed geological, biological and historical importance encircle Vatnajökull (Figure 32.3), many of them on the Nature Conservation Register which lists places worthy of protection (Nature Conservation Agency, 1996). There can be no doubt that, from a conservation viewpoint, a national park that included some or many of these areas would be considerably more valuable than a park that included only the glacier.

Originally introduced by a biologist and Member of Parliament, the idea of designating *all* of Iceland's major glaciers together with their immediate surroundings as national parks has been around for some years. For Vatnajökull, it gained momentum in 1999 with a parliamentary resolution, as mentioned above. This happened in the midst of a great controversy over proposed hydropower developments at the north-east corner of Vatnajökull, where the National Power Company intended to flood some wetlands of great biological importance. Eventually this project was discarded in favour of another and much larger one (cf. National Power Company, 2003), which would spare the wetlands but would instead open up to development a vast 'wilderness' area immediately to the north of the ice cap. This has now gone ahead, to the outrage of conservationists.

Figure 32.3 Hikers in the mountains southeast of Vatnajökull. The glacially-dammed marginal lake caused catastrophic floods in the early 20th century, but with the thinning of the glacier these floods have become much smaller. (Photo: Karl Benediktsson)

32.4 Communities and current livelihoods

Those who live in the vicinity of Vatnajökull belong to several distinct farming communities, with only one town of note in the area. The communities to the south of the glacier will be most directly affected by the designation of the National Park. Our research has thus concentrated mostly on the people living in these communities. However, the initial survey of opinions included people in the north-east and north, whose lands extend up to the glacier's edge although the actual settlements are mostly some distance away.

The southern communities belong to two municipalities (see Figure 32.1): Hornafjörður (2,300 inhabitants) and Skaftárhreppur (550). Both have been formed by a recent amalgamation of a number of smaller rural districts. The town of Höfn in Hornafjörður (1,750), in the far east, is the single urban node – an important service centre, but above all a fishing town. In Skaftárhreppur, the village of Kirkjubæjarklaustur (150) is the only non-agricultural settlement. The sparsely populated rural areas consist of a number of communities that in former times were rather distinct social entities. Sheep and dairy farming were the mainstays of the agricultural economy.

With rural depopulation, caused partly by broad social changes and partly by changed conditions for farming (Benediktsson, 2001), the social fabric has been altered and old boundaries partly erased. A search for new economic alternatives has been on the agenda for some years, yielding several new opportunities, such as small-scale fish farming but, above all, tourism. For one of the rural districts, Öræfi (where Skaftafell NP is located), tourism has become especially important (Benediktsson & Skaptadóttir, 2002; Jóhannesson *et al.*, 2003).

In the municipality of Hornafjörður in particular, tourist operators have already started to take advantage of Vatnajökull. Boat trips on the glacial lagoon Jökulsárlón have become some of the most popular tourism products in the whole country. Other forms of adventure tourism have also been developed. Glacier tours, utilising both snow-scooters and specially equipped four-wheel drive jeeps, are offered by operators from the town of Höfn. Such activities bring their own specific dilemmas related to conservation and tourism. While the traffic on the glacier has, in general, not reached any critical limits, there have been incidents where the 'wilderness experience' of hikers on Öræfajökull has been severely downgraded by swarms of noisy motor vehicles (Einarsson, 2002, cf. Figure 32.4). This will be one of the issues awaiting planners and managers of the new national park.

Figure 32.4 'Superjeep' tourism on Vatnajökull. The 'motorized outdoors' industry is well developed in Iceland, both as a recreational activity for local people and as a tourism attraction. (Photo: Edda Ruth Hlín Waage)

32.5 The National Park and local communities

During the recent conflict about hydropower development vs. conservation, the importance of national parks for tourism was increasingly used by conservationists as an argument for protecting this expanse of wilderness from the ever more voracious power industry. The Vatnajökull NP proposal, however, was also put forward by some as a 'win-win' solution, accommodating the wishes of conservationists for a new protected area while offering new local economic opportunities through tourism, and still allowing hydropower development to go ahead in the areas just outside the park. This seemed indeed to be a 'safe' and non-controversial park proposal.

A special preparatory committee was appointed by the government in February 2002 – the same year as the park was to be established. The committee held meetings with representatives of the seven local governments of the region, landowners, tour operators and various other local and non-local stakeholders. In addition, the committee consulted several associations and societies as well as the committee for public lands (see below). Finally, open meetings were held for the general public in the local communities.

In October 2002, the committee finally delivered a report (Preparatory Committee for the Establishment of Vatnajökull National Park, 2002), concluding that the project would be most easily carried out if the park boundary simply followed the outline of the glacier, with the exception of those parts of Skaftafell NP that are beyond the ice-limit, as well as the Lakagígar Natural Monument. The report highlighted some specific issues for resolution before going further, including detailed mapping of landowners' rights and opinions. The need for judicial resolution of various issues regarding land ownership on the glacier and adjacent to it was stressed. The committee also emphasised the importance of cooperation with local governments, landowners, tour operators and the planning committee for the central highlands. In addition, the report indicated the need to evaluate the economic benefits of the park's establishment as well as its value for scientific research.

Our research has endeavoured to track this consultation process, with participant observations at meetings and interviews with various people, both directly involved with the work and on the sidelines (Waage & Benediktsson, 2002; Benediktsson et al., 2003). An initial questionnaire survey was carried out in late 2001 and early 2002 among the populations in the districts to the south, east and north of the ice cap. Questionnaires were mailed out to a 10% random sample of the adult population, followed by phone calls a few days later. However, the response rate was rather low, 48%, with 187 questionnaires completed. That said, the survey revealed that about half of the people surveyed had a positive or very positive attitude towards the park's establishment. It was notable, however, that many stated that they did not know much about the project. This is not surprising: although it had by then been discussed for some time, very little had been decided upon. It was also noticeable that most respondents, including both park supporters and opponents, viewed the national park, and in fact nature as such, from a very utilitarian standpoint. A minority of around 20% stated that they were 'rather opposed' or 'very opposed' to the designation of the park.

A series of qualitative interviews was carried out in mid-2002 in order to clarify a number of issues. Nearly 30 local people in the two municipalities south of Vatnajökull were visited and asked about their individual views – their hopes and fears – regarding the proposed park. The group included people of both sexes from most parts of the occupational spectrum – farming people, tourism operators, local officials and others. The interviews were recorded and transcribed, and then analysed thematically. Below is a brief discussion of the major themes which emerged.

32.5.1 The National Park as a marketing device

It is obvious that local people perceive this new park from a mainly economic perspective. When asked about the reason for the park's establishment, a young farmer answered bluntly

"To attract more tourists. The uniqueness of the glacier – this is maybe what can be sold most easily."

This statement echoes the sentiments of many others. This same farmer answered a question about the most sensible way of deciding the park's boundaries thus:

"Of course it depends on what we are going to sell, I mean, are we only going to sell the glacier or are we going to sell its surroundings?"

While this farmer did not reflect on the basis for this market advantage, some others did so, echoing well-established arguments in both Icelandic and international discourse on conservation and tourism:

"... with the growth of cities in so many parts of the world, there is increasing shortage of places, both in Iceland and elsewhere, that can be marketed as unspoiled nature."

There are high hopes for increased employment in tourism, related to the downturn in farming over the past few years. Rural people are reassessing their options and they look, for example, to Skaftafell NP and what it has done for the district of Öræfi. That district has fared much better than many others, in part because of the re-invigoration of the local economy through tourism. Some go as far as stating that this is the 'only hope' for the rural areas. Not everybody believes in this, however. Some of the rural people interviewed thought that the local and national authorities have promised too much in this regard. They pointed to the seasonal nature of tourism employment; and to the fact that many of the existing summer jobs are currently filled not by local people, but by seasonal workers brought in from elsewhere. Thus tourism may not be a panacea for continued settlement of the rural districts.

32.5.2 Fears of an underfunded National Park

While many people base their economic expectations on the experience from Skaftafell NP, another aspect of the history of that park provides an example to be avoided: the chronic lack of funds for the construction of necessary facilities and even for basic park maintenance. In the past, the physical condition of some footpaths and other facilities has deteriorated at Skaftafell (Sæþórsdóttir *et al.*, 2001), as well as at other national parks, due to the significant increase in tourist numbers. Maintenance carried out by volunteers, mainly from the British Trust for Conservation Volunteers, has prevented serious damage. The story is similar for the nature reserve of Lónsöræfi to the east, which is now a popular hiking area (cf. Sæþórsdóttir *et al.*, 2003). This has created negative feelings among some landowners there towards the Nature Conservation Agency (from 2003 a department within a newly-created Environment and Food Agency), which they see as having avoided its responsibilities.

It is hardly an exaggeration to say that the state did not, in the past, provide sufficient money for the Agency to carry out its tasks in the best way possible, although park managers certainly made the best of what they received. Despite modest increases, it is debatable whether funding is sufficient today, The example of the new Snæfellsjökull NP is instructive. In 2001, a working group appointed by the Minister for the Environment estimated that well over 100 million Icelandic krónur (approx. £800,000 according to exchange rates in mid-2003) would be needed to establish this new park, set up a visitor centre, and ensure its proper day-to-day running for the first year (Working Group for the Establishment of National Park on Snæfellsnes, 2001). For the first full year of operation (2002), the Nature Conservation Agency asked for 33 million krónur for this park. The final allocation for 2002 was 10 million (a little less than £80,000), plus housing for the newly-appointed park manager. Since the park was designated, the limited financial means have restricted the capacity of the park manager to prepare the area for the anticipated influx of tourists.

To be fair, the situation has improved. Extra money was allocated early in 2003 for the hiring of a second full-time warden at three national parks managed by the Agency. Yet, given the numerous protected areas and the many issues which urgently need attention, it would still be difficult to state conclusively that either the current or previous governments have seen conservation as a priority when it comes to funding, despite considerable rhetoric.

Worries about adequate funding are reflected in remarks such as the following, made by a rural woman:

> *"I think that, here in Iceland, the national park idea is great for providing opportunities for politicians and dignitaries to sign some papers and have a drink … but what then? … There is money … for one round of drinks – to say toast for the new national park, but then there is no more money to continue. There is no policy so that we know where we are going … no funding at all! Nothing!"*

For this person and some others in the local communities, the disingenuous handling of conservation matters in general by the state does not bode well for the new park.

How then can parks be financed? For some years, Icelanders have discussed whether to charge entrance fees at some of the most popular tourist sites, including national parks. Although matching well with the neo-liberal orthodoxy which has pervaded the corridors of power for some time, charging entrance fees for national parks would in fact be quite contrary to the country's strong tradition of public rights of access.

32.5.3 A critique of top-down tactics

The third major theme found in the interviews is a familiar one in similar circumstances: a sense of a gap between the 'grass-roots' and the authorities, whether national or local. Local farming people criticise what they perceive as late and limited attempts at establishing a genuine dialogue between themselves and those pushing for the park's establishment:

> *"They started to talk about this park probably two years ago and they have held conferences and have been, as they say, interpreting the view of the locals. But it wasn't until late this winter [i.e. 2001] that they invited landowners in every district to a meeting. And I say they started at the wrong end. I feel they were not interpreting any views except for their own, these officials – they took it for granted that they were representing the locals."*

However, this criticism may not be fully justified. Many of those who were instrumental in having the park proposal taken up by the central authorities were, in fact, local tourism entrepreneurs. It is true that the approaches to consultation have been rather unimaginative, and have mainly consisted of open meetings where representatives from the Ministry for the Environment, the Nature Conservation Agency, the national NGO

Landvernd and the municipal governments have presented their views and asked for responses. Few local people have attended. Participatory observation by our research assistants at these meetings and subsequent interviews with some those attending revealed that many had the feeling that the Ministry had already made up its mind about the shape of the new park – despite assertions to the contrary.

Another highly charged issue concerns ownership claims, and this too has fuelled suspicions about the state's intentions. In 1998, legislation on 'public lands' was passed, aiming to sort out once and for all some long-standing, thorny questions regarding ownership of the central highlands (Anon., 1998). According to this new legislation, the state is to assume ownership over large parts of the highlands. Protracted legal proceedings are currently taking place in several regions, including those to the south of Vatnajökull. Most of the land in question has been used and managed as common property by the farmers in the adjacent lowland districts, but the state has also claimed considerable areas of land which have long been considered as privately owned. Farmers and landowners have had to legally defend their claims and, needless to say, the issue has aroused heated passions among the local population. Some people inevitably see the Vatnajökull NP proposal in this light: as yet another encroachment by the state upon local ownership rights. Even if the decision to draw the boundaries at the margins of the ice cap was intended to limit the risks of such conflicts, it is perhaps not surprising that they have arisen.

32.6 Conclusion: frozen opportunities?

"Nature protection? As far as I know there has been no discussion about it!"

Together with its surroundings, Vatnajökull has all the necessary attributes to become an important national park. However, the above remark, made by a local woman, indicates that the deliberations about Vatnajökull NP have taken a rather one-sided form, namely that of economic development. This is not only the main concern of local people: it is also central to the rationale put forward by the Ministry for the Environment for designating the new park (cf. Preparatory Committee for the Establishment of Vatnajökull National Park, 2002). Many people feel that the protection of nature is not at all the only issue at stake.

This comes as no particular surprise, given our earlier discussion of national parks as hybrid spaces. Our account has highlighted the various ways in which culture, ideology and politics are playing a role in the ongoing process of park creation in Iceland. Indeed, the establishment of protected areas is bound to be a juggling act in many different respects: between conservation and development; between local control of resources and central/national planning and management. The trick is not to 'freeze' the situation by getting off on the wrong foot.

And here we think that more imaginative ways for genuinely involving local communities have to be worked out. Only by doing so will it be possible to draw the boundaries of the park so that it truly serves the interests of conservation. While necessary, local input into the park designation process is not enough on its own for establishing a successful park. The other crucial input – one which seems not to be guaranteed in this case – is the financial means needed for the proper planning and construction of necessary facilities and the employment of adequate staff. Merely placing a National Park label on an expanse of ice is not enough – either to protect nature or to create sustainable economic opportunities for local people.

Postscript

Late in 2004, following an offical decision in the public lands dispute for the municipalities to the south of Vatnajökull, Skaftafell National Park was extended to include most of the southern half of Vatnajökull, together with Lakagígar (cf Figure 32.1). This is a significant step towards the realization of the Vatnajökull National Park concept.

Acknowledgements

This project is a collaborative effort of people at the University of Iceland, Hornafjörður Municipality, Kirkjubæjarstofa Research Centre, East Iceland Development Agency, and the Environment and Food Agency, led by the first author. The project has received support from the Icelandic Research Council, RANNÍS. The authors thank research assistants Steingerður Hreinsdóttir and Edda Ruth Hlín Waage for their invaluable inputs. Three referees made useful suggestions. The authors alone are responsible for the arguments contained in the chapter and these arguments do not necessarily represent the opinion of their institutions.

References

Aarsæther, N. & Bærenholdt, J.O. (2001). *The Reflexive North.* Nordic Council of Ministers, Copenhagen.

Alþingi (1999). *Þingsályktun um Vatnajökulsþjóðgarð [Parliamentary Resolution about Vatnajökull National Park].* Accepted by Alþingi 10 March 1999. (123. löggjafarþing 1998-1999, 16. mál, þskj. 1144). http://www.althingi.is/altext/123/s/1144.html.

Anonymous (1998). *Lög um þjóðlendur og ákvörðun marka eignarlanda, þjóðlendna og afrétta [Laws on National Lands and the Determination of the Boundaries of Private Lands, National Lands and Common Lands]* No. 58 (1998).

Benediktsson, K. (2001). Beyond productivism: regulatory changes and their outcomes in Icelandic farming. In *Developing Sustainable Rural Systems,* ed. by K.-H. Kim, I. Bowler & C. Bryant. Pusan National University Press, Pusan. pp. 75-87.

Benediktsson, K. & Skaptadóttir, U.D. (2002). *Coping Strategies and Regional Policies: Cases from Iceland (Working Paper 2002:5).* Nordregio, Stockholm.

Benediktsson, K., Waage, E.R.H. & Hreinsdóttir, S. (2003). *Þjóðgarðstal: Viðhorf heimamanna til Vatnajökulsþjóðgarðs [Talking of Parks: Attitudes of Local People Towards Vatnajökull National Park].* Háskólasetrið á Hornafirði, Höfn.

Björnsson, H. (1988). *Hydrology of Ice Caps in Volcanic Regions.* Societas Scientiarum Islandica, Reykjavík.

Butler, R.W. & Boyd, S.W. (2000). *Tourism and National Parks: Issues and Implications.* John Wiley, Chichester.

Castree, N. (2001). Socializing nature: theory, practice, and politics. In *Social Nature,* ed. by N. Castree & B. Braun. Blackwell, Oxford. pp.1-21.

Castree, N. & Braun, B. (Eds) (2001). *Social Nature.* Blackwell, Oxford.

Cronon, W. (1995). *Uncommon Ground.* W.W. Norton, New York.

Einarsson, E. (2002). Vélsleðaakstur á Hvannadalshnúk [Snow-scooter traffic on Hvannadalshnúkur]. *Morgunblaðið,* 26 May 2002, 49.

Ilbery, B. & Kneafsey, M. (1998). Product and place: promoting quality products and services in the lagging rural regions of the European Union. *European Urban and Regional Studies,* **5**, 329-341.

Jenkins, T.N. (2000). Putting postmodernity into practice: endogenous development and the role of traditional cultures in the rural development of marginal regions. *Ecological Economics*, **34**, 301-314.

Jóhannesson, G., Skaptadóttir, U.D. & Benediktsson, K. (2003). Coping with social capital? The cultural economy of tourism in the north. *Sociologia Ruralis*, **43**, 3-16.

Kneafsey, M. (2000). Tourism, place identities and social relations in the European rural periphery. *European Urban and Regional Studies*, **7**, 35-50.

Kneafsey, M. (2001). Rural cultural economy: tourism and social relations. *Annals of Tourism Research*, **28**, 762-783.

Lash, S. & Urry, J. (1994). *Economies of Signs and Space*. Sage Publications, London.

Latour, B. (1993). *We Have Never Been Modern*. Harvard University Press, Cambridge MA.

Machlis, G.E. & Field, D.R. (2000). *National Parks and Rural Development: Practice and Policy in the United States*. Island Press, Washington DC.

Macnaghten, P. & Urry, J. (1998). *Contested Natures*. Sage Publications, London.

Mels, T. (1999). *Wild Landscapes: The Cultural Nature of Swedish National Parks*. Lund University Press, Lund.

Minister for the Environment (2000). *Skýrsla umhverfisráðherra um möguleika á stofnun Vatnajökulsþjóðgarðs [Report of the Minister of the Environment on the possibilities for the designation of Vatnajökull National Park]*. (125. löggjafarþing 1999–2000, 642. mál, þskj. 1300). http://www.althingi.is/altext/125/s/1300.html (Accessed 20 June 2003).

Minister for the Environment (2003). Auglýsing um friðlýsingu Kringilsárrana, N-Múlasýslu [Notice about the Preservation of Kringilsárrani, N-Múlasýsla]. *Stjórnartíðindi*, **B 181**, 7 March 2003.

National Power Company (2003). *Kárahnjúkar Hydropower Project*. http://www.karahnjukar.is/en/.

Nature Conservation Agency (1996). *Náttúruminjaskrá [The Nature Conservation Register], 7th edition*. Náttúruverndarráð, Reykjavík.

Preparatory Committee for National Park Designation (1997). Þjóðgarður á utanverðu Snæfellsnesi Lokaskýrsla [National Park in Outer Snæfellsnes – Final Report]. Unpublished report.

Preparatory Committee for the Establishment of Vatnajökull National Park (2002). Unpublished report.

Ray, C. (1998). Culture, intellectual property and territorial rural development. *Sociologia Ruralis*, **38**, 3-20.

Ray, C. (1999). Endogenous development in the era of reflexive modernity. *Journal of Rural Studies*, **15**, 257-267.

Soper, K. (1995). *What is Nature?* Blackwell, Oxford.

Sæþórsdóttir, A.D., Gísladóttir, G., Ólafsson, A.M., Sigurjónsson, B.M. & Aradóttir, B. (2001). *Þolmörk ferðamennsku í þjóðgarðinum í Skaftafelli [Tourism Carrying Capacity in Skaftafell National Park]*. Icelandic Tourist Board, University of Iceland and University of Akureyri, Akureyri.

Sæþórsdóttir, A.D., Gísladóttir, G., Aradóttir, B., Ólafsson, A.M. & Ólafsdóttir, G. (2003). *Þolmörk ferðamennsku í friðlandi á Lónsöræfum [Tourism Carrying Capacity in the Lónsöræfi Nature Reserve]*. Icelandic Tourist Board, University of Iceland and University of Akureyri. Akureyri.

Waage, E.R.H. & Benediktsson, K. (2002). Vatnajökulsþjóðgarður: Sjónarmið úr grasrótinni [Vatnajökull National Park: A View from the Grass-roots]. *Landabréfið*, **18-19**, 41-57.

West, P.C. & Brechin, S.R. (1991). *Resident Peoples and National Parks*. University of Arizona Press, Tucson.

Working Group for the Establishment of National Park on Snæfellsnes (2001). Þjóðgarðurinn Snæfellsjökull [Snæfellsnes National Park]. Unpublished report.

33 Outdoor education and outdoor recreation in Scotland

Peter Higgins

Summary

1. The landscape and natural heritage are the stimuli for, and the space within which, outdoor recreation and outdoor education have developed in Scotland. The origins of this development are reviewed, case studies are presented and the socio-economic, educational and policy implications considered.

2. Outdoor education centres are significant employers in some rural areas. Outdoor recreation generates perhaps at least £600 to £800 million of Scotland's tourist income, much of which is in rural areas and extends the traditional tourist season.

3. Evidence from one area of Scotland (Lothian Region) suggests that the pattern of outdoor education provision has changed significantly in recent years.

4. Recent legislation (e.g. Land Reform Scotland (2003) Act) and other developments offer a sound basis for outdoor education to relate closely to the natural heritage.

5. There are significant implications for policy on rural development and education.

33.1 The landscape of Scotland: a recreational and educational space

At over 78,000 km², Scotland covers around a third of the UK and has a coastline of approximately 4,000 km (around half that of the UK). It contains most of the high mountains and, wildest coastline areas. Much of the 'wild land' is held in the form of 'sporting estates' and maintained and managed primarily for the purpose of hunting deer or grouse and/or for fishing. This form of outdoor recreation is usually tied to ownership of the land whereas, for example, walking, mountaineering and canoe-sports require access to the landscape without such management regimes.

The combination of variable climate, geological history and the resulting topography has had a major influence on the development of outdoor activities in Scotland. For example, the development of winter mountaineering required freeze-thaw cycles of the accumulated snow in the gullies of mountain crags. This particular form of climbing led to technical developments which have had an international impact (McNeish & Else, 1994). The opportunities offered by the natural physical heritage of Scotland have also

Higgins P. (2005). Outdoor education and outdoor recreation in Scotland In *Mountains of Northern Europe: Conservation, Management, People and Nature,* ed. by D.B.A. Thompson, M.F. Price & C.A. Galbraith. TSO Scotland, Edinburgh. pp. 349-358.

encouraged the development of river and sea kayaking, sailing, rock climbing and mountaineering, etc. (Mackay, this volume). Variable winter snowfall allowed the development of skiing from as early as 1892 (Simpson, 1982).

Clearly the Scottish countryside, wild areas and the surrounding seas have long been attractive to outdoor people from all over Britain and other parts of the world. Little wonder that Scotland provided an early venue for the development of formalised outdoor education. A number of factors led to a long and documented tradition of outdoor recreation in the UK. These include the influence of the 'Romantic' school of writers and artists (e.g. Blake, Wordsworth and Constable) and the 'age of exploration' (initially for political, economic and scientific reasons) as epitomised by the race to the Poles, Mount Everest, etc. (Higgins, 2002; Mackay, this volume).

33.2 Outdoor education - formalised adventure

A number of factors led to favourable conditions for the development of outdoor education. By the 1960s several Acts of Parliament had led to the protection of the natural heritage, and increased holiday opportunities for workers at all levels led to subsequent growth in interest in outdoor activities. The 1944 Education Act, the 1945 Education (Scotland) Act and other reports had encouraged the use of the outdoors for environmental and nature studies (Cook, 1999). Finally, the rise of the 'progressive' education movement through the 20th century until around the 1970s encouraged the use of experiential educational techniques. For detailed summaries see Parker & Meldrum (1973) and Higgins (2002).

The 1944 Education Act explicitly stated the value of direct experience of the outdoors and encouraged Local Education Authorities to establish 'camps' for the purpose. During the 1960s and 1970s, most Local Educational Authorities offered progressive outdoor educational opportunities for the school pupil, and many bought and converted old mansions as residential bases from which outdoor adventurous activities and field studies could take place. At this time Scotland, and in particular Lothian Region (Edinburgh and the surrounding area) had what was probably the most comprehensive formalised provision of outdoor education in the world (Cheesmond, 1979). Consequently, outdoor education courses designed to meet the demand throughout the UK for trained teachers and instructors were established in a number of colleges and universities (Higgins & Morgan, 1999).

Notably, 1948 saw the establishment of Glenmore National Forest Park (which in 2003 became part of the Cairngorms National Park) and Glenmore Lodge, Scotland's National Outdoor Training Centre (Loader, 1952). The progressive nature of these developments, acknowledging respectively the importance of outdoor recreational spaces and outdoor educational provision for field studies, physical activities and 'citizenship' in establishing templates for future developments, cannot be underestimated. That both were founded essentially on an 'ideal of liberal democracy' and a 'willingness to experiment', and were supported by the Scottish Office (Lorimer, 2003, p. 204) says much of post-war notions of 'national heritage' and the desire to 'instil a sense of citizenship'.

33.3 Theoretical perspectives on outdoor education

Many of the arguments put forward for the educational use of the outdoors are of course the same in Scotland as they are elsewhere in the UK and other parts of the world.

Theoretical perspectives on outdoor education in Scotland are notable because of its early development in the context of the natural and cultural heritage. For critiques see Hopkins & Putnam (1993), Higgins & Loynes (1997), and Nicol & Higgins (1998a,b).

The expectation is that an outdoor educator will work safely and professionally within the whole domain in Figure 33.1, shifting emphasis from one area to another as opportunities arise within the programme. In recent decades the emphasis has been on personal and social development or the outdoor activities themselves. However there is evidence that, in response to global environmental imperatives, environmental education is now becoming more fully integrated into outdoor educational practice (Cooper, 1991; Nicol & Higgins, 1998a; Crowther *et al.*, 1998).

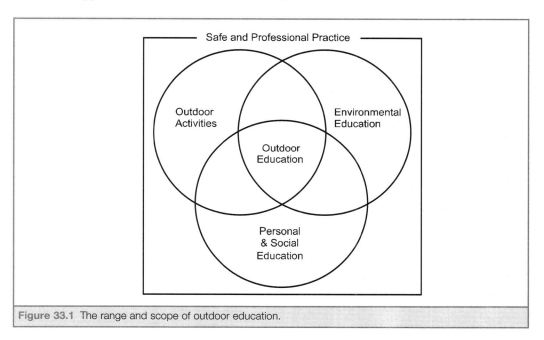

Figure 33.1 The range and scope of outdoor education.

33.4 Outdoor education - a changing pattern of provision

In 1983, the Countryside Commission for Scotland published the results of a survey (conducted in 1982) reporting the existence of some '163 residential centres providing for outdoor activities of a recreational or educational kind'. It was the view of the authors that the centres represented 'a resource of considerable value to both the education and leisure markets'. They presage the financial pressures which have led to a decline in provision in recent years, noting that 'in the public sector, outdoor education may be looked upon as one of the first extras to be cut, partly due to the high costs of transport and building maintenance'. They emphasise the importance of this form of educational provision and suggest ways to meet this challenge (Countryside Commission for Scotland, 1983). Little published evidence exists concerning changing patterns of provision other than for Lothian Region, and to a lesser extent Strathclyde Region.

The changes in provision in Strathclyde Region have recently been charted by Halls (1997a,b). He notes the growth and increasing 'organisation' of the subject area to a peak in the 1970s and early 1980s and identifies reasons (such as its non-statutory status) for the subsequent decline of robust, diverse and Region-wide provision.

Lothian Region was very successful in developing outdoor education during this period. In 1978 and 1979, Lothian Region and Dunfermline College, Edinburgh collaborated in a substantial research project to review the status of outdoor education within Lothian Region (Cheesmond, 1979). This report also considered ways in which Lothian Region might sustain and develop its provision. This review of provision is summarised in Table 33.1 and is compared with the current situation:

Table 33.1 Changing provision in outdoor education in Lothian Region.

Provision in 1978-1979 (Cheesmond, 1979)

- All secondary schools had some programme of outdoor education, either formal or informal. Almost all secondary schools (45) employed promoted staff responsible for outdoor education. There was a high number of other staff (estimated 500 to 600) who assisted in outdoor education provision, ranging from only one or two in some schools to over 50% in others. The majority of these were volunteers.

- Four outdoor education centres were operational, each fully staffed.

- Several 'resource bases' (centralised stores of equipment, etc.) were provided.

- Primary School provision was very limited.

Provision in 2000 (Chalmers Smith, Chair of Lothian Association of Outdoor Education Staff, pers. comm., 2000)

- Under Local Government Reorganisation legislation enacted in 1996, Lothian Region was split into four new Councils.

- Very few secondary schools had a formal or informal programme of outdoor education.

- Few secondary schools (six) employed any staff responsible for outdoor education, and several of these had only a part responsibility for the subject. The total school-based staff (including special schools) was about seven full-time equivalents.

- Relatively few other staff assisted in outdoor education provision.

- Two outdoor education centres were operational, each with a reduced staff complement.

- One 'resource base' existed.

- Prior to Local Government Reorganisation, a 'Disadvantaged Pupils Fund' existed to help with the cost of residential courses. Only two of the four new Councils provided this support.

- Primary school outdoor education involvement was somewhat more extensive, but this was rarely reflected in staff appointments and designated responsibilities.

A number of factors have led to a decline in formal outdoor education provision in the UK. For example, recent legislation (Activity Centres (Young Persons Safety) Act, 1995) has posed organisational problems which have been translated into financial and staffing pressures. Similarly, reductions in the Rate Support Grant from Central Government have led to substantial cuts (in real terms) in Local Authority provision for most services, and other non-core subjects such as music, art, drama and curriculum support have also suffered. In Scotland, the 1996 changes to the structure of Local Authorities were the 'last straw' for many Centres. This process, designed to reduce bureaucracy through the generation of 'Single Tier Authorities' left these Authorities searching for ways to balance their budgets.

Whilst there has been no recent survey of residential provision following these changes, there does appear to have been a substantial reduction. As part of a survey conducted in 1998, Nicol (1999) established that there were nine remaining Local Authority residential outdoor education centres compared to 15 identified by the Scottish Advisory Panel for Outdoor Education in 1996. There is some evidence that, whilst Local Authority provision has decreased, the commercial and charitable sectors have gained bookings from schools. The Adventure Activities Licensing Authority (AALA), who inspect providers to ensure safe practice under the Young Persons Safety (Activity Centres) Act 1995, estimated that, whilst there are still nine Local Authority centres, there are now 59 residential centres working within its scope (AALA Inspectors, Newtonmore, pers. comm., 2003). Provision seems to be more heterogeneous as the AALA inspectors report that there are a further 82 'licensed providers'. These include non-residential centres, Local Authorities, farmers seeking to diversify and small businesses of one or more individuals. Nonetheless, if the situation in the rest of Scotland mirrors the former Lothian Region, where so few schools now have a member of staff with a responsibility for outdoor education (Table 33.1), the long-term prospects of continued school commitment cannot be taken for granted.

This situation has been compounded by the recent outbreak of foot and mouth disease in the UK. As many providers are now more commercially orientated, the resulting closures led to financial difficulties. Whilst a number of compensation claims have been successful, it is clear that the sector is precariously balanced and its contribution to the rural economy difficult to predict.

In 2003, the New Opportunities Fund allocated £87 million over three years for 'PE, Sport and Out of School Hours Activities' (www.nof.org.uk). These funds can be accessed for outdoor activities and this may well provide a basis for development in the sector. Nonetheless it is clear that the reliance of an aspect of educational provision on such funds demonstrates a lack of political commitment and does not lead to long-term stability.

33.5 Economic impact of residential outdoor education centres

The outdoor centres built or converted in the 1960s and 1970s were generally in rural or mountainous areas. Consequently, many of these communities have benefited socially, economically and in employment terms. Higgins (2000) presents a case study which relates to one such area in Scotland where in 1996 (this is the last period for which a full year's data was available) a small town (population approximately 9,000) had three Local Authority outdoor centres (referred to as A, B and C) within a 15 mile range. The area was chosen because the Centre Principals were willing to provide budgetary information. Higgins (2000) gleaned data on numbers of employees, turnover, etc. and used standard 'economic multipliers' to estimate the employment generated in the local community as a result of these primary data. The 'turnover' represents the total income to the centres (including any support from Local Authorities) and this is spent on salaries and operational costs. The multipliers came from a survey commissioned by the Scottish Tourist Board (STB) which was carried out by the Surrey Research Group (Scottish Tourist Board & Surrey Research Group, 1992). At the time of the study for each of the three centres, this did not include expenditure on buildings maintenance and development.

The assumption behind the use of multipliers is that, for each person working (direct employment) in a given area, there will be local expenditure which will generate further employment (indirect employment). Those involved in this form of service industry will also spend money locally and will in turn generate jobs (induced employment). There will, of course, be regional and seasonal variations to apply to these 'multipliers'. The results of the study are summarised in Table 33.2.

Table 33.2 Calculation of employment resulting from Centres A, B and C (from Higgins, 2000).

Centre A	Turnover (1995/6) = £605,000	Direct Employment (FTE) = 24.0
Centre B	Turnover (1995/6) = £681,000	Direct Employment (FTE) = 35.5
Centre C	Turnover (1995/6) = £551,000	Direct Employment (FTE) = 19.0

Applying the multipliers provides the following estimates:

Centre A	Indirect Employment = 3.0	Induced Employment = 2.7
Centre B	Indirect Employment = 3.4	Induced Employment = 3.0
Centre C	Indirect Employment = 2.8	Induced Employment = 2.4

The total full time equivalent (FTE) employment resulting from these centres (direct + indirect + induced) is therefore 96 individuals. This represents a small but significant proportion of the employment opportunities in the area. The local Job Centre provided the number of those registered as unemployed in the area, a total of 443. Therefore, if the three centres were all closed and all staff sought re-employment in the area, the unemployment total would increase by around 20%. Almost all of those working in these centres at the time of the study were 'settled' in the area and, whilst closure would probably result in some individuals leaving the area, this would in turn only exacerbate the impact on the local service sector.

One additional consideration is the long-term benefits to a rural area arising as a result of the throughput of young people and adults through the centres. The estimated number visiting this area, probably for the first time because they were on a course at one of the centres, is c. 7,000 to 8,000 per annum (Centre Principals, 1997, pers. comm.). If even a small proportion of these develop an affection for the area and return later as an adult, perhaps on a day trip or holiday, this will bring additional economic benefits to the local economy.

33.6 Changing patterns of outdoor recreation

The few studies of changing patterns of outdoor recreation in Scotland are, at best, partial and tend to focus on the economic impact rather than social trends, though the latter have been partially reviewed by Higgins (2001). Nonetheless, attempts have been made to draw these studies together. These have primarily focused on the economic contribution of the sector, indicating that recreational use of the outdoors represents a significant source of revenue to the Scottish rural economy.

The two most recent reviews have been by Scottish Natural Heritage (1998) and Higgins (2000) and these come to similar conclusions. The SNH publication (based on 1992 data) estimates the contribution of open air recreation to the Scottish economy as about £730 million, supporting 29,000 (FTE) jobs (Scottish Natural Heritage, 1998). A further 1,700 (FTE) jobs are related to environmental education and tourism associated with the natural heritage (Scottish Natural Heritage, 1998).

Higgins (2000) considered studies published over a 15 year period and came to similar conclusions. This study addresses the economic significance of outdoor recreation in comparison with the hunting of deer, grouse, pheasants and fish which takes place on the majority of 'sporting estates' in Scotland. It concludes that, whilst hunting clearly makes a contribution to the rural economy, activities such as hiking, water sports, field nature study and mountaineering are of greater national significance and currently at least £600 million to £800 million (Higgins, 2000; Wightman & Higgins, 2001).

Higgins (2000) was able to include more recent material such as that from the Cairngorms Recreation Study which revealed that, in the period September 1997 to August 1998, visitors enjoyed 123,000 days out in the Cairngorm mountains and spent £2.2 million in the area (Taylor & MacGregor, 1999). Notably, much of the recreational activity took place outwith the traditional tourist season, with summer accounting for approximately 39%, autumn 27%, winter 13% and spring 20% (Taylor & MacGregor, 1999). This is in line with other studies cited by Higgins (2000) and a more recent (2003) empirical study of the economic impact of water-based recreation in the catchment of the River Spey (Riddington et al., 2004). This study estimated water-sports activity days at over 38,000 with a total expenditure by participants of £1.7 million. In comparison, angling activity involved over 54,000 participant days and £11.8 million expenditure; the higher figures reflecting the national significance of the river for salmon angling (Riddington et al., 2004, p. 2).

Outdoor activities and other forms of 'open air recreation' constitute a major proportion of the Scottish tourist industry. Perhaps not surprisingly, there appears to be an increasing number of candidates seeking the National Governing Body awards which qualify them to lead and instruct others (Principal of Glenmore Lodge, Scottish Sports Council National Mountaineering Centre, pers. comm., 1997). This is mirrored by the increasing number of freelance outdoor instructors now operating in the Cairngorms, Skye and Fort William areas (AALA Inspectors, Newtonmore, pers. comm., 2003).

33.7 Concluding comments

This chapter has not sought to address the issue of whether outdoor recreation influences outdoor education or *vice versa*. Rather, it has noted the common requirement for appropriate 'spaces' within which these activities take place, namely the countryside of Scotland. Furthermore, increasing commercialisation of outdoor centres and the growth in freelance staff working in both outdoor centres and in a guiding capacity points to increasing permeability between outdoor education and recreation (Barnes, 1998).

A number of comments can be made about the significance of these activities.

In economic terms, it is important to note that a high proportion of the activity noted above takes place in rural areas which often have few other forms of employment (Scottish Natural Heritage, 1998). For example, the closure of an outdoor centre in a small rural community which does not have a diverse and robust economy may have substantial consequences.

The majority of tourist income is seasonal in nature, however some of the outdoor activities which generate such a substantial proportion of the income of the Highland economy do so either throughout the year (e.g. mountaineering) or in the winter (e.g. skiing). For example, SNH (Scottish Natural Heritage, 1998, p. 13) estimated that '77% of hill walker/mountaineer visitor days in the Highlands and Islands are spent in October to June, compared with just 53% of all visitor days in the area'. The impact of such activities in maintaining year-round tourism jobs should not be underestimated as it reduces the seasonal nature (three months - July to September) of the more traditional forms of tourist employment. Similarly, a substantial proportion of the money spent by countryside users of this type is in 'low intensity' industry such as bed and breakfasts, garages, cafes and small hotels. The income therefore goes directly into the local economy (e.g. Hunt & Dearden, 1996).

Many changes relevant to the broader debate on land management are now taking place in Scotland. For example traditional use of 'hunting estates' is changing to a more diverse form of management (Higgins et al., 2002), and the Land Reform (Scotland) Act 2003 will have an impact on patterns of ownership and hence land management. Furthermore, by enshrining customary traditions of recreational access to countryside in law, this Act has made clear the significance the Scottish Parliament places on such activity. Two National Parks have now been established and growth in visitor numbers can be expected. This requires politicians and planners to give thought to any economic and environmental impact.

Raising awareness of the rights and responsibilities of access to the countryside is implicit in the Land Reform legislation. A recent estimate suggests that 300,000 to 500,000 Scottish school pupil days a year are spent in outdoor education (Higgins, 2002), and it seems clear that access to and management of the natural heritage of Scotland should be a central feature of this provision. To this end SNH has recently devoted considerable effort to raising awareness amongst the public and commissioned a report specifically to explore formal and informal curricular opportunities for the Scottish Outdoor Access Code (Higgins et al., 2004).

It is apparent that social and economic dimensions of rural areas are significant factors in the decision making process of the Scottish Parliament, and this chapter reflects a range of important issues to be considered in the long-term planning for outdoor recreation and outdoor education in Scotland. First, these closely related sectors have significant economic and hence social influence on rural areas. Second, whilst outdoor education provision has been in decline in recent decades, new funding and flexible approaches to provision have maintained a base for growth which, with encouragement from SNH and Government, might focus on relationships with the natural heritage.

References

Activity Centres (Young Persons Safety) Act (1995). HMSO, London.

Barnes, P. (1998). What is going on in the outdoor industry? *Horizons*, **1**, 10-13.

Cheesmond, J. (1979). *Outdoor Education Research Project*. Lothian Region and Dunfermline College of Education, Edinburgh.

Cook, L. (1999). The 1944 Education Act and outdoor education: from policy to practice. *History of Education*, **28**, 157-172.

Cooper, G. (1991). The role of outdoor and field study centres in educating for the environment. *Journal of Adventure Education and Outdoor Leadership*, **8**, 10-11.

Countryside Commission for Scotland (1983). *Outdoor Education Centres in Scotland: A Report on the Situation in 1982 and Developments Since 1970*. Countryside Commission for Scotland, Perth.

Crowther, N. Higgins, P. & Nicol, R. (1998). The contribution of outdoor education towards learning to sustain. In *Learning to Sustain*, ed. by J. Smyth. SEEC, Stirling and Scottish Natural Heritage, Perth. pp. 58–60.

Education Act (1944). HMSO, London.

Education (Scotland) Act (1945). HMSO, London.

Halls, N. (1997a). The development of outdoor education in Strathclyde Region (Part 1). *Scottish Journal of Physical Education*, **25**, 36-38.

Halls, N. (1997b). The development of outdoor education in Strathclyde Region (Part 2). *Scottish Journal of Physical Education*, **25**, 12-28.

Higgins, P. (2000). The contribution of outdoor recreation and outdoor education to the economy of Scotland: case studies and preliminary findings. *Journal of Adventure Education and Outdoor Learning*. **1**, 69-82.

Higgins, P. (2001). Experience without limits? Risk, reality and the role of outdoor education. In *Hemmungsslos erleben? Horizonte und Grenzen*, ed. by F. Paffrath & A. Ferstl. Zeil, Augsburg. pp. 120-134.

Higgins, P. (2002). Outdoor education in Scotland. *Journal of Adventure Education and Outdoor Learning*, **2**, 149-168.

Higgins, P. & Loynes, C. (1997). On the nature of outdoor education. In *A Guide for Outdoor Educators in Scotland*, ed. by P. Higgins, C. Loynes & N. Crowther. Scottish Natural Heritage, Perth. pp. 6-8.

Higgins, P. & Morgan, A. (1999). Training outdoor educators: integrating academic and professional demands. In *Outdoor Education and Experiential Learning in the UK*, ed. by P. Higgins & B. Humberstone. Luneburg University Press, Luneberg. pp. 7-15.

Higgins, P., Wightman, A. & McMillan, D. (2002). Sporting estates and recreational land use in the highlands and islands. ESRC Report R000223 163.

http://www.education.ed.ac.uk/outdoored/research/abstract_higgins.pdf.

Higgins, P., Ross, H., Lynch, J. & Newman, M. (2004). Building the Scottish Outdoor Access Code and responsible behaviour into formal education and other learning contexts. Unpublished report. Scottish Natural Heritage, Perth.

Hopkins, D. & Putnam, R. (1993). *Personal Growth Through Adventure*. David Fulton Publishers, London.

Hunt, J. & Dearden, M. (1996). *Ross and Cromarty Upland Footpath Survey: Annual Patterns of Use and Spending*. Scottish Natural Heritage, Perth.

Land Reform (Scotland) Act (2003). HMSO, London.

Loader, C. (1952). *Cairngorm Adventure at Glenmore Lodge: Scottish Centre for Outdoor Training*. William Brown, Edinburgh.

Lorimer, H. (2003). Telling small stories: spaces of knowledge and the practice of geography. *Transactions of the Institute of British Geographers*, **28**, 197-217.

McNeish, C. & Else, R. (1994). *The Edge: One Hundred Years of Scottish Mountaineering*. BBC Books, London.

Nicol, R. (1999). A survey of Scottish Outdoor Centres. *Horizons*, **6**, 14-16.

Nicol, R. & Higgins, P. (1998a). Perspectives on the philosophy and practice of outdoor education in Scotland. Paper presented at an international conference, Umeå, Sweden, August 1998.

Nicol, R. & Higgins, P. (1998b). A Sense of Place: a context for environmental outdoor education. In *Celebrating Diversity: Learning by Sharing Cultural Differences*, ed. by P. Higgins & B. Humberstone. European Institute for Outdoor Adventure Education and Experiential Learning, Buckinghamshire. pp. 50-55.

Parker, T. & Meldrum, K. (1973). *Outdoor Education*. Dent, London.

Riddington, G., Radford, A., Anderson, J. & Higgins, P. (2004). *An Assessment of the economic impact of water-related recreation and tourism in the catchment of the River Spey*. Spey Catchment Steering Group, Aviemore.

Scottish Natural Heritage (1998). *Jobs and the Natural Heritage: The Natural Heritage in Rural Development*. Scottish Natural Heritage, Perth.

Scottish Tourist Board & Surrey Research Group (1992). *Scottish Tourism Multiplier Study (Volume 1)*. Scottish Tourist Board, Edinburgh.

Simpson, M. (1982). *'Skisters': the Story of Scottish skiing*. Landmark Press, Carrbridge.

Taylor, J. & MacGregor, C. (1999). *Cairngorms Mountain Recreation Survey 1997-98*. Scottish Natural Heritage Research, Survey & Monitoring Report No. 162.

Wightman, A. & Higgins, P. (2001). Sporting estates and outdoor recreation in the Highlands and Islands of Scotland. In *Enjoyment and Understanding of the Natural Heritage*, ed. by M.B. Usher. The Stationery Office, Edinburgh. pp. 171-176.

34 SCANNET: a Scandinavian-North European network of terrestrial field bases

O.W. Heal, N. Bayfield, T.V. Callaghan, T.T. Høye,
A. Järvinen, M. Johansson, J. Kohler, B. Magnusson,
L. Mortensen, S. Neuvonen, M. Rasch & N.R. Saelthun

Summary

1. The development of SCANNET, supported by the European Commission, has emphasised the value of establishing links among Northern field sites in considering long-term ecological and environmental dynamics. It highlights the benefits of knowledge exchange, combining different types of information and approaches, technology transfer, enhanced data access, and increased communication with local and regional stakeholders.

2. SCANNET now provides an integrated facility enabling students, researchers, managers and policy-makers to access information and expertise at field sites throughout this highly sensitive region, i.e. it is a 'one-stop-shop'.

3. The distribution of SCANNET sites, rich in mountains, covers the broad range of predicted climate change in the region, and climate observations are well replicated across the network. However, each site tends to have particular subjects for intensive observations. This provides a diversity of subject coverage but little replication of long-term observations and experiments, a feature that provides the potential to extend future observations across the network.

4. Extension of observations has begun through application of automatic photography to determine timing and distribution of snow cover. A Memorandum of Understanding has been signed by SCANNET partners to formalize their commitment to continue co-operation

5. Formal and informal links have been established with related site networks and programmes in the north (e.g. ENVINET, DART, CALM), in the mountains (e.g. ALPNET, GLORIA), in the UK and Europe (e.g. ECN, NoLIMITS) and globally (e.g. ILTER, GTOS). The forthcoming Arctic Climate Impact Assessment (ACIA) provides detailed hypotheses of climate-induced future changes, which will be tested and refined through regional and circum-Arctic observation networks such as SCANNET and CEON.

Heal, O.W., Bayfield, N., Callaghan, T.V., Høye, T.T., Järvinen, A., Johansson, M., Kohler, J., Magnusson, B., Mortensen, L., Neuvonen, S., Rasch, M. & Saelthun, N.R. (2005). SCANNET: a Scandinavian-North European network of terrestrial field bases. In *Mountains of Northern Europe: Conservation, Management, People and Nature*, ed. by D.B.A. Thompson, M.F. Price & C.A. Galbraith. TSO Scotland, Edinburgh. pp. 359-364.

34.1 Why do we need a northern network of sites?

Northern mountains and tundra include some of Europe's last wilderness areas (Nagy *et al.*, 2003). They show great topographic and climatic diversity, strong environmental and ecological gradients, and have a specialised and diverse flora and fauna. The region also supports widely distributed human populations and localised industrial developments. However, it is experiencing rapid environmental and social changes and is particularly vulnerable to predicted climate changes which have important implications throughout the northern hemisphere (Weller, 2000).

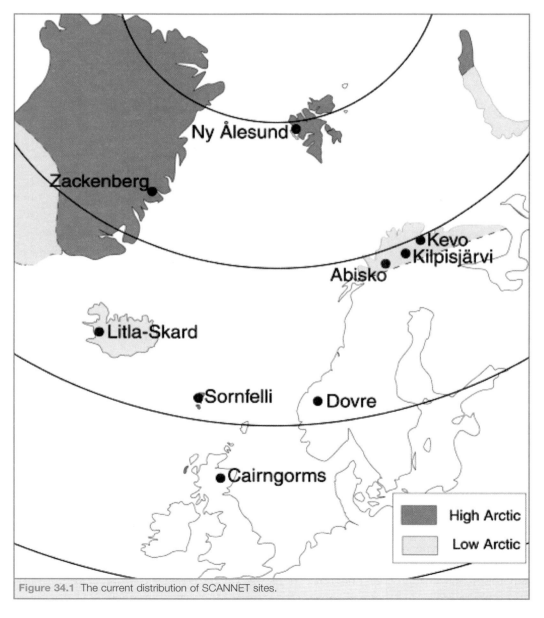

Figure 34.1 The current distribution of SCANNET sites.

Recent international discussions on climate change impacts in Arctic-alpine regions concluded that field stations are critical to the scientific infrastructure (Turunen *et al.*, 1999). These retain data on research publications, results, experiments and field plots generated by 'transient' researchers. This archival function, combined with local and historical environmental data, is increasingly important for understanding global change.

Unfortunately, interaction between field stations is very limited, thus comparative observations and experiments across the region are minimal. In response, the European Commission supports the SCANNET network of Field Sites and Research Stations (Figure 34.1), aiming to:

- provide data and information on environmental change to a wide range of users; and
- improve geographical and environmental coverage and comparability of long-term observations and experiments.

34.2 What progress have we made in SCANNET?

SCANNET activities focus on five topics.

34.2.1 Climate variability

Comprehensive and often comparable climate data are available from all sites, with systematic observations extending back to 1913 at Abisko, Sweden. Unfortunately, snow dynamics, soil microclimate and hydrology – critical and sensitive drivers of ecological change – are poorly represented in conventional observations. Importantly, tests at Cairngorms (Scotland), Zackenberg (Greenland) and Sornfelli (Faroes) (Figure 34.1) demonstrated that robust automatic digital cameras can monitor snow cover dynamics under extreme climatic conditions. This addition to climate observations will provide critical information for conservation, land use and hydrology in a changing climate.

34.2.2 Regional climate change scenarios

Early activities have been to access global and regional models and relate probable future climates to SCANNET sites. Despite strong local influences of topography, snow and ice and large variations in model projections, clear spatial patterns in projected climates emerge. Annual mean air temperatures are expected to rise by more than 3°C in high Arctic Svalbard, 2-3°C in northern Scandinavian and east Greenland sites and 1-2°C at sites in southern Scandinavia, Faroes, Iceland and Scotland. These increases are expected by 2050. Temperatures will increase mainly in winter, with less frequent extreme lows, and more frequent extreme highs. Winter wind speed will increase moderately. Precipitation is projected to increase by about 10%, mainly at high latitudes in autumn and winter (Saelthun & Barkved, 2003).

34.2.3 Land use and society interaction

Intensity and type of human impact vary greatly within the region. SCANNET is identifying key human drivers of change at the sites, defining indicators of social and environmental change, and undertaking risk assessments (Bayfield *et al.*, 2000). Initial information from stakeholders shows that climate dominates considerations of change across sites. However, each site has distinctive local concerns reflecting specific geographic and resource-based conditions, e.g. degree of maritime influence and associated fishing, compared with land use. So, whilst some observations are regionally applicable, local interests critically influence selection and interpretation of site observations.

34.2.4 Variation in trends of biodiversity among the sites

Low species diversity in northern sites reflects climatic severity, spatial isolation and recent deglaciation. Similarly, genetic diversity and active speciation are frequent phenomena. Documentation and assessment of the diverse baseline information on habitat richness, species diversity and intraspecific variability will precede improvement in standardisation and comparability of observations. Check lists, with taxonomic standardisation, for major taxa (birds, vascular plants, lichens) are well developed and selected invertebrate groups are now being addressed (Lepidoptera, Carabidae). SCANNET documentation will build on results from ALPNET (Nagy *et al.*, 2003).

34.2.5 Species performance and phenology

Species phenology is particularly sensitive to climatic perturbations, e.g. synchronous fluctuations of mammal populations and bird migration (Forchhammer *et al.*, 2002). Climate change has stimulated many studies on the potential of large-scale climatic phenomena to explain patterns in population dynamics. Importantly, a purely correlative approach is inefficient because individuals respond to short-term changes in resources, predation, or microclimate rather than large-scale climatic fluctuations (Post *et al.*, 2001). Also, correlation suggests that species inhabit 'temperature envelopes', moving to higher altitudes and latitudes as temperature rises. This neglects the importance of species interactions, behaviour and dispersal (Hodkinson, 1999) – species cannot move higher when they are on top of the mountain! Hence, population dynamics can only be understood through integrated analysis of climate and ecology data from several trophic levels at major field sites.

Species response to past climates helps our understanding of future events (Forchhammer *et al.*, 2002). This process involves monitoring species throughout northern mountains and tundra, generating large-scale data critical for modelling and prediction. We can also learn much from the past. Nineteenth and 20th century naturalists were devoted to documenting phenological events, publishing in local journals but not reaching the international scientific community. At least one rediscovered work has been used in a modern context (Lauscher & Lauscher, 1990). Tracking such data is difficult but provides valuable additions to modern monitoring, and will be available on our web-based database.

34.3 The future

The development of SCANNET, supported by the European Commission, has emphasised the value of stronger links between northern field sites in considering long-term environmental, social and ecological change. The SCANNET partners have acknowledged the value by signing a Memorandum of Understanding to continue co-operation in technology transfer, enhanced data access, and communication with local, regional and international stakeholders. SCANNET now represents an integrated facility which is linked to other short- and long-term programmes and networks in the European North (e.g. ENVINET, DART), and the Arctic (e.g. CALM), in the mountains (e.g. ALPNET, GLORIA), in individual countries (e.g. UK ECN), in Europe (e.g. NoLIMITS), and globally (e.g. ILTER, GTOS).

The details of data and information from across the SCANNET network and a summary of observed changes is being summarised in Callaghan *et al.* (2004). The results have also contributed regional and terrestrial information (Callaghan, 2004) to the major Arctic Climate Impact Assessment (ACIA, 2004), which predicts the climate-induced changes that are expected over the 21st century. Thus, ACIA now provides comprehensive and detailed hypotheses which must be tested against long-term observations and experiments in regional and circum-Arctic networks such as SCANNET and CEON respectively.

34.4 Further information

Further information on the programmes named in this chapter can be obtained from the following web sites.

- ALPNET: a network on Alpine Biodiversity which ran for a period of 3 years from July 1997 – http://www.esf.org/esf_article.php?activity=2&article=195&domain=3.
- CALM: Circumpolar Active Layer Monitoring – http://www.geography.uc.edu/~kenhinke/CALM/index.html.
- CEON: Circumarctic Environmental Observatories Network – http://www.ceoninfo.org/.
- DART: Dynamic response of the forest-tundra ecotone to environmental change – http://www.dur.ac.uk/DART/.
- ECN: Environmental Change Network – http://www.ecn.ac.uk/.
- ENVINET: a European Network for Arctic-Alpine Multidisciplinary Research – http://envinet.npolar.no/Default.htm.
- GLORIA: The Global Observation Research Initiative in Alpine Environments – http://www.gloria.ac.at/res/gloria_home/.
- GTOS: Global Terrestrial Observing System – http://www.fao.org/GTOS/.
- ILTER: International Long Term Ecological Research – http://www.ilternet.edu/.
- NoLIMITS: Networking of Long-term Integrated Monitoring In Terrestrial Systems – http://nolimits.nmw.ac.uk/.
- SCANNET: Scandinavian/North European Network of Terrestial Field Bases – http://www.envicat.com/scannet/Scannet.

References

ACIA (2004). *Impacts of a Warming Arctic: Arctic Climate Impact Assessment.* Cambridge University Press, Cambridge.

Bayfield, N.G., McGowan, G.M. & Fillat, F. (2000). Using specialists or stakeholders to select indicators of environmental change for mountain areas in Scotland and Spain. *Oecologia Montana*, **9**, 29-35.

Callaghan, T.V. (Ed) (2004). Climate Change and UV-B Impacts on Arctic Tundra and Polar Desert Ecosystems. *Ambio Special Issue*, **33**, 385-479.

Callaghan, T.V., Johansson, M., Heal, O.W., Sælthun, N.R., Barkved, L.J., Bayfield, N., Brandt, O., Brooker, R., Christiansen, H.H., Forchhammer, M., Høye, T.T., Humlum, O., Järvinen, A., Jonasson, C.,

Kohler, J., Magnusson, B., Meltofte, H., Mortensen, L., Neuvonen, S., Pearce, I., Rasch, M., Turner, L., Hasholt, B., Huhta, E., Leskinen, E., Nielsen, N. & Siikamäki, P. (2004). Environmental changes in the North Atlantic Region: SCANNET as a collaborative approach for documenting, understanding and predicting changes. *Ambio Special Report 13*, November 2004, 39-50.

Forchhammer, M.C., Post, E. & Stenseth, N.C. (2002). North Atlantic Oscillation timing of long- and short-distance migration. *Journal of Animal Ecology*, **71**, 1002-1014.

Hodkinson, I.D. (1999). Species response to global environmental change or why ecophysiological models are so important: a reply to Davis *et al. Journal of Animal Ecology*, **68**, 1259-1262.

Lauscher, A. & Lauscher, F. (1990). *Phanologie Norwegens, Teil IV.* Eigenverlag, Vienna.

Nagy, L., Grabherr, G., Korner, C.H. & Thompson, D.B.A. (Eds). (2003). *Alpine Biodiversity in Europe.* Springer Verlag, Berlin.

Post, E., Forchhammer, M.C., Stenseth, N.C. & Callaghan, T.V. (2001). The timing of life-history events in a changing climate. *Proceedings of the Royal Society of London B*, **268**, 15-23.

Saelthun, N.R. & Barkved, L. (2003). Climate Change Scenarios for the SCANNET Region. Report SNO 4663-2003. Norwegian Institute for Water Research, Oslo.

Turunen, M., Hukkinen, J., Heal, O.W., Saelthun, N.R. & Holten, J.I. (Eds). (1999). A terrestrial transect for Scandinavia/Northern Europe: Proceedings of the International Scantran Conference. Ecosystems Research Report 31. European Commission, Brussels.

Weller, G. (2000). The weather and climate of the Arctic. In *The Arctic. Environment, People, Policy*, ed. by M. Nuttall & T.V. Callaghan. Harwood Academic Publishers, Amsterdam. pp. 143-160.

Summary

1. This chapter describes the application of Geographical Information Systems (GIS) and public participation GIS (PPGIS) to the identification of wild land and targeting areas for wilderness restoration in mountain regions.
2. Two assumptions are made: (a) the best areas for restoration are those that are adjacent to and link together areas that already possess the best qualities of remoteness and naturalness; and (b) widespread support through public participation is essential to the successful implementation of policy objectives.
3. Existing wild land areas and possible restoration target areas are mapped.

35.1 Wild Land Policy

Scottish Natural Heritage (SNH) and The National Trust for Scotland (NTS) have published wild land policy statements (SNH, 2002; NTS, 2002). Both stress the importance of Scotland's wild land resource, the need to address the pressures leading to loss of wildness (e.g. hill track construction and hydro/wind power schemes) and support for National Planning Policy Guideline 14 on the Natural Heritage (Scottish Office, 1998). Similarly, both dwell in detail on what wild land means to the Scottish countryside, and draw up lists of wild land indicators.

"Wild land in Scotland is relatively remote and inaccessible, not noticeably affected by contemporary human activity, and offers high-quality opportunities to escape from the pressures of everyday living and find physical and spiritual refreshment." (NTS, January 2002, p.4).

Key indicators identified by SNH and NTS include: remoteness, inaccessibility, solitude, perceived naturalness, absence of modern human artefacts and land use, rugged terrain, scenic grandeur, physical challenge and risk. Their policy objectives are to safeguard wild land, enhance nature, promote responsible recreational use, restore past damage and promote awareness. Both stress the need to identify core areas of wild land as a precondition of management. Both are careful to avoid reference to the wilder areas of Scotland as 'wilderness' as, ecologically and historically speaking, there is no true wilderness left in the country, and the term can in some contexts be politically sensitive. Nonetheless,

Carver, S.J. (2005). Mountains and wilderness: identifying areas for restoration. In *Mountains of Northern Europe: Conservation, Management, People and Nature*, ed. by D.B.A. Thompson, M.F. Price & C.A. Galbraith. TSO Scotland, Edinburgh. pp. 365-370.

it is the qualities of wilderness or wildness (i.e. large and remote areas, devoid of human activities and influence and possessing pristine/unaltered ecosystems) that provide a benchmark against which the wild qualities of Scottish landscapes can be judged.

35.2 Identifying core wild land

The utility of GIS in mapping wilderness quality has been demonstrated elsewhere (see Lesslie, 1994; Carver *et al.*, 2002; Sanderson *et al.*, 2002). Briefly, it is based on the assumption that the more remote and unaltered a landscape is, the greater its wilderness quality. The indicators described by SNH and NTS for wild land can be mapped using surrogates such as distance from the nearest road and combined to calculate a map of overall wildness. This chapter demonstrates how these methods can be applied to identify target areas for restoration. It is suggested here that core wild lands can be mapped according to human perceptions using readily available datasets and the wilderness continuum concept identified by Nash (1981). Figure 35.1a shows results for Scotland using mapped indices describing remoteness from settlement and mechanised access, visibility of human artefacts in the landscape and a classification of land cover into naturalness classes (Tables 35.1 and 35.2).

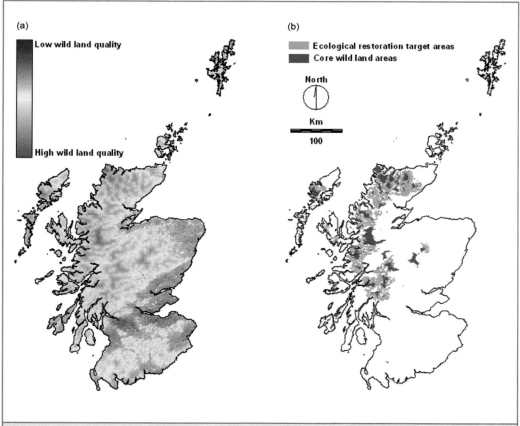

Figure 35.1 Example wild land maps. a) Wild land continuum using equally weighted factors. b) Core wild land and restoration target areas.

Table 35.1 Derivation of wild land mapping indices.

Index	Measure	Data source(s)
Remoteness from settlement	Distance from human settlement weighted by size class: city, town, village	Countryside Information System: Ordnance Survey topographic data, 1km^2 resolution
Remoteness from mechanised access	Distance from mechanised access weighted by size class: motorway, A road, B road, minor road	Countryside Information System: Ordnance Survey topographic data, 1km^2 resolution
Visibility of human artefacts	Number of visible artefacts weighted by distance from viewpoint	Countryside Information System: Ordnance Survey topographic data, 1km^2 resolution
Naturalness of land cover	Reclassification of land cover data into naturalness classes (see Table 35.2)	Countryside Information System: CEH (ITE) Land Classification System System, 1km^2 resolution

Table 35.2 Classification of CEH Land Classification (Fuller et al., 1994) into naturalness classes and associated map scores.

Land Classification	Naturalness Class	Score
Bog	Natural to semi-natural	4.5
Bracken	Natural to semi-natural	4.5
Coastal Bare	Natural	5.0
Coniferous Woodland	Semi-natural to non-intensively managed	3.5
Deciduous Woodland	Natural to intensively managed	3.0
Grass Shrub Heath	Semi-natural to non-intensively managed	3.5
Grass Heath	Natural to semi-natural	4.5
Inland Bare	Mainly natural with some elements of semi-natural, non-intensively managed and non-natural	4.5
Inland Water	Natural (Non-natural when impounded for Hydro-electric Power and/or water supply)	5.0 (0.0)
Managed Grassland	Managed to intensively managed	1.5
Marsh/Rough Grass	Mainly non-intensively managed to managed with elements of semi-natural and intensively managed	2.5
Saltmarsh	Natural	5.0
Sea/Estuary	Natural	5.0
Shrub Heath	Semi-natural	4.0
Suburban	Non-natural with some elements of intensively managed	0.5
Tilled Land	Intensively managed	1.0
Unclassified	Unclassified	0.0
Urban	Non-natural	0.0

35.3 Remoteness modelling in mountain areas

Remoteness in terms of the time taken to walk from the nearest point of mechanised access is an important wild land indicator in mountain areas. Straight-line distances are easy to calculate but give a poor indication of physical remoteness because of the difficulty of traversing the intervening terrain. Calculations of off-road travel times based on Naismith's Walker's Rule, a shortest path algorithm and information on local terrain, barrier features and ground cover, can be used as an index of remoteness (Carver & Fritz, 1999). The effect on access times if footbridges and hill tracks are removed in wild land restoration work can be explored using this approach, while results can be used for mapping management zones. Perceived remoteness is more difficult to assess, but is essentially controlled by physical remoteness and the visual intrusion (or lack of) from human artefacts (roads, dams, buildings, plantation forest, etc.). Levels of visual intrusion can be measured using GIS. Varied relief in mountain areas often screens human features from view, thereby enhancing perceived remoteness.

35.4 Identifying target areas for restoration

Wilderness restoration, or 're-wilding', is the planned enhancement of the natural or wild qualities of selected areas of countryside with the aim of restoring habitats and even whole landscapes to near-natural conditions (Council for National Parks, 1998). Determining where core wild land areas are, under current conditions, is central to identifying target areas for restoration. A network of core wild land areas linked by restoration target areas is suggested as a starting point for re-wilding mountain areas. Figure 35.1b shows an example of a possible network for Scotland based on identifying land with semi- to near-natural vegetation classes and low visual impact from human features that lie within 10 km of the core wild land areas (top 5%) from Figure 35.1a. Such areas will already possess these wild land attributes, but would benefit from targeted removal of human artefacts such as hill tracks, footbridges, shelters, plantation forestry, etc. to improve their remoteness and landscape value within a managed programme of re-wilding, the efficacy of which can be monitored through further mapping and analysis.

35.5 A role for the public?

The definition and mapping of wild land is heavily reliant on people's perceptions of wild land values. It makes sense therefore to include the public within the planning process based on the assumption that, without public support, policy on these issues may flounder. Methods of surveying the opinions of large numbers of people using Internet-based GIS have recently been developed (Carver et al., 2002) and can be seen online at www.ccg.leeds.ac.uk/teaching/wilderness/. Briefly, this includes an online version of the datasets and analysis used to produce Figure 35.1a that allows users to choose which factors they feel are relevant, weight them according to their perceived importance, produce a map of the wilderness continuum and top-slice the map to identify the wildest areas. These methods could be used to incorporate an element of public participation in the implementation of policy on wild land and re-wilding. This is lent extra weight by the fact that mountain communities are often dispersed and isolated, while other stakeholders are likewise dispersed and varied, making the Internet an ideal medium for this kind of consultation. It is suggested here that such an approach, based more on the principles of PPGIS, may better address the difficulties of taking present ownership, land use, cultural views and multi-agency/multi-user interests into account, and so be a useful way forward in implementing policy on wild land and re-wilding in mountain areas.

References

Carver, S. & Fritz, S. (1999). Mapping remote areas using GIS. In *Landscape Character: Perspectives on Management and Change*, ed. by M.B. Usher. The Stationery Office, Edinburgh. pp. 112-126.

Carver, S., Evans, A. & Fritz, S. (2002). Wilderness attribute mapping in the United Kingdom. *International Journal of Wilderness*, **8**, 24-29.

Council for National Parks (1998). *Wild by Design: a Guide to the Issues.* Council for National Parks, London.

Fuller, R.M., Groom, G.B. & Jones, A.R. (1994). The Land Cover Map of Great Britain: an automated classification of Landsat Thematic Mapper data. *Photogrammetric Engineering & Remote Sensing*, **60**, 553-562.

Lesslie, R. (1994). The Australian National Wilderness Inventory: wildland survey and assessment in Australia. In *Wilderness - The Spirit Lives: 6th National Wilderness Conference*, ed. by C. Sydoriak. Bureau of Land Management, Fish & Wildlife Service, Forest Service, National Biological Service, National Park Service, and Society of American Foresters. Santa Fe, NM. pp. 94-97.

Nash, R. (1981). *Wilderness and the American Mind.* Third Edition. Yale University Press, New Haven.

National Trust for Scotland (2002). *Wild Land Policy. January 2002.* National Trust for Scotland, Edinburgh.

Sanderson, E.W., Jaiteh, M., Levy, M.A., Redford, K.H., Wannebo, A.V. & Woolmer, G. (2002). The Human Footprint and the Last of the Wild. *Bioscience*, **52**, 891-904.

Scottish Natural Heritage (2002). *Wildness in Scotland's Countryside: a Policy Statement. November 2002.* Scottish Natural Heritage, Perth.

Scottish Office (1998). *National Planning Policy Guidelines. NPPG14: Natural Heritage.* Scottish Office, Edinburgh.

36 Priorities for the conservation and management of the natural heritage in Europe's high mountains

Georg Grabherr

Summary

1. High mountain regions have been and will continue to be affected by global change, in particular by land use and climate change.
2. The traditional rural mountain world has virtually disappeared during the past 50 years, leaving behind a few, important remnants.
3. Visions for the future need to be discussed, and then measures implemented to allow for a sustainable form of development which includes a set-aside strategy in leaving areas to the natural dynamic.
4. A better balance between preservation and development needs to be found, globally, in support of mountain environments.

36.1 Introduction

Mountains occur in all life zones, from the Equator to the Poles (WCMC, 2002). They are hotspots of biodiversity, both in absolute terms and relative to the surrounding lowlands (Barthlott *et al.*, 1996; Nagy *et al.*, 2003; Thompson *et al.*, this volume). Out of a total of 11,000 vascular plant species in Europe, more than 2,000 are estimated to occur in the alpine regions of Europe (Väre *et al.*, 2003). Mountains have evolved, providing in their harsh and stressful environments open areas for radiative speciation. Mountains, especially during the Ice Ages, acted as barriers for migrating floras and faunas, and by separating populations, which then developed into new taxa or ecotypes by way of genetic drift (see Agakhanjanz & Breckle (1995) for Middle Asia; Hedberg (1992) for tropical mountains). Conversely, during glacial retreats, reinvading populations of closely related species have intermixed, a process which induced speciation by means of hybridisation and polyploidisation. These speciation processes were set in highly diverses landscape with their characteristic habitats (e.g. heathlands, screes, rock faces, wetlands, glacial moraines, snow beds, different bedrock materials), resulting in a highly diverse alpine and nival flora (Grabherr, 1995).

Mountains are major contributors to overall global biodiversity (Wielgolaski, 1997; Körner & Spehn, 2002). As biodiversity hot spots, they merit special attention and conservation. The general objectives of maintaining native biodiversity, as laid out by Noss (cited in Hamilton, 2002), are to:

Grabherr G. (2005). Priorities for the conservation and management of the natural heritage in Europe's high mountains. In *Mountains of Northern Europe: Conservation, Management, People and Nature,* ed. by D.B.A. Thompson, M.F. Price & C.A. Galbraith. TSO Scotland, Edinburgh. pp. 371-376.

1. represent all ecosystem types and successional stages across their natural range;
2. secure viable populations of all native species in their natural patterns of abundance and distribution;
3. maintain ecological and evolutionary processes, such as disturbance, nutrient and water cycling, and predation; and
4. ensure that the biological diversity of each region can respond naturally to short-term and long-term changes.

In mountain environments, the best way to achieve these objectives is, simply, to keep them as natural as possible.

36.2 Mountain wilderness – towards a strategy

High mountain environments have remained in a relatively natural state, either as part of extensive wilderness regions, such as those in the Scandes, the Urals and the polar regions, or as islands surrounded by a sea of cultivated land (most European mountains: Alps, Pyrenees, Carpathians). The degree of naturalness increases rapidly with elevation. In Austria, for example, where cultivation since prehistoric times has largely transformed the land, up to half the mountain forests are natural or at least semi-natural today (Grabherr *et al.*, 1998). Above the treeline, human impact diminishes. The high alpine and nival zones are almost natural, and have remained as wilderness areas (in the Alps about 10-20% of the total area; Kaissl in WCMC, 2002); maintaining, or restoring, the wilderness character of large areas is undoubtedly the main conservation priority in mountains.

The real challenge for the near future is, in my view, not – as frequently stated – how to sustain agriculture in the mountains of Europe, and thereby conserve rural landscapes in the montane and alpine zones, but how to manage the remaining wilderness and recently abandoned agricultural land. A wilderness strategy for mountains is required to harmonise economic use and conservation objectives. The ongoing exploitation of high mountain environments for tourism, mining, hydroelectricity, wind and solar power generation, and industrial-scale farming should be controlled under the principles of sustainable land management. Extensive set-aside areas are one example of creating areas which are free of human disturbance and can provide for biodiversity and important ecosystem-derived services and goods, such as clean water for urban areas.

36.3 Land use and cultural change

Today's people perceive high mountain regions differently, each according to their interest. These include endless wilderness areas; Heidiland 'with Barry dogs and Swiss cheese'; the home of monsters; a skiing paradise; a challenging area for mountain biking, paragliding, or hunting. Often forgotten is the fact that people have survived for hundreds, in some parts even thousands, of years in the harsh environment of high mountain regions (e.g. in the Alps, Caucasus, Pyrenees, and the Scottish and Scandinavian mountains). The remains of these old cultures are disappearing rapidly. These cultures have developed a variety of responses to the challenges of high mountain living conditions (see Messerli & Ives, 1997; Körner & Spehn, 2002). An important legacy of these cultures is their characteristic landscapes and their natural heritage, modified and enriched through traditional land use (e.g. Thompson *et al.*, 1995; Erschbamer *et al.*, 2003).

Transhumance systems were probably the most ancient forms of exploiting the treeless alpine environments (see also M. Edwards, this volume; Thomson, this volume). The alpine-type transhumance system has made use of the availability of high altitude pastures in the summer, thereby freeing up the land around settlements for grain and winter fodder production. Some of these farming systems were of local importance (e.g. Alps, Scandes, Scottish Highlands). Others, such as the large numbers of transhumant grazing stock on the Iberian Peninsula in the Middles Ages, used a range of resources (including the alpine pastures of the Pyrenees) and prevailed as economic activities over centuries. The systems that developed in the Alps during medieval times were either of a subsistence type or of a feudal type. Fields were established up to the treeline, and the grasslands of the uppermost reaches were managed for cattle grazing. The pastures were cleared of boulders (and were even weeded sometimes). Summer shelters helped make full use of summer grazing land, and allowed for the manufacturing and preserving of dairy produce, such as cheese. Winter storage of forage for livestock was essential. This took various forms, such as coppicing trees for fodder for goats, and making hay on hay meadows, which extended up to the steep summit slopes. In so doing, the mountain farmer created a high variety of meadow types, pastures, and ruderal plant communities which resulted in an increased level of biological diversity.

It is this enrichment in community diversity and in the mosaic pattern of the landscape which has to be considered when looking at mountain landscapes. The impact of pastoralism on plant species diversity has been less important, although dispersal of ruderals in particular has been enhanced by transhumance and along trade routes. As all activities were farming-orientated, their impacts on the fauna were more controversial. The hunting of large carnivores to extinction in many countries has led to changes in native ungulate species distributions. Changing land use, conservation measures and the recent reintroduction of carnivores e.g. wolf (*Canis lupus*) from the Apennines to the S.W. Alps, and the brown bear (*Ursus arctos*) from Slovenia to the Austrian Alps, and some other species such as ibex (*Capra ibex*) and bearded vulture (*Gypaetus barbatus*) in the Alps is changing this pattern yet again.

This entire cultural heritage is on the verge of disappearance. Tourism and urban development (with commuter systems) offer attractive alternatives to the laborious life of a mountain farmer. The remaining farmers are engaging in large-scale agricultural production technologies, which require the use of chemicals (fertilizers, pesticides, food supplements for cattle) and machinery to fulfil economic criteria alone for optimal production. The dynamics of the mountain system, which have been rather stable for centuries, are changing, or have already changed. The challenge for today's generation of mountain dwellers is to find solutions to achieve a new harmony between their economic needs and capabilities on the one hand and nature on the other.

The conservation of traditional rural landscapes in the mountains requires diverse and locally relevant strategies; local people must become positively involved (see Ferguson & Forster, this volume). In addition, society at large needs to be educated to respect mountain life and recognise the values of traditional cultures and their roots. The neglect surrounding the conservation of the cultural landscapes often arises from a failure to recognise their value, as opposed to the lack of resources (see Thompson *et al.*, 1995); in some cases it is the result of depopulation (e.g. Andre, 1998). A set-aside-strategy, whereby some agricultural areas are abandoned to natural succession, might certainly be one choice but should only be applied where active land use in a modern, adapted form cannot be maintained in the long term.

36.4 Climate change – a major, new driver of change

Among the drivers of global change, such as land use, climate, nitrgen-deposition, biotic exchange, atmospheric carbon dioxide (e.g. Sala & Chapin, 2004; Fowler & Battarbee, this volume), climate change might become the most important in alpine environments. The exact mechanisms and impacts of abrupt climate warming are only now being predicted, and careful evaluation of regional and local characteristics needs to be made for the individual European high mountain systems (Alps, Carpathians, Pyrenees, Apennines, the mountains of the Balkan peninsula, Caucasus, Urals, the Scottish Highlands and the Scandes; see Fowler & Battarbee, this volume).

Pollen records have shown that the climate of N.W. Europe has been warming and the area of arctic-alpine vegetation has been shrinking for the last 15,000 years (see Nagy *et al.*, 2003). Shorter-term changes (over a time span of up to 150 years) in the vegetation have also been observed (e.g. Nagy *et al.*, 2003). The most striking change concerns plant species richness on 30 high mountain summits (>3,000 m), for which old precise records of the summit flora were available (oldest record from 1835). In the Alps, comparisons have shown higher species richness in 1992-1993 than in earlier times. An upward establishment of alpine species has been clearly documented (Grabherr *et al.*, 1994). This increase in species richness was not related to factors such as airborne nitrogen deposition or direct human impact (trampling or dispersal). Instead, detailed modelling has indicated that the increase in altitudinal range of some species is not likely to involve simple elevational displacement of today's vegetation belts (Gottfried *et al.*, 2002). Rather, the models have shown that plant species respond, individually, to migration pathways and local refugia. As a result, new types of species assemblages might appear. Some of today's plant assemblages might disappear if competitors become established in high numbers. Importantly, endemic alpine floras may become extinct under the predicted scenarios for climate change, such as has been predicted for the Sierra Nevada, Spain (Pauli *et al.*, 2003).

Climate change will affect the mountain environment with its biodiversity, and will modify its capacity to supply goods and services. Of the changes, those in the cryosphere are the most important (Haeberli & Beniston, 1998). Overall, the glaciers of the Alps are receding (they have lost 50% of their mass and 30% of their total area) and some smaller ones have disappeared (WCMC, 2002). The permafrost line is also moving upwards. As a consequence, much unstable ice-free moraine and soil material is being released, which increases the danger of floods and rock falls. In the longer term, the water supply from high mountains may decline as a result of the reduction in discharge from glaciers. This may have dramatic consequences for arid regions, in particular.

Acknowledgements

I am grateful to Laszlo Nagy, Des Thompson, Jo Newman and an anonymous reviewer for improving the manuscript.

References

Agakhanjanz, O. & Breckle, S.-W. (1995). Origin and evolution of the mountain flora in Middle Asia and neighbouring mountain regions. In *Arctic and Alpine Biodiversity: Patterns, Causes and Ecosystem Consequences (Ecological Studies Vol. 113)*, ed. by F.S. Chapin & C. Körner. Springer, Berlin. pp. 63-80.

Andre, M.F. (1998). Depopulation, land-use change and transformation in the French Massif Central. *Ambio*, **27**, 351-353.

Barthlott, W., Lauer, W., Placke, A. (1996). Global distribution of species diversity in vascular plants: towards a world map of phytodiversity *erkunde*, **50**, 317-327.

Erschbamer, B., Virtanen, R. & Nagy, L. (2003). The impact of vertebrate grazers on vegetation in European high mountains. In *Alpine Biodiversity in Europe*, ed. by L. Nagy, G. Grabherr, C. Körner & D.B.A. Thompson. Springer, Berlin. pp. 377-398.

Gottfried, M., Pauli, H., Reiter, K. & Grabherr, G. (2002). Potential effects of climate change on alpine and nival plants in the Alps. In *Mountain Biodiversity. A Global Assessment*, ed. by C. Körner & Spehn, E. Parthenon Publishing, Boca Raton. pp. 213-224.

Grabherr, G. (1995). Alpine vegetation in a global perspective. In *Vegetation Science in Forestry: Global Perspective based on Forest Ecosystems of East and Southeast Asia. Handbook of Vegetation Science 12/1*, ed. by E.O. Box, R.K. Peet, T. Masuzawa, I. Yamada, K. Fujiwara & P.F. Maycock, Kluwer Academic Publishers, Dordrecht. pp. 441-451.

Grabherr, G., Gottfried, M. & Pauli, H. (1994) Climate affect on mountain plants. *Nature*, **369**, 448.

Grabherr, G., Koch, G., Kirchmeir, H. & Reiter, K. (1998). *Hemerobie österreichischer Waldökosysteme*. Universitätsverlag Wagner, Innsbruck.

Haeberli, W. & Beniston, M. (1998). Climate change and its impacts on glacier and permafrost in the Alps. Ambio, 27, 258-265.

Hamilton, L.S. (2002). Conserving mountain biodiversity in protected areas. In *Mountain Biodiversity. A Global Assessment*, ed. by C. Körner & Spehn, E. Parthenon Publishing, Boca Raton. pp. 295-306.

Hedberg, O. (1992). Afroalpine vegetation compared to paramo: convergent adaptations and divergent influence. In *An Andean Ecosystem Under Human Influence*, ed. by H. Balslev & J.L. Luteyn. Academic Press, London. pp. 15-30.

Körner, C. & Spehn, E, (Eds) (2002). *Mountain Biodiversity. A Global Assessment*. Parthenon Publishing, Boca Raton.

Messerli, B. & Ives, J.D. (Eds) (1997). *Mountains of the World. A Global Priority*. Parthenon Publishing, Carnforth.

Nagy, L., Grabherr, G., Körner, C. & Thompson, D.B.A. (Eds) (2003). *Alpine Biodiversity in Europe*. Springer, Berlin.

Pauli, H., Gottfried, M., Dirnböck, T., Dullinger, S. & Grabherr, G. (2003). Assessing the long term dynamics of endemic plants at summit habitats. In *Alpine Biodiversity in Europe*, ed. by L. Nagy, G. Grabherr, C. Körner & D.B.A. Thompson. Springer, Berlin. pp. 195-208.

Sala, O.E. & Chapin, T. (2004). Scenarios of global biodiversity. *Global Change Newsletter*, **43**, 7-12, 19.

Thompson, D.B.A., Hester, A.J. & Usher, M.B. (Eds) (1995). *Heaths and Moorland: Cultural Landscapes*. HMSO, Edinburgh.

Väre, H., Lampinen, R., Humphries, C. & Williams, P. (2003). Taxonomic diversity of vascular plants in the European alpine area. In *Alpine Biodiversity in Europe*, ed. by L. Nagy, G. Grabherr, C. Körner & D.B.A. Thompson. Springer, Berlin. pp. 133-148.

WCMC (2002). Mountain-watch. www.unep-wcmc.org/mountains/mountain_watch/pdfs.

Wielgolaski, F.E. (Ed) (1997). *Polar and Alpine Tundra*. Elsevier, Amsterdam.

37 Personal reflections on the conference

Simon Pepper & Eli Moen

Society is rather muddled about mountains. Whole sections of our culture label them simplistically as good (a source of inspiration) or bad (a cause of hardship) – an institutionalised polarity which lies at the heart of many conflicts. Meanwhile the true values of mountains receive little recognition; and the environment on which we all depend continues to decline. Action is required.

This message emerged with vivid and alarming clarity from the conference which formed the basis of this book.

Two truths emerged also. Firstly – as the motto for the International Year of Mountains says – 'we are all mountain people'. Wherever we live, even on the flattest parts of the Earth, mountains affect our daily lives, and we affect the mountains, more than most of us begin to realise. Secondly, we will never do justice to mountains, and resolve their problems, until we recognise their huge importance to the social, environmental and economic needs of the whole of society.

So, we need to make the case for mountains, and for a more integrated approach to their protection, use and management. Importantly, this case should embrace the interests of the people – residents and users, as well as wider society at a regional, national and global level – who know, love, fear and need the mountains in so many different ways. This conference presented many of the biophysical parts of the jigsaw. The people element was less well covered by the content or represented in the audience, despite the title of the conference. Yet, it is the people themselves – especially those who live there – who are absolutely central. It is they who know the place best, who use it day in day out, and who will eventually be the main victims of decline or beneficiaries of recovery.

It is time to bind what we know scientifically about mountains into a debate with the people – insiders and outsiders – in search of solutions which meet everyone's needs. Local and national interests are too often at odds with each other. Instead, a partnership between these groups should be delivering outcomes for the benefit of both. There are deals to be done, concerning a wide range of uses and issues. Protected area designations (e.g. National Parks in Scotland) provide one good arena for such deals, with lessons for the wider scene. This could include innovative use of agricultural subsidy, for example, exploiting Common Agricultural Policy reforms to help deliver environmental and social objectives. Indeed, because of the scale of its funding, agriculture policy may be even more important than environment policy in dealing with the challenges faced here. A new deal could also be sought with the sporting estate sector, whose impact on the social, environmental and economic fabric of the Scottish Highlands is immense.

Pepper, S. & Moen, E. (2005). Personal reflections on the conference. In *Mountains of Northern Europe: Conservation, Management, People and Nature,* ed. by D.B.A. Thompson, M.F. Price & C.A. Galbraith. TSO Scotland, Edinburgh. pp. 377-378.

But we need also to address the underlying architecture of policy and policy making. Encrusted traditions and historical legacies should be confronted, not tiptoed around. Perverse agricultural policies, sectoral divisions, a disenfranchised rural population, centralised decision-making, archaic land tenure, and so on. These relics, obstructing change, can confound every well-intentioned measure. It is time for new vision, leadership and progress. Can National Parks show the way?

This is where investment is needed, easing the path for progress and engaging mountain communities in fresh initiatives, capitalising on their knowledge, motivation and commitment to the future, for the good of the whole of society. Of course, it is essential that such work is well informed. To this end, the conference has made a very welcome, invigorating and valuable contribution, crammed with enlightening insights from many perspectives. Let us not ignore its message. The content of this book should be compulsory reading for all involved.

38 Looking forward from Pitlochry and the International Year of Mountains

Colin A. Galbraith & Martin F. Price

The international conference 'Nature and People: Conservation and Management in the Mountains of Northern Europe' attracted nearly 300 participants from a wide range of backgrounds. While the majority came from the UK, particularly Scotland, 33 attended from overseas, both from Northern Europe (Faroes, Finland, Iceland, Ireland, Norway, and Sweden) and also from further afield (Australia, Austria, Belgium, Russia, Switzerland). The diversity of participants was reflected in the range of complex issues addressed during presentations, and in poster and discussion sessions.

The programme of the conference provided both overviews and detailed case studies with regard to the mountain environments of Northern Europe, their land use issues and challenges; management influences – particularly in relation to national parks; and relevant trends and prospects not only in mountain areas, but also at wider scales. The need to ensure effective mechanisms to share information was often mentioned. This is true at every spatial and temporal scale, from early geological times to the present day. By understanding history, we may have a better chance of influencing and successfully reacting to future trends.

Key issues

A number of important issues emerged from the conference. First, we need to value our common inheritance at the regional scale, but also to recognise its European context. Based on their recognition of this common heritage, people living in and visiting each part of the mountains of Northern Europe need to develop their own sense of place, which may differ between residents and visitors. Yet, while there is considerable diversity both within and between our countries, there is also a commonality of views on, and objectives for, many issues. Further development of such shared understanding is necessary to make better decisions at local and wider scales. This requires that the necessary data and information are obtained, compiled, analysed, and disseminated, at scales from the local to the national and international. To be useful at more than local scales, comparable methods for these activities are essential. In the long term, systematic monitoring is required to understand trends in processes and their impacts. This implies a need for methods to measure change objectively: a process which requires partnerships among all stakeholders to define and utilise clear and useful indicators.

The research agenda is huge. Some of the key issues include:
- how can the ecosystem approach be applied effectively?
- how do the effects of different ecological, economic, and social processes vary at different scales?

Galbraith, C. A. & Price, M. F. (2005). Looking forward from Pitlochry and the International Year of Mountains. In *Mountains of Northern Europe: Conservation, Management, People and Nature,* ed. by D.B.A. Thompson, M.F. Price & C.A. Galbraith. TSO Scotland, Edinburgh. pp. 379-380.

- what are the most appropriate levels of grazing for specific habitat outcomes?
- what are the impacts of pollution on different ecosystems and species, and how do these change and accumulate over time?
- how can local people have sustainable livelihoods and continue to maintain the cultural landscapes that typify the mountains of northern Europe?

A number of land management challenges emerged, some common across the region, others much more regionally and nationally specific. Public-private partnerships may represent particular opportunities for innovation, and there are many different examples. We have to recognise that the objectives of primary land management are moving away from the production of agricultural and forest products – as well as animals for fieldsports – to the production of wider ecosystem services. This implies the need to develop and clarify these services, including those related to the provision of clean water, functional ecosystems, and attractive landscapes which are especially important for tourism. New ways have to be found to maintain the environments and societies of mountain systems; tourism, now the greatest single employer and driver of economies in many parts of the mountains, may play a key role. Climate change is likely to bring new challenges in planning land management in these areas.

Despite the many commonalities across the mountains of Northern Europe, the real variations must also be recognised, particularly in the political sphere. Many countries are members of the European Union (EU) but Faroes, Iceand and Norway are not. Even within the EU Member States, there are significant differences of opinion, both between governments and among the public, as to the most desirable degree of EU support and integration and the extent to which policies should be aligned – for instance with regard to the reform of the Common Agricultural Policy. Often, the perspectives of people living in mountain areas on such issues are not the same as those of others on the margins of mountain areas.

Looking forward

By jointly hosting this conference, Scottish Natural Heritage (SNH) and the Centre for Mountain Studies (CMS) believe that we have taken important steps forward in recognising the similarities, complementarities and differences across the mountains of Northern Europe – and in identifying the resulting opportunities. This book captures the contributions to science, advice and policy. To move forward, we now need concerted action involving key people. The process begun at the conference should include the encouragement of contacts both between governments and with NGOs (e.g. IUCN, WWF, Birdlife International, recreational NGOs), providing a model of collaborative effort for Europe. Working together implies effective flows of information; the creation of a Northern European node of the European Mountain Forum may be a means for doing this.

In conclusion, we hope that the conference and this book have provided a valuable contribution from Scotland, in the International Year of Mountains, and, especially, to European discussions on the conservation and management of mountain regions. There is an urgent need for real integration, both across sectors and at national and regional scales. We need to monitor progress, and to return to these topics, possibly with another conference in 2010.

Index

Note: Page numbers in **bold** indicate chapters

AALA (Adventure Activities Licensing
  Authority) 352, 355-6
Aalbu, H.: on delimitation of mountain
  areas 305, **325-33**, 355-6
Aarnes, E. 157
Aarsæther, N. 337
Aas, B. 155
Abernethy Forest RSPB Nature Reserve 225,
  239-50
  aims of conservation management 241-2
  benefits of healthy environment 248
  biodiversity 246-8
  Caledonian pinewoods 242-4
  innovative management 244-5
Abisko Mountains (Sweden) 58-9, 60, 61-2,
  64, 65, 66, 360
Abruzzo National Park (Italy) 268
Access Forum 135
access and recreation in Scottish
  Highlands 135-6
Accipiter spp. 253
accommodation, tourist 206-7
ACIA (Arctic Climate Impact Assessment)
  363
acid deposition/acid rain 46, 72-6, 77,
  80, 94
Acidification of Mountain Lakes:
  Palaeolimnology and Ecology project
  76
Action Plan, Cairngorms 276-7, 287
Activity Centres (Young Persons Safety)
  Act (1995) 353
activity holidays *see* centres; outdoor
  education
activity-specific changes in tourism 204
Aðalsteinsson, S.182
adaptive ecosystem managememnt 257
Addison Committee 8
Adelboden Conference and Declaration
  (2002) 18, 20
advantage, environmental 230
Adventure Activities Licensing Authority
  352, 355-6
AEM (European Association of Elected
  Representatives from Mountain Areas)
  325
aeolian deposits 183
aerial photographs, orthorectification
  of 197, 199
aerosols 76
Aeschna caerulea 62
aesthetics *see* ethical and aesthetic
AEWA (African-Eurasian Migratory
  Waterbird Agreement) 49
afforestation 228
Africa/African 12, 25
 -Eurasian Migratory Waterbird
  Agreement 49
  International Year of Mountains 15,
  16-17, 20
  Ministerial Conference on Environment
  16-17

Agabus wasasjernae 248
Agakhanjanz, O. 371
`Agenda 21' 3, 11, 12, 28-9
Agrawal, A. 283
agribusiness 373
agriculture 9, 46, 253, 309
  dairy farming 254
  decline 277
  development of 142, 143-4, 145
  ecosystem approach 24, 28
  Europe from 1960 to 2000 158
  farms deserted 339, 340
  Iceland before 1900 179, 181
  Norway 152, 158
  pre-industrial 143-4
  subsidies 175
  in valleys 166
  see also grazing
Agrostion alpinae 61
Agustí-Panareda, A. 80
AL:PE project 76
Alatalo, J.M. 66
Albritton, D.L. 104
Alcamo, J. 27-8
Alces alces 245, 254, 266, 270
Alectoria ochroleuca 311
Allan, N.J.R. 152
allemannsretten 167
Alley, R.B. 104
Aløingo 336
Alopex lagopus 312, 313-14
alpine bioclimatological zone 45
alpine plants 26
alpine sulphur-tresses (*Alectoria
  ochroleuca*) 311
ALPNET 359, 362, 363
Alps 7, 26, 113, 331
  delimitation of mountain areas 326, 331
  environmental history 142, 145
  recreation 128, 129
  rivers of 25
Alta 318
ALTER-NET (A Long-Term Biodiversity,
  Ecosystem and Awareness Research
  Network) 221
altitude
  and bioclimatological altitude zones 45
  and ozone concentrations 82
aluminium smelting 10
Amhuinnsuidhe and North Harris Estate 296
Amphlett, A.: on Abernethy Forest **239-50**
An Teallach 186
Andes 17, 20, 27
Andre, M.F. 47, 373
Angelini, P. 16
Anser
  A. albifrons 314
  A. erythropus 314
ants 247
Apennines 373, 374
Aquila chrysaetos 253, 265, 312-13
Arctic 318

-alpine habitat *see* Highlands
  effect 311
Arctic charr (*Salvelinus alpinus*) 76
Arctic Climate Impact Assessment 363
Arctic fox (*Alopex lagopus*) 312, 313-14
Arctic saxifrage (*Saxifraga nivalis*) 62
Arctostaphylos uva-ursi 317
Areas of Outstanding Natural Beauty 169
Arft, A.M. 66
Argentina 18
Arnalds, O. 179, 320
Arnesson-Westerdahl, A. 267, 268
Artemesia norvegica 100
Ashein, V. 254
Asian Development bank 20
Asia-Pacific 12
  International Year of Mountains 15
Assynt Crofters Foundation 297
Atholl, Duke of 7
Atlantic 104
  climate 46
  North Atlantic Oscillation 106-7
atmosphere 73
auroch (*Bos primigenius*) 245
Australia 26, 379
Austrheim, G. 153, 156, 164, 169
Austria 236, 328, 372, 373, 374, 379
  Alps, pollution 78
  International Year of Mountains 15
  lakes contaminated 78, 80
  mountain municipalities 331-2, 333
  tourism 207, 229
autumnal moth (*Epirrita autumnata*) 312
avalanches 63, 65
Avena sativa 143
Averis, A. 242
Averis, A.B. 46
Averis, A.M. 186-7
Avon, Loch 240, 241
azalea, wild (*Loiseleuria procumbens*) 100
Azerbaijan 17, 314

Bac an Lochain, Lochan 93
BADC (British Atmospheric Data Centre)
  108
Baddeley, J.A. 84, 242, 320
Badenoch 248
Bærenholdt, J.O. 254, 337
Baetis rhodani 76
Baillie, G.: on geology and nitrogen
  deposition on vegetation **121-5**
Bain, D.C. 122
Baines, D. 243
Baldascini, A. 47
Baldock, D. 228
Bali 17
Ballantyne, C.L. 192
Ballantyne, C.K. 65
Balmoral 115
bamboo 26
BAP *see* Biodiversity Action Plan
Barkved, L. 107, 361

barnacle goose (*Branta leucopsis*) 314
Barnes, P. 355
barren land, mountains seen as 6
Barrett, J.E. 58
Barry, R.G. 46
Barth, E. 151
Barthlott, W. 371
Barton, H. 285
Batterbee, R.: on climate change and pollution 26, 41, 46, **71-87**, 89, 104, 105, 106, 121, 146, 374
Bayfield, N.
  on international environmental monitoring 140, **213-22**
  on SCANNET **359-64**
Beag, Loch 93
Bealach na h-Uidhe, Loch 93
bearberry (*Arctostaphylos uva-ursi*) 317
bearded vulture (*Gypaetus barbatus*) 373
bears 16
Beaumont, D.J.: on Abernethy Forest 225, **239-50**
beaver (*Castor fiber*) 245
Becker, A. 20, 105, 146
Behre, K.-E. 47
Bele, B. 153
Belgium 379
Ben Macdui 188
Ben mor Coigach 186
Ben Wyvis 186, 187, 189
Benedictow, A. 94
Benediktsson, K.: on Vatnajökull National Park 305, **335-47**
Beniston, M. 146, 374
Bennet, D. 319
Berge, E. 236
Berglund, B.E. 146
Bern Convention (1982) 51
Berry, P.M. 100, 101
berry-picking 209, 254, 260, 266, 321
Betula
  B. pendula 153-4
  B. pubescens ssp czerepanovii 166, 307, 312
Bhuic Moir, Loch 93
Bhuju, D.R. 320
Bhutan 15, 17, 18, 20
Bibby, J.S. 191
biking 209
bioclimatological, zones 45
biodiversity 51, 221, 252, 362, 371
  Abernethy Forest 246-8
  action *see* Biodiversity Action Plan
  changes in Norway 155-7
  decline 159
  ecosystem approach 26-7, 29
  Global Biodiversity Strategy (1992) 49
  Global Mountain Biodiversity Assessment 20
  grazing, responses to 155-8
  Norway and Wales compared 167-9
  see also geodiversity and biodiversity
Biodiversity Action Plan 311
  for Abernethy Forest 241, 243, 246
  for lichen 100

Biodiversity, Convention on 19, 24, 27, 29-33, 100
biological diversity *see* biodiversity
Biosphere Reserves 47
birch (*Betula*)
  altitudinal limit 312
  cleared 308, 312
  Iceland 338, 339
  mountain (*B. pubescens ssp czerepanovii*) 166, 307, 312
  replanting 316
  weeping (*B. pendula*) 153-4
birds
  Abernethy Forest 241, 247
  Birds of Conservation Concern 243
  Birds Directive (EU) 50, 51, 52, 230
  and climate change in Scotland and Wales 100
  natural heritage trends 310, 311, 312, 314, 320
  Northern Europe 7, 9
  of prey 318
  scavenging 320
  see also grouse
Birks, H.J.B. 90, 105, 106, 142, 143, 165, 174
Birnie, R.V. 191
  multi-purpose management 225, **227-38**
Bishkek Conference on Mountain Research (1996) 13, 17, 18
Bishkek Global Mountain Summit (2002)
  and Bishkek Mountain Platform 17, 18, 20
Björnsson, H. 146
Bjørse, C. 146
black grouse (*Tetrao tetrix*) 225, 241, 243, 311
blaeberry (*Vaccinium myrtilis*) 188, 244, 317, 321
Blake, William 10, 350
blanket bog/mire 172, 192, 241, 247, 309
Blatt, S.E. 316
Blechnum spicant 106
Blunier, T. 104
Blyth, S. 20, 26, 327
BMP (Bishkek Mountain Platform) 17, 18
boar, wild (*Sus scrofa*) 245
Bobbink, R. 121, 124
Bodén, B. 210
bog/mire 35, 61, 154
  blanket 172, 192, 241, 247, 309
  bog myrtle (*Myrica gale*) 317
  bog-moss (*Sphagnum skyense*) 311
  raised 237
  woodland 242, 247
Bonn Convention on Migratory Animals 48, 49, 380
Boreal climate 46
Boreo-Atlantic climate 46
Börgesson, M. 35
Borrini-Feyerabend, G. 282
Bos primigenius 245
Bos taurus see cattle
Boyd, S.W. 338
Bradshaw, R. 146

Brady, E. 291
Branta leucopsis 314
Braun, B. 337
Braun-Blanquet, J. 61
Brazier, V. 59
Brazil 158
  see also UNCED
Breckle, S.-W. 371
Britain *see* United Kingdom
British Atmospheric Data Centre 108
Britton, A.J.: on geology and nitrogen deposition on vegetation 42, **121-5**, 320
Brook, E.J. 104
Brooker, R.: on international environmental monitoring **213-22**
Brovold, E. 172
Brown, A. 46
Brown, K. 278
brown bear (*Ursus arctos*) 245, 253, 265, 373
brown trout (*Salmo trutta*) 76, 94
browsing index 316-17
Bruar Intake 110, 112
Brunsden, D. 65, 185
Bryn, A. 164
Bryophytes 247
Bubo bubo 253
Bugmann, H. 20, 105, 146
Bulgaria 329, 331-2, 333
Búmodel grazing simulation model 179, 181-4
Burnett, M.R. 65
burning *see* fire
Burt, D. 5
Bush Estate: ozone concentrations 82-3
bushmeat 16
Buteo lagopus 265
Butler, R.W. 338

Cadell, W.: on skiing in Scotland 140, **197-202**
cadmium pollution 76, 78
Caenlochan 115
Cairn Gorm 145, 186, 188, 239, 240
Cairngorms 115, 355
  climate change and pollution 6, 93
  Environmental Change Network 214, 216-18, 220, 221
  geodiversity and biodiversity 58, 59, 60, 61, 62-3, 65, 66
  Management Strategy 243
  Massif 239, 241
  multi-purpose management 227-8
  Partnership 276, 277-8, 281, 286
  recreation 130-3
  SCANNET 360, 361
  Working Party 8-9, 10
  see also Abernethy; Lochnagar
Cairngorms National Park 8, 9, 225, 239, 350
  establishing **275-89**
  future challenges 283-4
  housing 277, 285-6
  importance of 276-7

lessons learned 281-3
management initiatives 278
processes used 278-81
special management needs 277
sustainable communities 284-5
international environmental monitoring 216-17
Recreation Study 355
Calamagrostion arundinaceae 61
Caledonian folding 46
Caledonian pinewoods 242-4, 247
Callaghan, T.V. 221
on SCANNET **359-64**
Callander, R. 236
Callaway, R.M. 218
Calluna vulgaris see heather
CALM (Circumpolar Active Layer Monitoring) 359, 363
Caloplaca nivalis 100
Camarero, L.: on pollutant status of Scottish lochs **89-97**
Canada 49
Canis lupus 245, 253, 254, 265, 312-13, 373
CAP *see* Common Agriculture Policy
capercaillie (*Tetrao urogallus*) 225, 241, 243
Capra spp. 154, 160
C. ibex 373
Capreolus capreolus 242, 243, 254, 316
carbon
cycle 105
dissolved organic (DOC) 107
carbon dioxide 192
Carex spp. 61
C. bigeolowii 62, 187, 188
Caribbean *see* Latin America/Caribbean
Caricion rigidae 61
Carl XVI Gustaf, king of Sweden 272
Carlisle, A. 243
Carn a' Chaochain, Loch 93
Carn nan Gobhar 186
Carney, D. 284
Carpathians 16, 145, 372, 374
Carrera, G. 79
Carter, T.R. 107
Carver, S.J.: on wild land 306, **365-9**
Caspian Sea 314
Castor fiber 245
Castree, N. 337
Cat, Loch nan 93
Catalan, J. 80, 81
Catopyrenium psoromoides 100
cattle (*Bos taurus*) 154, 155, 166
wild 245
see also grazing
Caucasus 27, 142, 372, 374
CBD (Convention on Biological Diversity) 19, 24, 27, 29-33, 100
CCW (Countryside Council for Wales) 172
CEH (Centre for Ecology and Hydrology) 216, 367
Central America 26
Central Asia 16-17, 20
Regional Watershed project 17

see also Kyrgyzstan
Central Europe *see* Eastern and Central Europe
Centre for Ecology and Hydrology 216, 367
centres, outdoor 198-202, 248
educational 350-5 *passim*
employment at 354
CEON (Circumarctic Environmental Observation Network) 221, 359, 363
Cernusca, A. 146, 218
Cervus spp
C. elaphus see red deer
C. nipon see Sika deer
Cetraria islandica 188
Chalmers Smith 352
Chambers, F.M. 164, 172
Chandler, T.J. 73
Chapin, T. 374
Charadrius morinellus 100, 310, 320
charitable ownership of land 233-4
Chase, T.N. 107
Cheesmond, J. 350, 352
China 26
Chleirich, Loch a' 93
Choire Dhairg, Loch a' 93
Christensen, J.H. and O.B. 106, 107
Christensen, N.L. 66, 67
Chrysura hirsute 246
Circumarctic Environmental Observatories Network 221, 359, 363
Circumpolar Active Layer Monitoring (CALM) 359, 363
CITES (Convention in International Trade in Endangered Species) 49
Clark, R. 269
clearances 135
climate 46, 230, 260, 319, 326
change *see* climate change
contrast index 331
Global Climate Models 104, 107, 113, 115
Northern European mountains 46
variability 301, 361-2
climate change 58, 301, 311, 374
Convention 49
effects on Scottish and Welsh montane ecosystems 41-2, **99-102**
management knowledge required 101
temperature and precipitation changes 99
future *see* modelling future climates
Northern Europe, environmental history 142, 144
and pollution 41, **71-87**
atmosphere 73
cloud droplet deposition 74-5
orographic enhancement of wet deposition 73-4
ozone 81-5
toxic trace elements and persistent organic pollutants 76-8
recent 80-1
see also global warming
Climate Impact Programme model 107, 109, 115

climbing 209, 269, 277, 319, 355
cloud droplet deposition 25, 74-5, 121
cloudberry (*Rubus chamaemorus*) 62, 100, 317
Cloutman, E.W. 172
CMS (Convention on Migratory Species) 48, 49, 380
coal 28, 144
Coed Gallywd 166, 171, 173-4
Coir' a' Ghrunnda, Loch 91, 93
Coire an Lochain, Lochan 93
Coire Choille-rais, Lochan 93
Coire Mhic Fearchair, Loch 92, 93
Coire na Caime, Loch 93
Coleoptera 247, 248
Coll, J.: on modelling future climates in Scottish Highlands 42, **103-19**
combustion, fossil fuel 75-6
Common Agriculture Policy, EU 52, 158, 330, 377, 380
and multi-purpose management 229, 231, 235
common land 158
`Tragedies of Commons' 252
Common Property Resources 252
common property rights 234, 236
`Common Sense and Sustainability' 9
Community Councils Group in Cairngorms 278-9, 281-3, 284, 287
community land ownership in Scotland 226, 234, 236, **295-8**
Assynt Crofters Foundation 297
Eigg, Isle of 297
existing experience of 296-7
sustainable development 295
see also Land Reform (Scotland) Act
competition 10
conflicts 29
Connell, R. 114
conservation
management, and convention 47-52
see also Habitats and Species Directive; priorities for conservation
conservations, management, *see also* Abernethy Forest
Constable, J. 350
contaminants *see* pollution
context 3-37
see also ecosystems; International Year of Mountains; Northern Europe
Conventions
on Biological Diversity 19, 24, 27, 29-33, 100
concerning Protection of World Cultural and Natural Heritage 48
on Conservation of European wildlife and natural habitats 51
on Conservation of Migratory Species of Wild Animals 48, 49, 380
on International Trade in Endangered Species 49
on Wetlands 47, 48
Cook, L. 350
Cooper, G. 351

COP7 (Seventh Conference of Parties of CBD) 19
Copus, A.K. 277, 328
Cosgrove, P. 277
cotton grass (*Eriphorum spp.*) 311, 317
Countryside Act (1968) 134
Countryside Commission 351
Countryside Council for Wales 172
Countryside Information System 367
Countryside and Rights of Way Act (2000) 292
Countryside (Scotland) Act (1967) 134
Cowie, N. 245
CPR (Common Property Resources) 252
Craigmere 239
Cramb, A. 236, 252
Crawford, R.M.M. 105
Creag Fhiaclach 308
crested tit (*Parus cristatus*) 241
crofting 297
Cronon, W. 337
crossbill
  parrot (*Loxia pytyopstittacus*) 241
  Scottish (*Loxia scotica*) 241
cross-leaved heath (*Erica tetralix*) 317
crowberry (*Empetrum nigrum*) 317
crows 320
Crowther, N. 351
cryosphere *see* glaciated areas
Cuba 15
Cubasch, U.99 104
cuckoo wasp (*Chrysura hirsute*) 246
Cuillin Hills: Coir' a' Ghrunnda, Loch 91, 93
Culicidae 260
cultural change and land use 372-4
cultural landscape model 168
  see also wilderness
cultural services 28
Cusco meeting of Mountain Partnership (2004) 19
Cyclotella pseudostelligera 80-1
Czech Republic
  geodiversity and biodiversity 58, 59-61, 63, 64-5
  pollution from (in Scottish lochs) 94

dairy farming/produce 16, 143, 254, 373
Dalarna 205-8, 267
Danube, River 25
Darling, F.F. 243
Darmody, R.G. 60
DART 359, 363
Daugstad, K. 164, 169, 171
Davidson, D.A. 192
Dawson, C.P. 269
Dawson, T. 100, 101
DDO (Draft Designation Order) 281
De Moor, M. 256
Dearden, M. 356
Debinski, D.M. 321
debris flows 64, 192
Decisions, EU 51
deer 6, 254, 277
  Abernethy Forest 41, 242, 243-4

Commission for Scotland 243
  management 243
  natural heritage trends 307, 309, 314, 315-16
  see also red deer; roe deer; Sika
Deeside 243
definitions of mountains 44-7, 326-8
Defoe, D. 6
deforestation and forestry 9, 46, 144, 159, 191, 231, 277
  ecosystems and 25-6, 28
  Forestry Authority 241, 243
  Forestry Commission 292
  natural heritage trends 308, 309, 319, 320
  Norway and Wales compared 166, 170
  see also forests; timber
delimitation of mountain areas for EU policy development 305, **325-33**
  definition of mountain region 326-8
  eligible areas identified 328-9
  mountain municipalities 331-3
  Nordregio study 329-31
  regional policies needed 325-6
DEM (Digital Elevation Model) 198, 331
Dennis, R. 245
Denniston, D. 146
Denstadli, J.M. 301
depopulation, rural 254, 373
deposition
  acid 46, 72-6, 77, 80, 94
  cloud droplet 25, 74-5, 121
  dry 76, 78
  occult 74-5, 121
  wet 73-8
Deschampsia flexuosa 61, 121, 188
Designations, EU 52
diatom, planktonic (*Cyclotella pseudostelligera*) 80-1
diatoms and pH reconstruction 77, 80
Digital Elevation Model 198, 331
Diphasiastrum alpinum 188
Diptera 247
Directives *see under* Europe
Disney, R.H.L. 248
dissolved organic carbon (DOC) 107, 192
DIVERSITAS 47
diversity, topographical, climate modelling and 107-8
DOC (dissolved organic carbon) 107, 192
Dodgshon, R.A. 153, 158
Domack, E. 305
domesticated animals 142
  see also cattle; sheep
dotterel (*Charadrius morinellus*) 100, 310, 320
Dougherty, A.P.: on ethical and aesthetic issues in management policy 226, **291-4**
Dovre 360
Downie, A. 279
Draft Designation Order 281
dragonflies, mountain (*Aeschna caerulea* and *Somatochlora alpestris*) 62
Drumochter Hills 186

dry deposition 76, 78
Dryas
  D. octopetala 100
  D. oreopteris 105
Dullinger, S. 153
Dunfermline Collage 352
Dunlap, R.E. 228
Dunslair Heights 82
Dvorák, I.J. 57-70
dwarf pine (*Pinus mugo*) 60
dwarf willow (*Salix herbacea*) 188
dynamic entities, ecosystems as 66

eagle owl (*Bubo bubo*) 253
East Asia 16, 17
Eastern and Central Europe 329, 333
  Carpathians 16, 145, 372, 374
  see also Poland
ECN *see* Environmental Change Network
eCognition 198
ecology 8
  deep 292
  ownership in Northern Scandinavia 252-4
  restoration areas 366
  see also ecosystems
ECOMONT 217-18
Economic and Social Cohesion, Second Report on 325
ECOSOC (UN Economic and Social Council) 13
Ecosystem Assessment, Millennium 27-8
ecosystems 232
  approach to meet human needs 3, **23-38**
    CBD's Ecosystem Approach 3, 29-33
    goods and services 24-7
    see also issues
  defined 24
  montane 65-6
  Scottish and Welsh, climate change and 41-2, 46, **99-102**, 105, 106, 107, 146
ecotourism 16
Ecuador 17, 18
Edgell, M.C.R. 164
Edinburgh 82
education
  Education Acts (UK, 1944 and 1945) 350, 353
  `progressive' 350
  see also outdoor education
Edvardsen, H. 253
Edwards, M. 246
  on Norway and Wales compared 139, 155, **163-77**, 373
Edwards, S.R.: on ecosystem approach **23-38**, 43, 46, 257
EEA (European Environment Agency) 51
EIA Directive (1985) 51
Eigg, Isle of, Heritage Trust 297
Einarsson, E. 341
electrification 142, 144
Elgersma, A. 254
Ellis, N.E.: on climate change and effects on Scottish and Welsh montane ecosystems 41-2, 46, **99-102**, 105, 106, 107, 146
Else, R. 349

EMEP (European Monitoring and Evaluation Programme) 94
emigration 144
Emmelin, L. 301
Empetrum
   E. nigrum 317
   E. nigrum spp. hermaphroditum 188
employment 248, 354, 356
Endangered Species Convention 49
endocrine disruption 94-5
Endresen, M. 155, 156
engagement, authentic 293
Engen, A. 152
England 142, 158, 215
   natural heritage trends 309, 310, 311, 318, 321
ENVINET 359, 363
Environmental Action Programme (EU) 76
environmental advantage 230
'Environmental Backlash' 253-4
Environmental Change Network (ECN) of United Kingdom 214-18, 219, 320
   Cairngorms 216-18
   database and users 216
   objectives 214
   protocols 214-15
   and SCANNET 359, 363
   sites 214
   strengths and weaknesses 216
environmental monitoring *see* international environmental monitoring
environmental philosophy 291
Environmental Protection Agency 204, 205
environments, mountain 41-136
   see also climate change; geodiversity and biodiversity; lochs, Scottish; modelling future climates; nature of mountains; nitrogen *under* geology; Scottish Highlands *under* tourism
EPA (Environmental Protection Agency) 204, 205
Epirrita autumnata 312
ErdasOrthoBASE 198
Erica tetralix 317
Ericsson, S. 144
Eriophorum spp. 317
   E. vaginatum 311
erosion
   high plateaus 179, 183, 185
   see also under soil
Erschbamer, B. 372
Erskine of Marr 8
Esjufjöll nature reserve 336, 338
Esseen, P.-A. 144
ethical and aesthetic issues in management policy 226, **291-4**
   differing park models 293-4
   environmental philosophy as bridge between science and policy 291
   Hume's is/ought dictum 292
   see also wilderness
ETOUR (European Tourism Research Institute) 204, 272

EU-EMERGE (European Mountain Lake Ecossystems: Regionalisation) 89
Eun, Loch nan 93
Euromontana 318, 325
Europarc 287
Europe/European Union 25, 44-5, 52, 72, 380
   CAP *see* Common Agriculture Policy
   Charter for Sustainable Tourism in Protected Areas 287
   Commission 44, 326-8, 359, 362
      Directorate-General for Regional Policy (DG Regio) 329
   conservation policies, directives and regulations listed 50-2, 76
   Directives 76
      Birds 50, 51, 52, 230
      Fish 50
      Habitats and Species 50, 51, 52, 90, 230
      Water 51, 230
   Environmental Action Programme 76
   International Year of Mountains 12, 13, 15, 16, 20
   mountain policy development *see* delimitation
   multi-purpose management 229, 231, 235, 237
   National Parks 269, 271
   policies 228-9
   recreation 128, 129
   Water Framework 230
   see also delimitation of mountain areas; geodiversity and biodiversity in European Mountains; Scandinavia
European Association of Elected Representatives from Mountain Areas 325
European Community Biodiversity Strategy (1998) 52
European Environment Agency (EEA) 51
European Environment Information and Observation Network, Regulation on (1990) 51
European Monitoring and Evaluation Programme 94
European Tourism Research Institute 204, 272
eutrophication 46, 72, 320
Evans, C. 76
Evans, R. 179, 192
Everest 320
extreme events in European Mountains
   episodic 63-5
   planning for 58
   weather, more frequent 106, 107

Faarland, T. 155
Faegri, K. 168
Falco
   F. columbarius 318
   F. rusticolus 318
Falkanger, T. 256
FAO *see* Food and Agriculture Organization

farming *see* agriculture
Faroes 218, 361, 379, 380
Faskally 110, 112
Ferguson, M.: on Cairngorms National Park 225, **275-89**
Fernandez, P. 76, 78, 89
ferns 105, 106, 164, 311
Feshie 218
Festuca vivipara 188
Fhraoich-choire, Loch an 93
Fhuar-thuill Mhoir, Loch an 93
Fiedler, P.L. 58, 66
Field, D.R. 338
field sports *see* hunting and shooting
Finland 80, 218, 379
   mountain municipalities 331-2, 333
   natural heritage trends 308, 313-19 *passim*, 321
   snowmobile damage 318
   tourism comparison with Sweden 204, 208,-10
   see also Saami; Scandinavia
Fionn Bheinn 186
Firbank, L.G. 46
fire 46, 192, 245
   controlled 51, 142, 143
fish 41, 76, 94
   acid stress 76
   contaminated 78, 79
   endocrine disruption 94-5
Fish Directive (EU) 50
Fisher, J.: on geology and nitrogen deposition on vegetation **121-5**
fishing 144, 209, 254, 260, 266, 270, 308
Fjellheim, A. 76
floods 144
flora *see* vegetation/plants
Folland, C.K. 99
food
   production, in Norway, future 139, **151-62**
   see also agriculture
Food and Agriculture Organization (UN) 12, 17, 19, 44
   see also International Year of Mountains
Forchhammer, M.C. 362
forestry *see* deforestation and forestry
forests 35
   clearance *see* deforestation
   cloud *see* cloud droplet deposition
   ecosystem approach 25-6, 28
   European Mountains 61
   major species *see* birch; pine
   Norway 152, 153, 155
   and Wales compared 164, 165-6, 173
   protection against fire regulation (1992) 51
   protection against pollutants regulation (1986) 51
   reindeer (*Rangifer t. fennicus*) 314-15
   remnant, deciduous 165
   tree-line 26, 308, 312
   see also Abernethy Forest
Forster, J.A.: on Cairngorms National Park 225, **275-89**

Forsyth, B. 279
Fort William 355
fossil fuel combustion 75-6
four-by-four vehicles *see* off-road
Fowler, D.: on climate change and
   pollution 26, 41, 46, **71-87**, 89,
   104-7 *passim*, 121, 146
foxes 107
France 47, 48
   agricultural decline 228
   lakes contaminated 78
   see also Pyrenees
Fredman, P. 272, 301
   on tourism in Northern Europe 140,
   203-12, 326
Freeman, C. 107
Frei, C. 106, 113
Freitag, R.D. 204
Frenzel, B. 142
freshwater
   acidification 84
   mussel (*Margaritifera margaritifera*)
   107
   rivers 15, 25
   see also lakes; lochs
Friðleifsdóttir, S. 5
Fritz, S. 368
Fryday, A.M. 311
Fukker, R.M. 367
Fuligula clangula 241
Fulufjället National Park (Sweden) 225,
   263-73
   planning process and conflicts 266-7
   planning process improved 267-8
   realization of project 268-72
   future 272-3
fungi 247, 311
future
   Cairngorms National Park 283-4
   climate *see* modelling future climates
   food production in Norway 139, **151-62**
   Fulufjället 272-3
   geodiversity and biodiversity 66-7
   management in Norway and Wales 174-5
   nature of mountains 52-3
   temperature changes 105-6, 110-12
Fyffe, M. 297

Galbraith, C.A. 306, **379-80**
Galium saxatile 188
Garcia, A. 47
García-Reyero, N. 95
Garten, Loch 239, 240, 248
Gaston, K.J. 167
GCMs (Global Climate Models) 104, 107,
   113, 115
GDP (Gross Domestic Product) 328
geese 314
GEF (Global Environment Facility) 20
genetics 26, 28, 362
   genetic drift 371
geodiversity and biodiversity in
   European Mountains **57-70**
   extreme episodic events 63-5
   future 66-7

geomorphological processes, active 62-3
geomorphological sensitivity,
   landscape stability and montane
   ecosystems 65-6
   landforms and habitat diversity 58-62
   species distributions and past events
   62
Geographical Information System *see* GIS
geology 46, 93
   and nitrogen deposition on vegetation
   42, **121-5**
   site differences 123-4
   species differences 122-3
geomorphology
   processes 62-3
   sensitivity 65-6
   value 264
geophysics of Northern Europe 141-2
geothermal areas 336, 338
Germany
   agricultural decline 228
   nature of mountains 48, 49
   pollution from (in Scottish lochs) 94
   tourism 229
Gerrard, A.J. 44
Geyer's whorl snail (*Vertigo geyeri*) 311
Ghana 16
Giant Mountains: geodiversity and
   biodiversity 58, 59-61, 63, 64-5, 66
Gibb, S.W.: on modelling future climates
   in Scottish Highlands **103-19**
Gibson, C. 283
Gigha, Isle of 296
Gilbert, J. 240
Gilbert, O. 100
Gimingham, C. 159, 277
Giorgi, F. 99
Giraldus Cambrensis 172
GIS (Geographic Information System) 90,
   193, 200, 216
   for Commission (GISCO) 330
   wild land 366, 368
Gjaerevoll, O. 169
glaciated areas 25, 142, 338-40, 374
   Quaternary, geodiversity and
   biodiversity in European Mountains
   and 58-9, 62
   sub-glacial chute channels 264
Glas Mael 186
Glen Cova 308
Glen Shiel 130-3
Glenmore Lodge 277
Glenmore National Forest Park and Centre
   350, 355
Glensaugh 321
Glenshee Ski Centre 198-202
Global Biodiversity Strategy (1992) 49
Global Business Network 35
Global Climate Models 104, 107, 113, 115
global conventions, strategies and
   agreements 48-9
Global Ecosystems Database 45
Global Environment Facility 20
Global Mountain Biodiversity Assessment
   20

Global Observation Research Initiative
   in Alpine Environments *see* GLORIA
Global Terrestrial Observing System 363
global warming 26, 43, 83, 146, 361, 374
   climate change and effects on Scottish
   and Welsh montane ecosystems 99,
   101
   modelling future climates in Scottish
   Highlands 104-6
   natural heritage trends 320-1
   see also climate change
GLOBE model 45
GLORIA (Global Observation Research
   Initiative in Alpine Environments)
   47, 214, 216-18, 219-20, 221, 359,
   363
Glück, P. 236
GNTEM (Peru) 16
goats (*Capra spp.*) 154, 160, 166
Godde, P.M. 203
golden eagle (*Aquila chrysaetos*) 253,
   265, 312-13
goldeneye (*Fuligula clangula*) 241
Göljådalen Valley 264
Gommenginger, C. 104
Gonzales-Rouco, H.F. 107
Good, J.E.G.: on climate change and
   effects on Scottish and Welsh
   montane ecosystems 41, 46, **99-102**,
   105, 106, 146
Goodin, R.E. 256
`goods', mountain 252
goods and services 27-8
   ecosystem approach 24-7
goose
   lesser white-fronted (*Anser
   erythropus*) 314
   white-fronted (*Anser albifrons*) 314
Gordon, J.E. 105, 185
   on geodiversity and biodiversity in
   European Mountains **57-70**
   on soil erosion in Scottish uplands
   191-5
Gorm, Loch 93
Gottfried, M. 105
Goulding, K.W.T. 84
Govan, H. 278, 281
Grabherr, G.G. 58
   on priorities for conservation 47, 306,
   371-5
Grace, J. 105
Grampians
   lochs, pollutant status of 90
   recreation 130-3
   soil erosion in Scottish uplands 191,
   192
grass 26, 61, 121, 154, 188
   aerial photographs digitally evaluated
   198-202
   natural heritage trends 311, 316, 317
   see also grazing; heath
Gray, J.M. 57
grazing/overgrazing and livestock 28, 66, 134
   natural heritage trends 308, 309,
   314-16, 317, 319-20

nature of mountains 46, 47
in Norway 152, 153-60
    biodiversity responses at community
        and population levels 155-7
    cessation and biodiversity changes
        157-8
    future 159-60
    sustainable future food production
        159-60
Wales compared with 166, 169, 173
simulation model for Iceland 179, 181-4
see also cattle; deer; goats; grass;
    horses; reindeer; sheep
great wood rush (*Luzula sylvatica*) 105-6
Greece 331-2
greenhouse gas emissions 49
Greenland 218, 321, 360, 361
Gregory, R.D. 243
Gregory, S. 73
Grieve, I.C. 66, 106, 191, 192
Grimalt, J.O. 78-9
    on pollutant status of Scottish lochs
        89-97
Grímsvötn geothermal area 336, 338
Griningsdalen 155
Gross Domestic Product 328
grouse (*Lagopus lagopus*) 159, 255, 266
    black (*Tetrao tetrix*) 225, 241, 243, 311
    red (*Lagopus lagopus scoticus*) 6, 46,
        130, 307, 309-10
Grove, J.M. 142, 146
GTOs (Global Terrestrial Observing
    System) 363
Gude, M. 64
Gudmunsson, G. 142
Guenther, A.B. 83
Guignon, C. 293
Guisan, A. 146
gullies 192
gulls 320
Gulo gulo see wolverine
Gunslay, N.: on Saami reindeer herders
    225, **259-61**
Gyllenhal carabid beetle (*Nebria
    gyllenhai*) 62
Gypaetus barbatus 373
gyrfalcon (*Falco rusticolus*) 318

habitat
    degradation 228
    see also soil; vegetation; water
Habitats and Species Directive 50, 51,
    52, 90, 230, 242, 247
HADRM3 (Hadley Centre Regional Climate
    Model) 105-16 *passim*
Haeberli, W. 374
Hafsten, U. 152
Hagen, S.E. 256
Hahnimäki, S. 312
Hall, A. 104
Hallanaro, E.-L. 46
    on natural heritage trends 143, 144,
        168, 305, **307-24**
Halls, N. 351
Hamilton, L.S. 371

Hancock, M. 245
Hannah, L. 287
Hansen, B. 104
hard fern (*Blechnum spicant*) 106
hare's-tail cotton grass (*Eriophorum
    vaginatum*) 311
Harkness, C.E. 144
HARM (Hull Acid Rain Model) 94
Harris 296
Harrison, P.A. 100
Harrison, S.J. 101
    on modelling future climates in
        Scottish Highlands **103-19**
Hastie, L.C. 107
hawk (*Accipiter spp.*) 253
Hay, A. 242
Haynes, V.M. 61, 66, 185
Heal, O.W. 144, 214, 217
    SCANNET 306, **359-64**
heath 6, 241
    dry 247, 317
    see also heather
    high plateaux and terrain sensitivity
        186-7
    wet 247, 317
    see also blanket bog
heather (*Calluna vulgaris*) 6, 42, 90,
    145, 154, 244
    aerial photographs digitally evaluated
        198-202
    burning 192
    geodiversity and biodiversity 61, 62
    geology and nitrogen deposition 121,
        122-4
    natural heritage trends 309, 316, 317,
        320
heavy metal deposition/pollution 46, 76,
    78, 90, 92, 95, 96, 106
Heberlein, T.H. 299, 301
    on tourism in Northern Europe 140,
        203-12, 326
Hebrides 296, 316
Hedberg, O. 371
Hedmark 152
Helander, E. 259
Hemiptera 247
Hendee, J.C. 269
herding *see* grazing; Saami
Heritage Trust for Isle of Eigg 297
Herries, J. 130-1, 132, 133
Hester, A.J. 316
Hetch-Hetchy dam (USA) 135
Heywood, D.I. 146
Higgins, P. 309
    on outdoor education in Scotland 305-6,
        349-58
High Summit (2002) 18
Highland cattle 245
Highlands and Islands Community Land
    Unit 296
Highlands and Islands Enterprise 328
    Community Land Fund 296
Highlands, Scottish 8, 377
    clearances 135
    conservation priorities 373, 374

high plateaux and terrain sensitivity
    140, **185-90**
    overgrazing 159
    see also Cairngorms; modelling future
        climates *and under* recreation
hiking 255, 260, 277, 319, 340, 355
    tourism in Northern Europe 206, 207,
        208, 210
Hill, A.R. 65
Hill, M.O. 105
Himalaya-Hindu Kush 17, 18, 20
Hipkin, J.A. 106
history
    of Northern European mountains 142-6
    of recreation in Scottish Highlands
        128-30
    see also land use change
Hjellbrekke, A.G. 74
Hobbs, R.J. 169
Hodkinson, I.D. 362
Hofer, T. 3, 29, 43
    International Year of Mountains **11-22**
Höfn 336, 340, 341
Hofsjökull 336
Holling, C.S. 57, 65, 66, 67, 257
Holloway, C.W. 243
Holocene 63, 105, 142, 143, 169
    forests 90, 164, 172
Hopkins, D. 351
Hordaland 152
Hori, T. 75
Hornafjörður 336, 340, 341
Horne, P.: on soil erosion in Scottish
    uplands **191-5**
Horrill, A.D. 244
horses 154, 166
Hörsten, L. 272
Hossel, J.E. 102
Houghton, J.T. 80
Housden, S.D.: on Abernethy Forest **239-50**
housing in Cairngorms National Park 277,
    285-6
Høye, T.T.: on SCANNET **359-64**
Huaraz Declaration (2002) 18
Huddleston, B. 12, 43, 44
Hudson, S. 210
Huenneke, L.F. 169
Hughes, R.E. 172, 173
Hull Acid Rain Model 94
Hulme, M. 99, 107, 109, 113, 116
Hulme, P.D. 191
human activities 58
    visibility in wild land 366, 367
    see also agriculture; recreation
Hume, D. 292
Hunt, J. 9, 356
hunting for food 254, 255, 260
    primitive 139, 143, 151, 152
hunting and shooting as sport 9, 46, 144,
    167, 209
    moose 270, 271
    natural heritage trends 308, 309
    Scottish Highlands 130, 134
Huntingford, C. 106, 113, 114
Huntley, B. 146

Huston, M. 153
hybridisation 372
Hydrology, Institute of 217
hydropower in Iceland 339
Hylocomium splendens 188
Hymenoptera 247
Hypnum jutlandicum 188

IAGM (Inter-Agency Group on Mountains) 12
ibex (*Capra ibex*) 373
Iceland 5, 142, 218, 379, 380
  climate 46
  high plateaux 181-3
  natural heritage trends 308, 312, 315,
    318, 319
  SCANNET 360, 361
  snowmobile damage 318
  tourism compared with Sweden 204,
    208-10
  transhumance and vegetation
    degradation in 139-40, **179-84**
  Búmodel grazing simulation model 179,
    181-4
  pre-modern agriculture 179, 181
  see also Vatnajökull National Park
Icelandic Sagas 308
Ilbery, B. 337
ILTER (International Long Term
  Ecological Research) 359, 363
image segmentation 197
India 17, 18, 322
industrialisation 142, 144-5
Initiatives, Mountain Partnership 20
Innerdalen 152
insects 311, 312
institutional scale and ecosystem
  approach 33-5
Inter-Agency Group on Mountains 12
interest-based management challenged 255
Intergovernmental Panel for Climate
  Change 80
International Development, UK Department
  for 284
international environmental monitoring
  140, **213-22**
  strengths and weaknesses 219, 220
  see also Environmental Change Network;
    GLORIA; SCANNET
International Long Term Ecological
  Research 359, 363
International Mountain Day (December
  11th 2002) 17
International Partnership for
  Sustainable Development in Mountain
  Regions 11, 29
  Mountain Partnership 19-21
International Trade in Endangered
  Species Convention (1973) 49
International Tundra Experiment 66
International Year of Mountains 3, 10,
  13-21, 287, 336, 377-80 *passim*
  Mountain Partnership 19-21
  objectives and activities 13-19
  websites 13, 14, 15, 27
invertebrates 246

investment and need for 29, 35-6
Ireland 379
  lakes contaminated 78
  pollution from (in Scottish lochs) 94
irrigation 228
issues in ecosystem approach 27-33
  goods and services, defining 27-8
  Principles of Ecosystem Approach 31-3
  supply and demand, balancing 28-31
Italy 18, 19, 268, 373, 374
  International Year of Mountains 16, 18
  lakes contaminated 78
  mountain municipalities 331-2
ITEX (International Tundra Experiment) 66
IUCN (World Conservation Union) 3, 48,
  49, 282
Ives, J.D. 12, 43, 372
IYM *see* International Year of Mountains

Jäckli, H. 63
Jackson, D. 216
Jakarta 12
Jämtland 205-8
Jämtlandsfjällen 265
Järvinen, A.: on SCANNET **259-64**
Jeník, J. 60, 61, 62, 65
Jenkins, G. 104, 107, 114
Jenkins, T.N. 337
Jentoft, S. 252
Jentsch, A. 153, 157
Jersletten, J.L. 260
Jevne, O.E. 152
Jia, X. 197
Johannesburg *see* World Summit on
  Sustainable Development
Jóhannesson, G. 340
Johansson, M.: on SCANNET **359-64**
Johansson, R. 265
Johns, T. 116
Johnson, R. 101
Johnson, S. 6
  on nature of mountains **43-55**
Jökulsárlón 336, 341
Jonasson, C.: on geodiversity and
  biodiversity in European Mountains
  45, 47, **57-70**
Jones, E.W. 62
Jones, M. 167, 169, 170
Jones, P.D. 107, 114
Jones, V.J. 76, 89
Josefsson, M. 57-70
Jotunheimen mountains 155
Juncus spp 61
  J. squarrosus 188
  J. trifidus 62, 188
Juniperus communis 155, 241

Kaland, P.E. 152
Kaltenborn, B.P. 301
Kampala Declaration on Environment for
  Development 16-17
Kapos, V. 12, 23, 44, 326, 327
Karjalainen, L. 246
Kärkevagge 64
Kazakhstan 17, 314

Kenya 18
Kernan, M.: on pollutant status of
  Scottish lochs 41, **89-97**
Kerr, R.A. 105, 107
Kiefer, R.W. 202
Kilpisjärvi 360
Kilsby, C.G. 107
Kim, E.G. 14, 146, 232
Kirby, K. 174
Kirby, P. 246
Kirkham, F.W. 320
Kirkjubæjarklaustur 336, 340
Kirkpatrick, A.H. 104
Kirunafjällen 265
Klanderud, K. 20, 105, 106
Klemen, J. 59
Kley, D. 83
Kneafsey, M. 337
Kociánová, M. 57-70
Koerselman, W. 122
Köhler, B. 58
  on SCANNET **359-64**
Koinig, K.A. 80
Kola peninsula 259, 314-15, 318
Konstantin, K. 260
Korea 16
Korhola, A. 80
Körner, C. 20, 47, 311, 371, 372
Kozlowska 58, 61
Kringilsárrani nature reserve 336, 338
Kristinsson, H. 308, 312
krummholz pine (*Pinus montana*) 61
Kuala Lumpur Conference of CBD (2004) 19
Kullman, L. 105, 312
Kuuluvainen, T. 246
Kvaalen, I. 253
Kvamme, M. 152, 169
Kyoto Protocol (1997) 49
Kyrgyzstan 11, 16
  Regional Watershed project 17
  see also Bishkek

Lagopus
  L. lagopus 159, 255, 266
  L. lagopus scoticus 46, 130, 307, 309
  L. mutus 100
lakes, mountain 72
  climate change and 79-81
  contaminated 76-9
  Scottish *see* lochs
Lambert, R. 227
Land Capability for Agriculture 191
Land Reform Policy Group 295
Land Reform (Scotland) Act (2003) 292,
  295, 356
  community's right to buy 296, 297
land rights 233-6
land use change 139-222
  and cultural change 372-4
  future food production in Norway 139,
    151-62
  in Northern Europe 46
  see also under tourism
  in Norway *see* biodiversity *under*
    Norway; environmental history

under Northern Europe; Wales *under*
Norway
in Scotland *see* high plateaux *under*
Scotland; skiing *under* Scotland;
soil erosion *under* Scotland
see also biodiversity and future food;
international environmental
monitoring; transhumance *under*
Iceland
landscape
and `people' 167
and scenery 6-7, 167
Landvernd 345
Lane, A.M.J. 214, 215
Langmuir, E. 9
Lapland 321
Lash, S. 337
Latin America/Caribbean 25, 26, 27
International Year of Mountains 12-13,
15, 16, 17, 18, 20
see also UNCED
latitude 45
Latour, B. 337
Lauscher, A. and F. 362
law *see* legislation
Lawers, Ben 93
Laxness, H. 5
lead deposition 76, 78, 90, 92, 95
Legg, C. 308
legislation 134-5, 168-9
countryside 134, 292
education 350, 352, 353
Iceland 345
local government reorganisation 352
National Parks 278, 283
pollution control 230
safety for young people 353
Scotland 134
education 350
land reform 292, 295, 296, 297, 356
see also Directives *under* Europe
Lehtonen, J. 312
Leith, I.D. 84
lemon scented fern (*Dryas oreopteris*) 105
Leopold, A. 292-3
Lepidoptera 247, 311, 312
Lepus timidus 100
Lesotho 16
lesser white-fronted goose (*Anser
erythropus*) 314
Lesslie, R. 366
Leys, K.F. 57, 65
lichens 61, 100, 247, 265, 311
Lillesand, T.M. 202
Lilly, A. 66
on soil erosion in Scottish uplands
140, **191-5**
Lima meeting (1995) 12-13
Lind, B. 300
Lindbladh, M. 144
Lindenmayer, D. 244
lingonberry (*Vaccinium vitis-idaea*) 311,
317
Liniger, H. 16, 25, 29
Linnard, W. 172, 173

liverworts 311
livestock *see* grazing
Lloydia serotina 100, 101
Loader, C. 350
Local Development Groups (Fulufjället) 268
Local Government Reorganisation Acts
(1996) 352
Loch Lomond and Trossachs National Park
275, 321
Lochaber 9
Lochnagar
diatoms and pH reconstruction 77
lead in water 78
pollution 89, 90, 93, 94-5
lochs, Scottish 275, 321
Abernethy Forest 239, 240, 241, 248
pollutant status of 41, **89-97**
results and discussion 92-6
sites and methods 90
logging, illegal 16
Loiseleuria procumbens 100
Lomond, Loch 275, 321
longhorn beetle (*Rhagium inquisitor*) 246
Long-Term Biodiversity, Ecosystem and
Awareness Research Network, A 221
long-term monitoring *see* international
environmental monitoring
Lónsöræfi nature reserve 336, 338, 343
Loomis, J.B. 204
Lorimer, H. 350
Lothian Region 350, 351-2
lousewort (*Pedicularis sudetica*) 62
Lowe, K. 107, 114
Lowenthal, D. 168
lowlands 29, 45
Loxia
L. *pytyopstittacus* 241
L. *scotia* 241
Loynes, C. 351
Luing cattle 245
Lundberg, A. 168, 172
Lundkvist, M. 58, 64
Lundquist, R. 264, 272
Luzula sylvatica 105-6
lynx (*Lynx lynx*) 245, 253, 254, 265

MA (Millennium Ecosystem Assessment) 27-8
Maastricht traty 52
MAB (Man and Biosphere Programme, 1971)
47, 48
Macaulay Land Use Research Institute
193-4, 216, 217
MacCaig, N. 6-7
McComb, W. 244
McConnell, J. 308
Macdonald, R. 8
McGibbon, J. 207
MacGregor, B.D. 236
MacGregor, C. 130-1, 132, 133, 277, 355
McGuire, D. 28
Machlis, G.E. 338
McHugh, M. 192
Mackay, J. 42
on recreation and Scottish Highlands
127-36, 145, 157, 208, 319, 326, 350

Mackenzie, J. 297
Mackey, E.C. 243, 309, 310, 311
MacKinnon, K. 20
Macnaghten, P. 337
McNeish, C. 349
McVean, D.N. 46, 106, 242
Madagascar 16
Madsen, J. 314
Magnússon, A. 182
Magnusson, B.: on SCANNET **359-64**
Magnusson, M.: on Northern Europe 3, **5-10**
Malmer, N. 142
Malung 267
mammals 247
see also grazing
Mamores 186
Man and Biosphere Programme (1971) 47, 48
management
Abernethy Forest 241-2, 244-5
Cairngorms 227-8, 243
National Park 277-8
conservation 47-52, 241-2
in Norway and Wales, future 174-5
management influences, practices and
conflicts 101, 225-302
see also Abernethy Forest; Cairngorms
National Park; community land
ownership in Scotland; ethical and
aesthetic issues; Fulufjället;
multi-purpose management;
ownership in Northern Scandinavia;
Saami; Sápmi
March, B.G.E. 318
Maree, Loch 93
Margaritifera margaritifera 107
maritime upland sensitivity to climate
change 104-8
Markland, J. 280
Marshall, J. 104
mason bee (*Osmia uncinata*) 246
Mather, A. 130-1, 132, 133
Matthews, K.B.: on skiing in Scotland
140, **197-202**
Maxwell, J.: on multi-purpose management
in Northern Europe 225, **227-38**
meadows, montane 47
Meagaidh, Creag 93
media, and International Year of
Mountains 14
Meehl, G.A. 46, 106
Megaselia
M. *abernethae* 248
M. *gartenis* 248
Melander, O. 60, 63
Melanitta nigra 265
Meldrum, K. 350
Mels, T. 337
Merano meeting of Mountain Partnership
(2003) 19
mercury pollution 79, 96
merlin (*Falco columbarius*) 318
Messerli, B. 12, 43, 372
metal pollution *see* heavy metal
Meuleman, A.F.M. 122
Mexico 26

International Year of Mountains 16
Mexico City 12
Meybeck, M. 12
Mhadaidh, Loch a' 93
'middle countryside' 231
Migon, P. 59
Migratory Species of Wild Animals,
Convention on Conservation of 48
Migratory Waterbird Agreement, African-
Eurasian 49
Miles, L. 185
milk quotas traded 254
Millennium Ecosystem Assessment (MA) 27-8
Milvus milvus 309
mining 28, 144
Ministry for Environment (Iceland) 343,
344, 345
mire *see* bog
Mitchell, F.J.G. 174
mites 247
Moberg, A. 107, 114
modelling future climates in Scottish
Highlands 42, **103-19**
maritime upland sensitivity 104-8
methods 108-10
objectives and approach 108
results and discussion 110-14
Moe, D. 152
Moen, A. 166, 169, 170, 306, 312
on conference **377-8**
Molau, U. 66
Molinia caerulea (purple moor grass) 121,
316, 320
Monadhliath Mountains 191
Mondhuie 246
monitoring *see* international
environmental monitoring
Montaigne, M.E. de 10
montane bioclimatological zone 45
montane ecosystems in European Mountains
65-6
Montieth, D.T. 76
Moor House 321
Moore, P.D. 172
moorlands 6
moose (*Alces alces*) 245, 254, 266, 270,
271
Morellet, N. 316
Morgan, A. 350
Morocco 16
Morrocco, S.M.: on Scottish Highlands,
high plateaux and terrain
sensitivity in 140, **185-90**, 191
Mortenen, L.: on SCANNET **359-64**
mosquitoes (*Culicidae*) 260
moss campion (*Silene acaulis*) 100, 101
moss, Skye-bog (*Sphagnum skyense*) 311
moss, woolly hair (*Racomitrium
lanuginosum*) 42, 61, 62, 84, 121,
122-4, 187, 188, 242, 320
moss-heath (*Carex bigeolowii*) 62, 187, 188
moths 311, 312
motorcycles forbidden 269
mountain avens (*Dryas octopetala*) 100
Mountain Biosphere Reserves 221

mountain birch (*Betula pubescens*) 166,
307, 312
Mountain Children, International
Conference of (2002) 18
Mountain Councils, future 256-7
Mountain Ecosystems, International
Meeting of (2002) 18
Mountain Forum 12-13, 17, 231
mountain hare (*Lepus timidus*) 100
Mountain Law (Norway) 256
Mountain Partnership 19-21
Mountain Populations, World Meeting of
(2002) 18
Mountain Research, Partnership
Initiative on 20
'Mountain Watch' 20
Mountain Women, Celebrating 18, 20
mountaineering *see* climbing; hiking
mountains *see* environments; land use
change; management; prospects
Muir, J. 144-5, 287
muirburn *see* fire
Müller, D. 299, 300
multi-purpose management in Northern
Europe 225, **227-38**
perspectives and policies, changing 228-31
property rights and implications 233-7
sustainable development 232-3, 237
municipalities, mountain 331-3
Munros (Scottish summits) 129, 276, 319
Murmansk 312
Murphy, J.M. 107
mushrooms 247
music and mountains 6-7
Myrica gale 317
myrtle, bog (*Myrica gale*) 317
Mývatn-Laxá 336

Naess, A. 292
Nagy, L 58, 66, 360, 362, 371, 374
nature of mountains **43-55**
Naismith's Walker's Rule 368
Nardus stricta 61, 121, 188
Nash, R. 135, 366
National Emissions Ceilings Directive of
EU 76
National Heritage Zone 239
National Infrastructure for Catchment
Hydrology Experiments 218
National Nature Reserves 169
National Parks 146, 377
Authority and Plan 275, 278
Iceland *see* Vatnajökull
multi-purpose management 230, 234,
236-7
Norway and Wales 168, 169, 172, 175
Sweden *see* Fulufjället
United States 168
see also Cairngorms National Park
National Parks (Scotland) Act 2000 278,
283
National Planning Policy Guidelines 365
National Trust for Scotland 365
National Working Group on Mountain
Ecosystems (Peru) 16

Natura 2000 50, 52
natural heritage trends in northern
Europe 305, **307-24**
contemporary pressures 316-19
grazing effects 314-16
Scandinavia 311-14
Scotland 308-11, 315-19 *passim*, 321
'naturalness' 165, 306-7
Nature Conservation Agency and Register
(Iceland) 339, 343-4
nature of mountains 41, **43-55**
conservation management and convention
47-52
definition of mountains 44-7
future 52-3
nature observation 209
Nature Reserves 169, 336, 338, 343
see also Abernethy Forest
NAU (North Atlantic Oscillation) 106-7
Nebria gyllenhai 62
Negros Committee (Philippines) 16
Nemitz, E. 78
Nepal 18, 26, 320
Neshatayev, V. and V. 312
Nesje, A. 146
Netherlands 121
Networking of Long-Term Integrated
Monitoring in Terrestrial Systems 259, 363
Neuvonen, S.: on SCANNET **359-64**
Newtonmore 353, 355
NGOs *see* non-governmental organisations
niche markets 127
NICHE (National Infrastructure for
Catchment Hydrology Experiments) 218
Nichols, W.F. 58, 65
Nicholson, J. 94
Nicol, R. 351, 353
Nilsson, S.G. 144, 146
nitric oxide pollution 74, 75-6, 82
nitrogen
emissions/pollution 76, 84, 106
-phosphorus ratio and plants 321
nival bioclimatological zone 45
NNRs (National Nature Reserves) 169
NoLIMITS (Networking of Long-Term
Integrated Monitoring in Terrestrial
Systems) 359, 363
nomadic pastoralists *see* Saami; Sápmi
non-governmental organisations 12, 14,
134, 233-4, 344-5, 380
see also World Conservation Union;
World Wildlife Fund
Nordregio study 329-31
Norrbotten 205-8, 259
North America 292
International Year of Mountains 15
see also Canada; United States
North Atlantic Oscillation 106-7
Northern Europe 3, **5-10**, 46, 53
environmental history of mountains 139,
141-9
change 142-6
geophysics 141-2
Northern Ireland 215, 310
Norway 14, 135, 236, 379, 380

agriculture compared with Sweden 158
biodiversity and future food
    production 139, **151-62**
    see also under Norway
climate 46
environmental history 141-2, 145
lakes contaminated 78
mountain municipalities 329, 331-2, 333
natural heritage trends 308, 311-12,
    313-15, 316, 318, 319
SCANNET 360
snowmobile damage 318
tourism compared with Sweden 204,
    208-10
and Wales compared 139, **163-77**
    biodiversity 167-9
    future management 174-5
    people and land 166-7
    see also ownership in Northern
    Scandinavia; Saami; Scandinavia
Norway spruce (*Picea abies*) 152, 166, 312
Norwegian mugwort (*Artemesia norvegica*) 100
Noss 371-2
NTS (National Trust for Scotland) 365
NUTS (Nomenclature of Territorial
    Statistical Units) levels 0 to 5
    328-9, 330
Ny Ålesund 360
Nyberg, R. 64
Nygård, T. 318

oats (*Avena sativa*) 142, 143
occult deposition 74-5, 121
Odonata 62, 247, 248
Oestlund, L. 147
off-road vehicles and snowmobiles
    damage by 318, 319
    forbidden 270
    Fulufjället National Park 266, 270, 271
    tourism in Northern Europe 206, 208,
    209, 210
Oksanen, L. 312
olive mayfly, large (*Baetis rhodani*) 76
Olsson, A. 164
    on Norwegian biodiversity and future
    food 143, **151-62**, 166, 169, 170
Olwig, K.R. 164, 166-7
opinions of local communities in
    Vatnajökull 341-5
    marketing device 342-3
    top-down tactics criticised 344-5
    underfunding fears 343-4
Öræfajökull 336, 338, 339
Öræfi 340, 343
Ordnance Survey 367
organic pollutants 46
    persistent 76-9, 89, 318
organic/organo-mineral soils eroded 192-3
O'Riordan, T. 232, 281, 284
ornithology *see* birds
orogenesis 46
orographic enhancement of wet deposition
    73-4
orthorectification of aerial photographs
    197, 199

Osborn, T.J. 106-7, 113
Osmia uncinata 246
Osprey Centre (Abernethy Forest) 248
Ostrom, E. 252
O'Sullivan, P.E. 241
outdoor education in Scotland 305-6,
    349-58
    changing pattern 354-5
    changing priorities 351-3
    as formalised adventure 350-1
    residential centres, economic impact
    of 353-4
overgrazing 265
overland flow risk and soil erosion 193-4
Ovis aries see sheep
Owen, J.A. 248
ownership of land 134
ownership in Northern Scandinavia 225,
    251-8
    adaptive ecosystem management 257
    dynamics, social and ecological 254-5
    ecologies 252-4
    'goods', mountain 252
    interest-based management challenged
    255
    resources 256-7
    rights-based management, claims for
    255-6
Ozenda, P. 46
ozone and climate change 81-5

Pacific *see* Asia-Pacific
PAHs (polycyclic aromatic hydrocarbons)
    76, 78
Palacios, D. 58
palaeolimnology 80
PAN Parks 272
Parent Teacher Associations 279
Parker, T. 350
parrot crossbill (*Loxia pytyopsittacus*) 241
Parus cristatus 241
pastoralism *see* grazing
Pauli, H. 47, 105, 374
Paus, A. 152
PCBs (polychlorinate biphenyls) 90, 92
Pearce, I.S.K. 124
Pearsall, W.H. 46
peat 309
    aerial photographs digitally evaluated
    198-202
    erosion/oxidation 191, 192
    see also blanket bog
Pedicularis sudetica 62
Pennines 73
Pepper, S. 306
    on conference **377-8**
Perkins, D.F. 144
permafrost 374
Perring, M. 218
persistent organic pollutants *see* POPs
Peru 26
    International Year of Mountains 12-13,
    16, 17, 18, 19
Peterken, G. 241
Peterman, U. 116

Petri, M.: on soil erosion in Scottish
    uplands **191-5**
Pettersson, B. 269
Pettersson, R.: on Sápmi, tourism in 226,
    299-302
Philippines 16
philosophy, environmental 291
    see also ethical and aesthetic
Phoenix, G.K. 321
Picea abies 152, 166, 312
Pickett, S.T.A. 153
Pillai, A.: on community land ownership
    in Scotland 226, **295-8**
Pilous, V 64
Piña, B.: on pollutant status of
    Scottish lochs **89-97**
pine juniper (*Juniperus communis*) 241
pinewoods 60, 264, 276
    Caledonian 242-4, 247
    geodiversity and biodiversity in
    European Mountains 60, 61
    Norway
    biodiversity in 153-4
    and Wales compared 166, 171, 172
    see also Abernethy Forest; Scots pine
Pinus
    P. montana 61, 171
    P. mugo 60
    P. sylvestris 153-4, 166, 172, 225,
    242, 246, 312
Pitcairn, C.E.R. 84, 123
Pitlochry 110
Pitt, D.C. 146
PlaNet Finance 20
planktonic diatom (*Cyclotella
    pseudostelligera*) 80-1
plants *see* vegetation
Plectrophenax nivalis 100, 276
Pleistocene 142
Pleurozium schreberi 188
Po, River 25
Poff, N.L. 58
Poiani, K.A. 58
Poland 333
    geodiversity and biodiversity 58,
    59-61, 63, 64
    lakes contaminated 78
    pollution from (in Scottish lochs) 94
policy
    development *see* delimitation of
    mountain areas
    issues *see* ethical and aesthetic
    principles and components of Ecosystem
    Approach 31, 32
    wild land 365-6
Pollard, S. 147
pollution 72-6, 312, 320-1
    control legislation 230
    Northern European mountains 46
    see also organic pollutants *and under*
    lochs, Scottish
polychlorinate biphenyls 90, 92
polycyclic aromatic hydrocarbons 76
polyploidisation 371
Polytrichum alpinum 188

Poore, M.E.D. 106
POPs (persistent organic pollutants)
76-9, 89, 318
population of mountains 23-4
Portugal 78, 331-2
Post, E. 362
post-industrial era 139, 145-6
poverty reduction 16
Power, T.M. 203
Prather, M. 83
Pratt, V. 291, 292
pre-Cambrian shields 46
precipitation
changes, future 106-7, 112-13
increase and montane ecosystems 99, 101
orographic effect on 73-4
rainstorm events 182
predators 167, 260
persecuted 253, 312-13
regenerated 254
pre-industrial agriculture 139
Preston, C.D. 310, 311
Price, M.F. 3, 232, 306, 379-80
ecosystem approach 23, 25, 26, 27, 29
on International Year of Mountains
11-22, 328
nature of mountains 43, 47
Northern Europe, environmental history
141, 146
Primula scandinavica 157
Principles of Ecosystem Approach 31-3
see also ecosystem approach
priorities for conservation and
management of natural heritage 306,
371-5
climate change 374
land use and cultural change 372-4
wilderness strategy 372
private goods 252
private ownership of land 234-6, 277
property rights in Northern Europe 251-8
and multi-purpose management 233-7
prospects 303-80
see also delimitation of mountain
areas; natural heritage trends;
outdoor education; priorities for
conservation; SCANNET; Vatnajökull
National Park; wild land
Protected Area Network Parks 272
Protected Landscapes (Norway) 169
Psenner, R. 80, 89
psychology of leisure 129-30
ptarmigan (*Lagopus mutus*) 100
Pteridium aquilinum 164, 311
public goods 252
pulp and paper industry 10
purple moor grass (*Molinia caerulea*) 121,
316, 320
purple saxifrage (*Saxifraga
oppositifolia*) 100
Putnam, R. 351
Pykala, J. 146
Pylvänäinen, M. 46, 315, 318
Pyrenees 142
conservation priorities 372, 373, 374

lakes contaminated 80, 81
quarrying 144
quartzite and soils 123
Quercus petraea 173
Quito Declaration (2002) 18

Rackham, O. 170
Racomitrium lanuginosum see woolly hair
moss
Raddum, G.G. 76
Rahmstorf, S. 104
railways 145
rain *see* precipitation
raised bog, active 237
Ramakrishnan, P.S. 322
Ramsar Convention on Wetlands and
Waterfowl Habitat 47, 48
Rana, P. 320
Rangifer spp see reindeer
Rapp, A. 58, 60, 63
Rasch, M.: on SCANNET **359-64**
Rasenan, S. 142
Ratcliffe, D.A. 44, 46, 47, 172, 173, 242
Ray, C. 337
Razkowska 58, 61
RCMs (Regional Centre Models) 107, 115
HADRM3 105-16 *passim*
UKCIP02 (Climate Impact Programme) 107,
109, 115
see also modelling future climates
RDB (Red Data Book) species 246
recreation *see* tourism/recreation
Red Data Book species 246
red deer (*Cervus elaphus*) 6, 254, 277
Abernethy Forest 242, 243-4
field sports 46, 159
management 243
natural heritage trends 307, 309, 314,
315-16
red grouse (*Lagopus lagopus scoticus*) 6,
46, 130, 307, 309-10
Red Hills 186
red kite (*Milvus milvus*) 309
Redo, Lake (Pyrenees) 80
Refjell mountain 156
Regional Centre Models *see* RCMs
regional policies needed 325-6
regulation services 28
Regulations, EU 51
Reid, P.A. 107
reindeer (*Rangifer tarandus*) 143, 166,
225, 245, 254, 265
forbidden 269
forest (*Rangifer t. fennicus*) 314-15, 317
herders *see* Saami; Sápmi
natural heritage trends 307, 308, 312,
314, 315, 317
Norway: biodiversity and future food
155-6, 157
ranges 315
Svalbard (*Rangifer t. platyrhynchus*)
317
tundra (*Rangifer t. tarandus*) 314, 315
remote sensing of effect of skiing in
Scotland 140, **197-202**

remoteness modelling 367, 368
renewable energy 46
Rennie, S.C. 216
research agenda 379-80
Reserves
Biosphere 47
Mountain Biosphere 221
see also Nature Reserves
residential centres *see* centres
resource ownership 256-7
Rettie, K. 280
`re-wilding' 368
Rhagium inquisitor 246
Rhine and Rhone rivers 25
Rhytidiadelphus loreus 188
Richards, J.A. 197
Riddington, G. 355
rights
-based management 255-6
land 233-6
of way 7
rime 75
Rio Conference *see* UNCED
rivers 15, 25
roads 145
walking distance from 366, 367, 368
Robertson, P.: on nature of mountains
43-55
rock climbing *see* climbing
Rodwell, J.S. 122, 187, 188
roe deer (*Capreolus capreolus*) 242, 243,
254, 316
Rognerud, S. 76, 79
Romania 16, 329, 333
romanticism 6-7, 10, 350
ROS (Recreation Opportunity Spectrum)
263, 269, 270
Rose, N.L. 76, 78, 146
on pollutant status of Scottish lochs
89-97
Rosenberg, L. 210
Rosseland, B.O. 78
rough-legged buzzard (*Buteo lagopus*) 265
Rouvinen, S. 246
rowan (*Sorbus aucuparia*) 316
Rowell, D.P. 107, 114
Royal Society for the Protection of
Birds 9
RSPB *see* Abernethy Forest; Royal Society
for the Protection of Birds
Rubensdotter, L. 64
Rubus chamaemorus 62, 100, 317
Rum 316
Rundgren, M. 146
Ruong, I. 259
Rural Development Plans 229
Rural Development Regulation 237
rushes (*Juncus spp.*) 61, 62, 186-7, 188
Russfjell mountain 156
Russia 312, 379
lakes contaminated 78
natural heritage trends 308, 319
pinewoods 246
Saami in 259
rye (*Secate cereale*) 143

Rynning, L. 256

Stursa, J. 60, 62
Saami reindeer herders 225, **259-61**, 308, 314
  ownership in Northern Scandinavia 253, 256, 257
  summer grazing 259-61
  see also Sápmi, tourism in
SACs (Special Areas of Conservation) 47, 50, 90
Saebø, A. 320
Saelthun, N.R. 107
  on SCANNET **359-64**
Saetersdal, M. 105
Sæþórsdóttir, A.D. 343
Sala, O.E. 374
Salicion herbaceae 61
Salix spp. 155
  S. herbacea 188
  S. lanata 308, 311
  S. repens ssp. argentea 61
Salmo trutta 76, 94
Salten 254
Salvelinus alpinus 76
Sameby district 259
Sámi/Sámi *see* Saami
Sánchez-Colomer, M.G. 58
Sandberg, A. 159
  ownership in Northern Scandinavia 225, **251-8**
Sandell, K. 208
Sanders, G. 94
Sanderson, E.W. 366
Sanderson, S. 257
sandstone and soils 123
Sápmi, tourism in 226, **299-302**
SARD *see* Sustainable Agriculture and Rural Development
Saussure, H.B. de 7
sawflies 247
Saxifraga
  S. cespitosa 100
  S. nivalis 62
  S. oppositifolia 100
scale and ecosystem approach 33-5
Scandinavia 20, 44
  climate 46, 331
  conservation priorities 372, 373
  delimitation of mountain areas for EU policy development 329, 331-2, 333
  natural heritage trends 311-14
  North European Network of Terrestrial Field Bases *see* SCANNET
  orogenesis 46
  pinewoods 246
  recreation 128, 135
  see also Finland; Norway; ownership in Northern Scandinavia; SCANNET; Sweden
SCANNET (Scandinavian-North European Network of Field Sites and Research Stations) 306, **359-64**
  climate variability 361-2
  future 362-3

  international environmental monitoring 214, 216-19, 220, 221
  land use and social interaction 361-2
  reasons for sites 360-1
Scherer, D. 64
Schiermeier, Q. 104, 107, 115
schist and soils 123
Schmidt, R. 80, 89
Schmittner, A. 104
Schor, J. 204
Schwarze ob Solden (lake) 78
scoter, common (*Melanitta nigra*) 265
Scotland 5
  climate 46
    change and montane ecosystems 99-102
  conservation priorities 372
  erosion *see* soil erosion in Scottish uplands
  geodiversity and biodiversity *see under* Cairngorms
  Highlands *see* Highlands, Scottish
  international environmental monitoring 215, 216-18
  lakes contaminated 78
  lochs *see* lochs, Scottish
  multi-purpose management 230, 231, 236
  natural heritage trends 308-11, 315-19 *passim*, 321
  occult deposition 75
  orogenesis 46
  ozone concentrations 82-3
  reindeer in 315
  SCANNET 360
  skiing in 140, **197-202**
  tourism 14, 204, 208, 229
  see also Abernethy; Cairngorms; climate change and effects; geology and nitrogen deposition; Grampians; lochs; soil erosion in Scottish uplands; wild land
Scots pine (*Pinus sylvestris*) 153-4, 166, 172, 225, 242, 246, 312
  see also Abernethy Forest
Scott, M. 309
Scott, Sir W. 6
Scottish Advisory Board for Outdoor Education 353
Scottish crossbill (*Loxia scotica*) 241
Scottish Land Fund 296
Scottish Mountaineering Club Journal 7
Scottish Natural Heritage 7, 8, 14, 135, 216, 380
  Cairngorms National Park advice 278-81, 282
  Futures framework 193
  National Heritage Zone 239
  outdoor education in Scotland 355, 356
  wild land 365
Scottish Office 310, 365
Scottish Outdoor Access Code 135, 356
Scottish Rights of Way Society 7
Scottish Tourist Board 354
SCP (spheroidal carbonaceous particles) 76, 92, 94, 95

screes, aerial photographs digitally evaluated 198-202
scuttle flies (*Megaselia abernethae* and *M. gartenis*) 248
secondary forest and scrub 47
sedge (*Carex spp.*) 61
seeder-feeder enhancement 121
Segerström, U. 146
Sekyra, J. 59, 60, 61
Sene, E.H. 28
SEPA (Scottish Environment Protection Agency) 216
SEPA (Swedish Environmental Protection Agency) 266-73 *passim*
Seppa, H. 142
Sernander, R. 168
sessile oak (*Quercus petraea*) 173
set-aside strategy 373-4
seter landscape 171-2, 174-5
sheep (*Ovis aries*) 6, 42, 166, 183, 254
  Abernethy Forest 244-5
  natural heritage trends 308, 309
  Northern Europe, environmental history 144-5
  Norway: on biodiversity and future food 154, 155
  see also grazing
Shepherd, G. 29
Sheppard, L.J. 84
Shetland 191
Shiel, Glen 130-3
Shirt, D.B. 246
shooting 177
  see also hunting
Sierra Nevada 374
Sievänen, T. 209
Siitonen, J. 246
Sika deer (*Cervus nipon*) 243, 307, 315-16
Silene acaulis 100, 101
Simmons, I.G.: on Northern Europe, environmental history 46, 139, **141-9**, 152, 174
Simpson, I.A. 181, 182, 184
Simpson, M. 350
Sissons, J.B. 46
Sixth Environmental Action Programme, EU (2001-10) 50
Sjodal mountain 154
Sjodalen 166, 170, 171-2
Skaftafell National Park 336, 338, 339, 340, 342, 343, 346
Skaftárhreppur 336, 340
Skaptadóttir, U.D. 340
skiing
  cross-country 206, 207, 208, 210, 277
  downhill 7-8, 145, 206, 207-8, 209, 277
  in Scotland 127, 140, **197-202**
Skye 91, 93, 186, 355
  bog-moss (*Sphagnum skyense*) 311
slender Scotch burnet moth (*Zygaena loti scotica*) 311
Sletten, K. 63, 64
Slovakia and Slovenia 331-2
slush torrents 64
SMD (sustainable mountain development) 11, 13

Smidt, J.T. 320
Smith, A.G. 172
Smith, B.F.L. 122
Smith, M.A.F. 285
Smith, S. 240
Smout, C. 8
Smout, T.C. 147, 309
Snæfellsjökull 336, 343-4
snails 311
SNH *see* Scottish Natural Heritage
Snow, D.W. 309
snow 25, 62, 63, 65
snow bunting (*Plectrophenax nivalis*) 100,
    276
Snowdon (Yr Wyddfa) 142, 145, 321
Snowdon lily (*Lloydia serotina*) 100, 101
snowmobiles *see* off-road vehicles
Snowy Mountains 26
social changes and tourism 204
social principles and components of
    Ecosystem Approach 31, 32
social psychology of leisure 129-30
Soderstrom, L. 246
Södra Jämtlandsfjällen, tourism in 299,
    301-2
Sogn og Fjordane 152
soil
    degradation 228
    erosion in Scottish uplands 140, **191-5**
        geographical extent 191
        mapping overland flow risk 193-4
        organic and organo-mineral soils
            192-3
        trends 192
    and geology 123
    and global warming 101
    good 230
    Northern European mountains 46
Somatochlora alpestris 62
Soper, K. 337
Sorbus aucuparia 316
Sornfelli 360, 361
Soukupová, L. 60, 61, 62
Sourhope 321
South Africa, World Summit in 11, 29
South America *see* Latin America
South Georgia reindeer 315
South-East Asia 25
    International Year of Mountains 16
Southern Uplands (Scotland) 191
Sovari, S. 80
SP (Stated Preference) method 300-1
Spain 16, 47, 328, 374
    lakes contaminated 78, 81
    mountain municipalities 331-2
    rural depopulation 229
    see also Pyrenees
SPAs (Special Protection Areas) 47, 50
spatial scale and ecosystem approach 33-4
Special Areas of Conservation 47, 50, 90
Special Protection Areas 47, 50
specialisation 231
speciation 371
Spehn, E.M. 20, 47, 371, 372
Sphagnum skyense 311

spheroidal carbonaceous particles *see* SCP
spiders 247
spirituality 27, 28, 29
sport 27, 145, 355
    see also hunting and shooting;
        recreation; skiing
Sports Council, Scottish 355
spruce 152, 265
Spusta, V. 65
Srokosz, J. 104
stability of landscape in European
    Mountains 65-6
Stankey, G.H. 269
state ownership of land 233-4
Stated Preference (SP) method 300-1
Statskog 256
steam power 142
Stenmark, I. 207, 210
Stenseke, M. 158
Steven, H.M. 243
Stewart, F.E. 309
stock *see* grazing
Stocker, T.F. 104, 107
Stoll-Kleemann, S. 281, 284
Stone, P.B. 12, 146
Størseth, K. 155, 156
Stott, P.A. 106
Stouffer, R.J. 104
Strathclyde Region 351-2
Strathspey 248
Strathglass Complex 93
strengths and weaknesses of
    international environmental
    monitoring 219, 220
Stroeven, A.P. 59
Structural Funds 325, 330
subsidies 158
    agricultural 175
Sudetic lousewort (*Pedicularis sudetica*)
    62
sulphur/sulphur dioxide pollution 74,
    75-6, 94, 121, 122
summer farming and transhumance 373
    decline 170, 171-2
    Northern Europe, environmental history
        143-4
    Norway 152-3
    Norway and Wales compared 164, 172
    see also under Iceland
summer grazing, Saami reindeer herders
    259-61
Summers, R.W. 243
supply and demand, balancing 28-31
supporting services 28
Surrey Research group 354
Surrounding Project, Fulufjället 267,
    268, 271, 272
Sus scrofa 245
sustainability 27
    Cairngorms National Park 284-6, 287
Sustainable Agriculture and Rural
    Development 29
in Mountain Regions, International
    Conference on (2002) 18
sustainable development 232-3, 237, 295

sustainable future food production 159-60
sustainable mountain development *see* SMD
Svalbard 79, 142, 218
    natural heritage trends 315, 321
    reindeer (*Rangifer t. platyrhynchus*) 317
    SCANNET 360, 361
    snowmobile damage 318
Swanson, F.J. 58
Sweden 20, 142, 379
    agriculture comparison with Norway 158
    geodiversity and biodiversity *see*
        Abisko Mountains
    mountain municipalities 331-2, 333
    National Park Plan 266-7
    see also Fulufjället
    natural heritage trends 308, 313-19
        passim, 321
    reindeer herders *see* Saami
    rural depopulation 229
    SCANNET 360, 361
    snowmobile damage 318
    tourism changes in 203-10
        comparison with Finland, Norway and
            Iceland 208-10, 218
        reasons for 207-8
        Sápmi **299-302**
    see also Scandinavia
Swedish Environmental Protection Agency,
    and Fulufjället 266-73 *passim*
Switsur, V.R. 143
Switzerland 25, 48, 49, 236, 379
    International Year of Mountains 17, 18,
        19, 20
    mountain municipalities 329, 331-2, 333
Sykes, J.M. 214, 215, 244

Tallis, J.H. 143, 187
Taranger, A. 256
Tarrason, L. 74
taxonomic diversity, Northern European
    mountains 46
Taylor, D.A. 13
Taylor, J. 130-1, 132, 133, 172, 277, 355
Tebaldi, C. 46, 106
Teesdale 321
Teigland, J. 209, 210
temperature
    changes, future 105-6, 110-12
    increase *see* global warming
    lapse rate 111-12
temporal scale and ecosystem approach 33-4
Ten Nation Scoping Study 229, 231, 237
Tetrao
    T. tetrix 225, 241, 243, 311
    T. urogallus 225, 241, 243
Tewnion, S. 100
THC (thermohaline circulation) 104
Thimphu Conference and Declaration
    (2002) 18, 20
Thomas, C.W. 59
Thompson, D. 100
Thompson, D.B.A. 7, 26, 41, 135, 185, 336
    conservation priorities 371, 372, 373
    geodiversity and biodiversity in
        European Mountains 57-70, 116

nature of mountains **43-55**, 191
Northern Europe, environmental history 145, 146
Norway and Wales compared 164, 169, 173
Thompson, R. 80
Thomson, A.: on Iceland, transhumance and vegetation 139, **179-84**
Thorsteinsson, I. 179, 184
ticks 247
Tilt, Glen 7
timber *see* deforestation
toadstools 247
Toll Lochan, Loch 93
topography, diversity, modelling climate difficult in 107-8
Torneträsk valley 59, 60
Torridon, Glen 92, 93
tourism/recreation 10, 14, 167, 229, 326, 373
  Cairngorms National Park 277, 287
  ecosystem approach 26-7, 29
  Fulufjället National Park 268-71
  natural heritage trends 308, 318, 319, 320
  in Northern Europe 140, **203-12**
    changes in Sweden 205-7
      reasons for 207-8
    comparison with Finland, Norway and Iceland 208-10
    environmental history 145-6
    Norway 157-8
    Sápmi **299-302**
  outdoor education in Scotland 354, 355
  Recreation Opportunity Spectrum 263, 269, 270
  Scottish Highlands 42, **127-36**
    access 135-6
    history of 128-30
    land 134-5
    participants 130-3
  skiing in Scotland 140, **197-202**
  Vatnajökull National Park 337-8, 340-1, 342-3
toxaphene 78
toxicity 76-8
  see also pollution
'Tragedies of Commons' 252
trampling 46, 66, 191, 192
transhumance *see* summer farming and transhumance
travel literature 129
tree-line 26, 308, 312
  see also forests
Trichophorum cespitosum 188
Trossachs 275, 321
Trotternish 186
truth, mountain 5
tufted saxifrage (*Saxifraga cespitosa*) 100
Tulostoma niveum 311
tundra 166, 360
  arctic-alpine 60-1
  International Experiment 66
tundra reindeer (*Rangifer t. tarandus*) 314
Turkey 16, 17

Turnbull Jeffrey Partnership 276
Turner, L.: on international environmental monitoring **213-22**
Turunen, M. 361

Uaine, Lochan 93
Uganda 16-17
UKCIP02 (Climate Impact Programme) model 107, 109, 115
Ukraine 16
UNCED (United Nations Conference on Environment and Development, 1992) 3, 12, 17, 28-9, 284
unemployment 266
UNEP (United Nations Environment Programme) 25, 48, 49, 257
  and International Year of Mountains 17, 19, 20
  World Conservation Monitoring Centre 20, 44, 326-7
UNESCO 13, 44
  Man and Biosphere Programme 47, 48
United Kingdom 47
  acid deposition 72-3, 75-6, 81-2
  agricultural decline 228
  climate 46
  Climate Impact Programme 107, 109, 115
  hunting and field sports 46
  mountain municipalities 328, 331-2
  and United States compared *see* ethical and aesthetic issues
  see also England; Scotland; Wales
United Nations
  Conference on Environment and Development *see* UNCED
  Convention on Biodiversity (Rio Convention, 1992) 49, 168-9
  Economic and Social Council *see* UNESCO
  Environment Programme *see* UNEP
  FAO *see* Food and Agriculture Organization
  Framework Convention on Climate Change (1992) 49
  Inter-agency Committee on Sustainable Development 12
  World Charter for Nature 48
  see also International Year of Mountains
United States 135
  environmental history 141
  Geological Survey's EROS Data Center 45
  National Geophysical Data Center 45
  national parks 168
  and United Kindom compared *see* ethical and aesthetic issues
Upská jáma corrie 63
Urals 372, 374
Urry, J. 208, 337
Ursus arctos 245, 253, 265, 373
Usher, M.B.: on natural heritage trends 143, 144, 168, 305, **307-24**

Vaccinium
  V. spp. 61, 62, 90
  V. myrtilis 188, 244, 317, 321

V. uliginosum 188
V. vitis-idaea 311, 317
Vågå 154
Vaisänen, R. 46
van der Wal, R. 121, 124
Van Liere, K.D. 228
Vandvik, V. 164, 169, 174
Vanek, J. 63
Väre, H. 371
vascular plants 311
  see also forests
Västerbotten 205-8, 259
Vatnajökull National Park plans 305, **335-47**
  area described 338-9
  communities and livelihoods 340-1
  nature protection not covered 345-6
  social nature, national parks as 337-8
  see also opinions of local communities
  vegetation/plants
    arctic-alpine 60-1
    and climate change in Scotland and Wales 100-1
    conservation priorities 371, 374
    mat strength 187-8
    mean change index map 310
    medicinal 26
    and pollution 72
    see also forests; grass; heath; heather
Vellinga, M. 104
Vera, F.W.M. 175
Vertigo geyeri 311
Vestur-Eyjafjallahreppur 181-3
Victorian attitudes to mountains 6-7
Vídalín, P. 182
Village Hall Committees 279
Villeneuve, A. 13
Virtanen, R. 316
visualization 201
VOCs (volatile organic compounds) 83
volcanoes 142, 338
voluntary principle 8-9
Volz, A 83
von Storch, H. 107
Vorkinn, M. 209
Vuorio, T.: on Sápmi, tourism in 226, **299-302**

Waage, E.R.H. 341, 342
Wales 6, 142, 145, 172-3, 215
  climate *see* effects *under* climate change
  natural heritage trends 309, 310, 311, 318, 321
  see also under Norway
Walker, M.F. 172
walking *see* hiking
Wallen, B. 142
Wallis de Vries, M.F. 146
Wallsten, P.: on Fulufjället National Park 225, **263-73**
Walsh, S. 20
Warren, C. 236
Washington Convention 49
wasps 246, 247

water 28
  beetle (*Agabus wasasjernae*) 248
  deforestation and 26
  ecosystem approach 25, 29
  pollution 228
  sports 355
  supply 135, 144
  see also lakes; rivers
Water Framework Directive (EU) 51, 230
Waterfowl Habitat Convention 47, 48
Watershed Management, Partnership
  Initiative on 20
Wathne, B. 76
Watson, A. 197, 283
Watt, A.S. 62
WCMC *see* World Conservation Monitoring
  Centre
Weiermair, K. 203
Weingartner, R. 16
Welch, D. 185, 187, 197
Weller, G. 360
Werritty, A. 65
West, P.C. 338
wet deposition 73-8
Wet Woods Restoration Project 242
Wetlands Convention 47, 48
WHC (World Heritage Convention) 48
Whightman, A. 309, 355
White, P.S. 153, 157
white-fronted goose (*Anser albifrons*) 314
white-stalk puffball (*Tulostoma niveum*)
  311
Whittow, J. 46
wild azalea (*Loiseeuria procumbens*) 100
wild forest reindeer (*Rangifer t.
  fennicus*) 314-15
wild land 306, **365-9**
  core areas 366-7
  policy 365-6
  public, role for 368-9
  remoteness modelling 367, 368
  restoration, target areas for 368
  see also wilderness
wild spruce (*Picea abies*) 152
wilderness concept 168, 170-2, 360, 365,
  372
  and UK's cultural lansdcape model
    compared 292-3
  see also SCANNET; wild land
Willhauck, G. 197
Willows, R. 114
willows (*Salix spp*) 61, 155, 308-9, 311
Wilson, A. 287
wind 46
winter tourism 204, 349
  see also skiing
Winterbottom, A. 101
Wirlgolaski, F.E. 371
wolf (*Canis lupus*) 245, 253, 254, 265,
  312-13, 373
wolverine (*Gulo gulo*) 253, 265, 312
  maps of range (1850-1998) 313
wood *see* deforestation
Woodin, S.J. 84, 242
woodland *see* forests

Woodland Assurance Scheme 246
Woodwell, G.M. 66
Woolgrove, C.E. 242
woolly hair moss (*Racomitrium
  lanuginosum*) 42, 84, 242, 320
  geodiversity and biodiversity in
    European Mountains 61, 62
  geology and nitrogen deposition on
    vegetation 121, 122-4
  Highlands, Scottish, high plateaux and
    terrain sensitivity 187, 188
woolly willow (*Salix lanata*) 308, 311
Wordsworth, W. 350
Working Party, Cairngorms 8-9
World Bank 20
World Charter for Nature (1982) 48
World Conservation
  Monitoring Centre (UNEP) 20, 44, 326-7,
    328
  Strategy (1980) 48
  Union (IUCN) 3, 48, 49, 282
World Food Summit; Five Years Later 17
World Heritage
  Convention and Centre 48
  Site 276, 320
World Mountain Peoples Association 18
World Mountain Symposium (2001) 18
World Resources Institute 49
World Summit on Sustainable Development
  (Bali) 17
World Summit on Sustainable Development
  (Johannesburg, 2002) 11, 29
  Mountain Partnership 19-21
World Wildlife Fund 16, 48
Worldwide Fund for Nature 272
Wright, G.G. 197
WSSD *see* World Summit on Sustainable
  Development
WWF (World Wildlife Fund) 16, 48

Yang, H.D. 76, 94
Yr Wyddfa (Snowdon) 142, 145, 321

Þórsmörk National Park 180
Þorvarðardóttir, G.: on Vatnajökull
  National Park 305, **335-47**

Zackenberg 360, 361
zones, bioclimatological 45
Zorita, E. 107
Zygaena loti scotica 311